BEHAVIORAL RESEARCH AND ANALYSIS

An Introduction to Statistics Within the Context of Experimental Design

FOURTH EDITION

BEHAVIORAL RESEARCH AND ANALYSIS

An Introduction to Statistics Within the Context of Experimental Design

MAX VERCRUYSSEN • HAL W. HENDRICK

CRC Press
Taylor & Francis Group
Boca Raton   London   New York

CRC Press is an imprint of the
Taylor & Francis Group, an **informa** business

CRC Press
Taylor & Francis Group
6000 Broken Sound Parkway NW, Suite 300
Boca Raton, FL 33487-2742

First issued in paperback 2018

© 2012 by Taylor & Francis Group, LLC
CRC Press is an imprint of Taylor & Francis Group, an Informa business

No claim to original U.S. Government works

ISBN 13: 978-1-138-07318-0 (pbk)
ISBN 13: 978-1-4398-1802-2 (hbk)

This book contains information obtained from authentic and highly regarded sources. Reasonable efforts have been made to publish reliable data and information, but the author and publisher cannot assume responsibility for the validity of all materials or the consequences of their use. The authors and publishers have attempted to trace the copyright holders of all material reproduced in this publication and apologize to copyright holders if permission to publish in this form has not been obtained. If any copyright material has not been acknowledged please write and let us know so we may rectify in any future reprint.

Except as permitted under U.S. Copyright Law, no part of this book may be reprinted, reproduced, transmitted, or utilized in any form by any electronic, mechanical, or other means, now known or hereafter invented, including photocopying, microfilming, and recording, or in any information storage or retrieval system, without written permission from the publishers.

For permission to photocopy or use material electronically from this work, please access www.copyright.com (http://www.copyright.com/) or contact the Copyright Clearance Center, Inc. (CCC), 222 Rosewood Drive, Danvers, MA 01923, 978-750-8400. CCC is a not-for-profit organization that provides licenses and registration for a variety of users. For organizations that have been granted a photocopy license by the CCC, a separate system of payment has been arranged.

Trademark Notice: Product or corporate names may be trademarks or registered trademarks, and are used only for identification and explanation without intent to infringe.

Library of Congress Cataloging-in-Publication Data

Vercruyssen, M. (Max)
    Behavioral research and analysis : an introduction to statistics within the context of experimental design / Max Vercruyssen, Hal W. Hendrick. -- 4th ed.
        p. cm.
    Rev. ed. of: Behavioral research and analysis / Hal W. Hendrick. 1981.
    Includes bibliographical references and index.
    ISBN 978-1-4398-1802-2 (hardcover : alk. paper)
    1. Psychometrics. 2. Experimental design. I. Hendrick, Hal W. II. Hendrick, Hal W. Behavioral reseach and analysis. III. Title.

BF39.H394 2012
150.72'7--dc23   2011036196

Visit the Taylor & Francis Web site at
http://www.taylorandfrancis.com

and the CRC Press Web site at
http://www.crcpress.com

In Memoriam

Hal W. Hendrick
1933–2011
Creator and Co-Author of This Book
International Ergonomics Expert

On May 13, 2011, while editing the page proofs of this fourth edition, the book's founder and co-author peacefully took his last breath. During his gallant 18-month battle with cancer he was able to meet and say goodbye to friends and relatives. He was disappointed not to be able to see this book in its final published form but was satisfied to know his last piece of scholarly work was completed. Professor Hendrick touched the lives of many and will be dearly missed. However, his memory and inspiration live on in those who still picture his constant smile, hear his frequent laughter, and remember his giving nature, friendly demeanor, nurturing collaborations, and gentle kindness. His author biography in this book relates some of his great achievements but he was much more on a personal level as a husband, father, grandfather, university professor and dean, international professional organization leader, consultant, expert witness, mentor, and friend to many. His individual successes were usually a result of his expending a lot of energy inspiring everyone around him to be the best they could be. He was a remarkable leader who profoundly changed the lives of many. Farewell and thank you …

Contents

Preface

This book was created by Dr. Hendrick (1981) before Dr. Vercruyssen joined him in writing the second edition (1989) and then became the lead author for the third edition (1990) and this one (2012). Its primary use was for graduate students in the behavioral sciences and education as a component of the training for completing a thesis. However, it has recently been expanded to a broader audience including undergraduates.

Since publication of the first edition of this text in 1981, we have received many suggestions and comments from readers for improving this book. Nearly all reader suggestions and some new ideas have been included in the current edition. We hope you find this revised edition useful and that you will send us comments for changes in the next edition.

PURPOSE AND AUDIENCE

As with the previous editions, the purpose of this book is to present an overview of statistical methods within the context of fundamental topics relevant to career professionals, including experimental design, data collection, data analysis, interpretation of results, and communication of findings. We envision this text as being particularly useful in four types of applications: first, as the introductory text in research-oriented statistical methods courses in the behavioral sciences, the natural sciences, education, engineering, business, and multidisciplinary applied fields; second, as a companion in advanced research methods or content-oriented experimental courses in these same disciplines; third, as a text to introduce graduate and honor undergraduate students to experimental research, statistical tools, and scientific communications (e.g., theses, refereed publications, research proposals, research presentations); and finally, as a resource for purposes of review and refamiliarization for professionals in a variety of research fields.

Because readers of this text are likely to be users rather than producers of statistical equations, the emphasis has been placed on explaining statistical procedures and interpreting obtained results without discussing the derivation of equations or history of the method. Probability theory is scarcely mentioned in the appendices; this is not a purpose of this text. Also, the mathematical demands are minimal—no college-level mathematics background is required or assumed. We want this text to become a single resource that will assist students and seasoned professionals in conducting scientific research and reporting it to the scientific community.

ORIGIN AND UNIQUENESS

The authors realized while teaching courses in experimental psychology and research methodologies that although the students had taken prerequisite courses in statistics and research methods, their actual knowledge of the basics of research design and statistical analyses varied greatly and was generally poorer than expected. In looking for a comprehensive text that could be used to bring the students up to speed, none was found totally suitable. The books either reviewed the basics of behavioral research and experimental design but provided only cursory coverage of statistical methods or they provided comprehensive coverage of statistical methods with very little coverage of the research context within which these methods are used. Nowhere could we find a resource that provided the methodology, statistics, and coverage of communication skills, so we set out to write our own.

Graduate students using this text in research or statistics courses required for degrees in psychology, communications, systems management, neuroscience, safety science, kinesiology and sport sciences, engineering, and human factors have preferred this book to widely utilized conventional texts because presenting statistical methods within the context of research design made the methods much more meaningful and easier to remember. This appeared to be of benefit from both motivational and learning standpoints. We found that when students are assigned experiments at the beginning of a course, for which they must collect and analyze data and formally write the results of their study, they have a specific purpose for learning the contents of this book and the whole learning process is facilitated. Therefore, we strongly recommend that readers have an experiment planned (e.g., thesis, dissertation, term project) before starting this text.

TEXT ORGANIZATION

Chapter 1 contains a general review of scientific research, including various approaches to empirical investigations. Chapters 2 and 3 provide an overview of descriptive statistics (univariate and bivariate). Chapter 4 presents statistical hypotheses testing, basic simple experimental designs, and commonly used parametric and nonparametric tests. Chapter 5 introduces simple univariate analysis of variance (ANOVA). Chapter 6 differentiates multifactor univariate ANOVA designs (between, within, and mixed). Chapter 7 prepares the reader for planning, conducting, and reporting a behavioral research study. Each chapter in the body now contains relevant keywords, chapter summaries, keyword definitions, and end of chapter exercises (with answers).

Appendix A includes statistical reference tables. Appendix B contains a glossary to facilitate access to information on statistical terms, symbols, and equations. Appendix C contains statistical equations.

Acknowledgments

Over the years, many individuals have contributed to the continued development of this book. Students in the authors' classes continue to make helpful comments each semester, but over the years major efforts have come from graduate students George Brogmus, J.C. Edwards, Gretchen Greatorex, Tina Mihaly, Olu Olofinboba, Sara Reynolds, and Kim Siu. Desktop publishing support was provided by Michelle Agustin and Gia Macabeo-Shahn. We especially appreciate assistance provided by high school student Nani Vercruyssen who helped change the wording so this book is more understandable to entry-level students and by doctoral student Yasuhiro Ueyama who made suggestions for wording that made this book easier for beginning researchers whose native language is not English. Mia Vercruyssen, a technical writer, provided valuable revisions to this edition.

We are grateful to the Literary Executor of the late Sir Ronald A. Fisher, F.R.S.; to Dr. Frank Yates, F.R.S.; and to the Longman Group Ltd., London for permission to reprint Tables C, D, and E in Appendix A from *Statistical Tables for Biology, Agriculture and Medical Research* (6th ed., 1974). Of course, we are also grateful to those who first introduced us to the field of statistics: Paul A. Games (MV), Thomas Pyle (MV), Robert Perloff (HWH), and Ben Winer (HWH). Thank you for supporting continued development of this instructional tool.

Max Vercruyssen
Honolulu, Hawaii

Hal W. Hendrick
Denver, Colorado

About the Authors

Max Vercruyssen, PhD, is the director of Hawaii Academy, a private school for lifetime fitness, gymnastics, and human sciences, where he also serves as chair of the research department and as head coach of the school's elite-level trampoline gymnastics teams.

Dr. Vercruyssen holds a bachelor's degree in experimental psychology; master's degrees in experimental and physiological psychology, exercise and sport sciences, and public health; a PhD in neuromuscular control, and pursued postdoctoral training to earn advanced certificates in ergonomics and gerontology. Most of his advanced statistics training was under Paul A. Games (Pennsylvania State University) and as required for experimental psychology, biostatistics, and longitudinal studies of aging.

At the University of Southern California (USC), Dr. Vercruyssen served as assistant professor of human factors and ergonomics, director of the Human Factors Laboratory, and codirector of the Laboratory of Attention and Motor Performance in the Andrus Gerontology Center. He also helped develop the university's ergonomics graduate degree programs and mentored the first ergonomics majors in safety science. At the University of Hawaii, he was an associate professor in psychology, gerontology, and geriatric medicine.

During the 1990s, Dr. Vercruyssen was also a research associate at the University of Minnesota's Center for Transportation Research, Institute of Intelligent Transportation Systems, and a distinguished fellow of gerontechnology at the Technical University of Eindhoven, Netherlands. Dr. Vercruyssen has authored or coauthored over 200 refereed publications and presented papers at international scientific and technical conferences. Nine years of his university appointments involved teaching experimental research and statistics courses and mentoring thesis students (the practical need that resulted in developing this book). His students have received national research awards and his athletes have held many national and world championship titles. His current challenge is to make required statistics training palatable for students who may not enjoy or want to study mathematics—even those with statistophobia—as well as for those with great ambitions for academic achievement.

Hal W. Hendrick, PhD, was emeritus professor of human factors and ergonomics at the University of Southern California. He held a BA in psychology from Ohio Wesleyan University, an MS in human factors, and a PhD in industrial psychology from Purdue University. Dr. Hendrick studied statistics and experimental design with APA Past President Professor Robert Perloff and Professor Ben Winer, the "guru" of behavioral research and analysis of his time. For six years, Dr. Hendrick developed and taught a course in statistics and experimental design at the United States Air Force Academy and for 19 years, a similar graduate-level course at USC.

Dr. Hendrick was a past chair of USC's Human Factors Department, former executive director of the university's Institute of Safety and Systems Management, and a former college dean at the University of Denver. He was a certified professional ergonomist, a diplomate of the American Board of Forensic Examiners, and a fellow of the American Psychological Society, the Human Factors and Ergonomics Society (HFES), the International Ergonomics Association (IEA), the American College of Forensic Examiners, and a charter member and fellow of the Association for Psychological Science. He was a past president of the Human Factors and Ergonomics Society, the International Ergonomics Association, the Foundation for Professional Ergonomics, and was a founding member and past president of the Board of Certification in Profession Ergonomics.

Dr. Hendrick was a recipient of USC's highest award for teaching excellence, the IEA Distinguished Service Award, the HFES Arnold M. Small President's Distinguished Service Award, the Jack A. Kraft Innovator Award, and the Alexander C. Williams, Jr. Design Award. Dr. Hendrick had over 45 years of experience as a human factors and ergonomics and industrial and organizational psychologist practitioner, educator, program administrator, and consultant. He was the author or coauthor of over 200 professional publications and three textbooks, and has edited or coedited 10 books, including the *Handbook of Human Factors and Ergonomics Methods* (CRC Press, 2005). Dr. Hendrick conceptualized and initiated the human factors sub-discipline of macroergonomics. He held a regular commission in the U.S. Air Force (Lt. Col., USAF ret).

1 Overview of Scientific Research

This chapter begins the book with an overview of science featuring a description of (1) the scientific method, (2) the goals, principles and assumptions of science, and (3) the approaches to conducting scientific research. Thus, we start with the big picture (the forest) before examining detailed procedures (the trees). This chapter covers these topics:

What Is Science?
Scientific Method
Goals, Principles, and Assumptions of Science
Five Basic Approaches to Scientific Research
Summary
Keyword Definitions
References
Exercises
Exercise Answers

KEYWORDS

Case History Approach
Caterus Parabus
Concurrent Validity
Confirmatory Experiment
Contiguity of Events
Control Group
Controlled Observation
Correlation Approach
Dependent Measure (DM)
Dependent Variable (DV)
Determinism
Empirical
Empirical Confirmation
Empirical Verification

Experimental Approach
Experimental Group
Exploratory Experiment
External Validity
Field Study Approach
Hawthorne Effect
Independent Variable (IV)
Internal Consistency
Internal Validity
Laws of Science
Limited Causality
Meaningful Problem
Nonequivalent Control Group
Operational Definition

Pilot Study
Predictive Validity
Predictor
Quasi-Experimental Study
Random Sampling
Realism
Reliability
Sampling With Replacement
Science
Scientific Method
Statistical Generalization
Stratified Random Sampling
Validity

WHAT IS SCIENCE?

As often as we use the term *science*, it is surprising how lacking in precision is our concept of this term. If you were to ask a dozen prominent researchers to define science, you probably would receive a dozen different answers. In spite of these apparent differences, there are at least several definitional characteristics upon which there is fairly universal agreement.

First, science is concerned with the study of meaningful problems. This means that the problem is capable of being answered directly through empirical investigation, that is, through direct observation. By way of contrast, a meaningless problem is unanswerable through empirical investigation. Meaningless problems often are concerned with ultimate causes, master motives, or supernatural phenomena. Typical of such questions are "what is God?" or "does the libido motivate all human behavior?" Because neither God nor the libido is directly observable, asking these questions is empirically meaningless. It should be emphasized that the *meaningful* and *meaningless* terms in no way imply degree of importance. Some of the most important questions asked by humankind are empirically meaningless, and our assumptions and hypotheses concerning their answers often have influenced the direction of scientific investigation. One merely needs to look at the historical impact of philosophical and religious thought on the scientific disciplines to gain an appreciation of this fact.

A second common characteristic of the sciences is that they attempt to answer empirically meaningful problems by use of the same general approach. This approach is referred to as the *scientific method*. The scientific method is a systematic process involving a sequential series of steps that a researcher follows in identifying and answering a problem. Putting the above two general characteristics together, we can define science as at least being *the study of meaningful problems through use of the scientific method* (McGuigan, 1960). We already have noted that such disciplines as religion and philosophy deal with non-meaningful problems and, strictly speaking, do not qualify as science. Other disciplines, such as music, art, and literature study meaningful problems. However, because they usually do not use the scientific method as their basic approach, they too normally are not thought of as sciences.

SCIENTIFIC METHOD

We already have noted that the scientific method is a process involving a sequential series of identifiable steps. Let us now take a look at this process. Although (as we shall see later in this chapter) a variety of approaches utilize the scientific method, the approach of primary interest in this text is the experimental method. The discussion of the scientific method therefore will center on its application in experimentation and focus on seven steps: (1) identify problem, (2) formulate hypothesis, (3) conduct pilot study, (4) collect data, (5) test hypothesis, (6) generalize results, and (7) replicate experiment.

IDENTIFY PROBLEM

An experiment typically begins with the formulation of a question or problem that is meaningful. This is not always easy to do. History is full of useless research that resulted from failure to ask the right question—to identify the real problem. Sometimes this failure results from imperfectly stating the problem. In other cases, the problem may have arisen from imperfect or inadequate observations. Failure to ask such questions as "is this problem important enough to warrant investigation?" or "if I solve this problem, will the results be used?" also can lead to wasted research. By taking care and investigating the background of the problem, stating the problem in various ways, defining its limits, and asking the above questions regarding its importance and utility, a researcher can save many hours of wasted effort.

FORMULATE HYPOTHESIS

Having identified the problem, the next step is to formulate a guess as to what is the most probable answer to the problem. This is stated in the form of an empirical hypothesis. An empirical hypothesis can be defined as a meaningful proposition—one that can be supported or refuted as being probably true or probably false through scientific investigation.

Typically, a hypothesis is a statement of a potential empirical relationship between two variables. If we ask the question, "what is the effect of ambient noise on performance?" the hypothesis might be "if ambient noise increases, then performance decreases." This is an empirical hypothesis because we can increase noise and observe what happens to the subjects' performance. It also is a statement of a possible relationship because we are proposing what will happen to one thing if the other thing changes—that the change in one variable is related to change in the other. Note that the manner in which we state our experimental hypothesis is an "if… then…" proposition or what formally is known as the logical form of the general implication. In reading research reports in professional journals, you often will see the experimental hypothesis state something like "the purpose of this study was to investigate the effect of environmental noise on performance," or "…to investigate performance decrement as a function of environmental noise." Both statements could easily be restated as "If… then…" hypotheses: "If environmental noise increases, then performance decreases."

While an experimental hypothesis is a statement of a potential relationship, it does not necessarily imply that one thing causes the other. Suppose we were to find that as the number of doctors in a given geographical area increases, the number of cows in that same area decreases. It would be difficult to explain this by assuming that increasing the number of doctors causes the decreases in cows, or vice versa! Rather, there probably is some underlying factor such as urbanization that affects both the number of doctors and cows in some systematic fashion.

Keep in mind also that the experimental hypothesis begins with "if." This caution is important because what happens in a research situation may not necessarily occur in real life.

Conduct Pilot Study

This may be the most important step in preventing wasted effort. Conducting a mini-experiment using a small number of subjects (participants) may serve as a preliminary test of the experimental hypotheses. It provides a means for testing the instructions to be sure they are understood by the participants as intended, taking "bugs" out of the data collection system, ensuring that the statistical procedures are going to work as expected, and may also prevent unexpected problems from becoming major limiting factors. Results from pilot studies can aid a scientist in refining his or her approach to a problem, increasing statistical power of the tests, and providing data trends useful in soliciting research funds to conduct the actual experiment.

Data from a pilot study can reveal expected results, determine the number of subjects needed for finding significance in the actual experiment, and even indicate the expected directional outcome (e.g., one-tailed versus two-tailed test). Generally, only six steps are attributed to the scientific method but the authors of this text feel so strongly that a pilot study is essential that it has been introduced here as a formal step of the scientific method. Note that many academic disciplines hold steadfast to the notion of only six steps in the scientific method and do not recognize conducting a pilot study as a true step.

Collect Data

After stating the experimental hypothesis and conducting a pilot study, the next step is to collect data that can be used to support or refute the hypothesis. Introducing a few concepts here may be useful to better understanding the value of traditional methods used in data collection. Most relevant may be (1) participant sampling, (2) treatment groups, (3) experimental variables, and (4) description of data.

Participant (Subject) Sampling

In behavioral research, collecting data usually begins with the selection of a group of subjects (participants) who are representative of the population of interest. To a large extent, the population of interest will be determined by the problem. For instance, if the problem concerns some aspect of performance of grade school students in the United States, then grade school students in the United States would be the population from which the subjects would be drawn. If, instead, high school students were used, the experimental results might tell us something about high school students, but not necessarily answer the question about grade school students.

The two primary methods for selecting a representative sample of subjects are (a) random sampling and (b) stratified random sampling.

Random Sampling

Random sampling is a method of selecting subjects that ensures every person in the population has an equal and constant chance of being selected. One way of doing this is to place all names on slips of paper and place the slips in a hat. Then, a name is drawn, the hat is shaken and another name is drawn. To keep the chances of selection constant, a drawn name must be placed back in the hat after each drawing; this is referred to as sampling with replacement. With very large or infinite populations, sampling may be performed without replacement because failure to replace a name has no practical effect on the chances of the remaining subjects to be drawn.

Another common technique is to arrange the population alphabetically and then select every fifth, tenth, thirtieth, etc., name, depending on the proportion of the population you want in the sample. This procedure assumes alphabetical arrangement is a random factor. To assist researchers, tables of random numbers have been generated, such as Table K in Appendix A of this text. By consecutively numbering each subject in the population of interest, we can turn to the table of random numbers and arbitrarily select the corresponding numbered subject until we obtain the desired sample size.

Stratified Random Sampling

In stratified random sampling, a variable that may affect the outcome of the study is identified and the population is stratified or classified with respect to that variable. Subjects for the sample then are drawn from these classes so that they are represented in the sample in the same proportion in which they occur in the population. For example, if living in a rural, suburban, or urban neighborhood were considered likely to affect people's attitudes in an opinion survey and 20% of the population was rural, then 20% of the sample would randomly be selected from the rural population. If religion also was considered an important variable, and one-third of the rural population was Catholic, then one-third of the rural sample would randomly be selected from the Catholic portion of the rural population.

A stratified random sampling most frequently is utilized when a very small percentage of a very large population that is heterogeneous with respect to important extraneous variables is to be used as the sample. Samples used in nationwide public opinion polls perhaps are the best known examples of this kind of sampling. By carefully stratifying the population on a number of important demographic variables, very accurate nationwide public opinion surveys are possible using very small samples of the total population.

Before going on, it may be very useful at this point to discuss other considerations in typical data collection, namely treatment groups, variables, and methods of describing the data collected. So important are these topics that they are major components of the method section in published papers, reports, and presentations. As will be discussed in later chapters, the method section typically presents a descriptions of participants (including demographics and selection method), experimental design (often simply the assignment of participants into treatment and control groups), experimental variables (usually independent and dependent variables), procedures (how the participants were treated), and the treatment of data (often the statistical procedures used to describe and interpret the data collected). The next sections briefly discuss some of these components as they apply to data collection.

Experimental and Control Groups

In a simple experiment, a researcher typically will assign the sample of subjects to two groups. Assignment of the subjects is done in such a way as to ensure that the groups are fairly equivalent with respect to some a priori matching criterion at the start of the experiment (procedures for assigning subjects to groups will be discussed in conjunction with the simple experiment designs in Chapter 4). One of the two subject groups is called the *control group* and usually receives the normal or standard amount of the treatment variable. In answer to the previously stated hypothesis regarding the effect of ambient noise on performance, the control group might be subjected to the amount of ambient noise normally present in their classrooms while they perform a designated task.

The second group is designated as the *experimental group* and receives the experimental level of the treatment variable. In our hypothetical study, the experimental group of grade school students might be subjected to a high ambient noise level while performing the designated task. Because the environment of the two equivalent subject groups is kept the same (controlled) except for the ambient noise level, any difference in task performance between the two groups can be attributed to the difference in noise level. This describes the *caterus parabus* principle of scientific research (i.e., all variables are controlled except the one being manipulated by the investigator).

Independent and Dependent Variables

We attempted in the above experiment to establish a relationship between two variables, one representing an aspect of the environment and the other an aspect of behavior. The aspect of environment (noise level) that was manipulated in the study is referred to as the independent variable (IV). It is the variable in the experiment that was deliberately changed or made different for the two subject groups. The resulting change in behavior (performance of designated tasks) is referred to as the dependent variable (DV) or dependent measure (DM). That is, we were studying the extent to which its value or amount appeared to be dependent on the value or amount of the independent variable.

Describing Collected Data

The data gathered may appear as a jumble of numbers unless we can reduce and simplify them in some way by "getting our arms around them" to make them meaningful. To do this researchers often turn to graphical and statistical methods for describing data. Referring to our noise study, we might compute the mean or average performance of the experimental group and compare it with the mean performance of the control group to see whether the results of the two groups differed. Graphical and statistical methods of description will be covered in Chapter 2.

Test Hypothesis

Let us assume that in our noise study we found a difference of four points between the experimental and control group means, with the control group having performed better than the experimental group. Is this a "real" difference, i.e., a difference due to the experimental manipulation that would have appeared if we tested our entire population? Or does this difference merely represent a fluctuation in performance of our sample group due to chance factors? In other words, is this difference merely the result of failure to control adequately for all possible extraneous variables such as differences between the groups' abilities or motivations?

To answer this type of question, a researcher conducts a statistical test of the difference between the means. As we shall see in Chapter 4, the particular statistical test the researcher uses largely is determined by (a) the specific nature of the data, and (b) the particular experimental design employed in the study. The outcome of this statistical test enables the researcher to state whether a difference is or is not significant, i.e., whether it is probably real rather than chance difference. Based on this outcome, the researcher can decide whether the experimental hypothesis was correct.

If the statistical test in our noise study were to indicate that the difference between the mean performance scores was probably a real one, we would conclude that increasing the ambient noise level in the experimental situation resulted in a decrease in the performance of the subjects and that this decrease would have shown up had we used the entire population represented by our sample.

Generalize Results

In concluding that the experimental hypothesis was correct, we have generalized the results to a population. This step of the scientific method, however, is fraught with many dangers and frustrations. The first question is "how far can we generalize our results?" Assuming the sample of grade schoolers was a representative one, we probably would feel safe in generalizing our findings to the grade school population from which the sample was drawn. What happens when we extend our results to other populations such as high school students or adults? Can we safely say, for example, that performance of adults on their jobs also will be affected adversely by high ambient noise levels? The broader we attempt to generalize our findings, the less confidence we can have in our generalizations.

Even with the grade schoolers we must be very careful how we generalize the results. In the experiment we used performance of a specific designated task only as the dependent variable. Can we generalize these results to similar tasks? What about other kinds of tasks? Here again, the broader the generalization, the less safe it becomes. For this reason, experimenters make it a point to emphasize the specific conditions under which their hypotheses were tested and to caution their colleagues regarding the extent to which they believe their results may be generalized safely.

REPLICATE EXPERIMENT

Having generalized the data, we should now be able to predict how a new sample of subjects will perform and conduct another experiment or series of experiments. This process is referred to as replication. If the predictions based on the original experiment are confirmed, the probability that the hypothesis is true is greatly increased. Only through extensive replication in a variety of new situations can hypotheses be so highly confirmed that they are regarded as laws of science.

Laws of science are experimental hypotheses that have been confirmed so many times that they are generally regarded as true. Examples include Newton's laws, Fitts' law, Hick-Hyman law, Weber-Fechner law, Hooke's law, Graham's law, and Boyle's law.

GOALS, PRINCIPLES, AND ASSUMPTIONS OF SCIENCE

GOALS OF SCIENCE

As with the definition of the term *science*, there is no universal agreement on just what are the goals of science. Fairly universal agreement has been achieved for at least two of the basic definitional characteristics of science and also for two of the basic goals of science. These goals can be stated briefly as *description* and *explanation*. Through these goals behavioral science tries to increase our understanding of ourselves and the world around us.

Description

Anderson (1971) noted that description can be regarded as the empirical goal of science and explanation as the theoretical goal. In the course of empirical investigation, a scientist attempts to describe events by reporting what is observed. In the previous hypothetical experiment, we observed and described the task performance of two presumably equivalent groups (experimental and control) under different levels of ambient noise. This enabled us to come to a conclusion regarding a hypothesis. Having determined that increasing ambient noise leads to a decrement in task performance, the next step in increasing our understanding is to explain why.

Explanation

In attempting to explain a dynamic that we as scientists have described, we try to relate our research results and perhaps those of others to some general principle or principles known or presumed to operate in the situation. The particular explanation accepted will likely influence both the direction of future research and the nature of actions taken in the real-life situation. For instance, the performance decrement noted in our noise experiment may be explained as the result of some direct physiological effect of high sonic vibration on the human organism, by the interference of noise with normal speech communication (if verbal coordination with others was essential to task performance), or to more psychological aspects such as the distraction characteristics of noise.

If the belief were that the first explanation was correct, future research effort probably would be directed toward learning more about the physiological effects of high sonic vibration. From a practical standpoint, if this physiological effect was thought to be potentially harmful, we might suggest steps to take to ensure that high ambient noise levels were not present in any classroom or work situation even if the performance decrement from high noise was not considered serious. If, on the other hand, the second or third explanations were considered more likely, our research effort might be directed toward studying ways to improve communication under high ambient noise or the differential effects of selected frequency bands on speech communication and/or distraction. In addition to attempting to reduce the ambient noise level, we might instigate training programs to improve student or worker performance under high ambient noise.

As can be seen, pursuing the goals of description and explanation leads to a developmental chaining of research activity. Once we described our observations, we attempted to explain them. The particular explanations proposed then led to new hypotheses that could be empirically investigated, described, and explained. In this way we increase our knowledge of ourselves and of the world around us.

PRINCIPLES OF SCIENCE

Empirical Verification

The first principle of science is that a descriptive statement can be regarded as true only if it is found to correspond with reality. This test of correspondence is made by observation of reality. *Observed reality* can have many meanings including divine revelation, astrological knowledge, and philosophical insight. For our purposes, observed reality refers to sensed data on which observers agree. This public aspect of observation rules out hallucinations and other idiosyncratic phenomena. While we may not make the necessary observations ourselves, we can require evidence that the correspondence between the statement and observation has been adequately examined. As a practical matter, this type of empirical verification is carried out frequently because no one can observe everything.

In performing an observation or in evaluating a report of a test performed by others, adherence to the four principles described is essential.

Operational Definition

All terms in a descriptive statement must be defined relative to the operations involved in manipulating or observing them. This means that we must precisely define our IV and DV in terms of how each is to be *measured*. For example, in our noise study, we must define *performance* by providing a specific description of the task and how one measures student performance.

The value of this principle is twofold. First, it requires us to confine our descriptions to those things that can in fact be empirically observed. Second, it ensures that we can communicate precisely what we mean by our descriptive terms to others. Two persons could conduct a study of the effect of ambient noise on performance and obtain very different results, simply because *ambient noise* and/or *performance* had different meanings in their studies. The principle of operational definition ensures that these differences would be readily recognizable and understood.

Controlled Observation

To make the descriptive statement that "a change in variable X produces a change in variable Y," a researcher must control or otherwise discount the effects of all other variables that may affect Y. Otherwise, he or she has no way to know whether the change in Y is attributable to variations in variable X or to some extraneous variable or variables. In our noise experiment, this was the reason for using a control group. The experimental and control groups were exposed to the same conditions or amounts of the extraneous variables and were allowed to differ only in the level of the IV (noise). Hence, a difference in the DV (performance) could be attributed only to the IV.

Statistical Generalization

Before attempting to generalize the results of an experiment to a given population, the results must be shown to hold for an adequate sample of that population. An *adequate* sample must (a) have been drawn by a random procedure and (b) consist of a significant number of members of that population to be considered representative. For example, if a researcher were to select John Doe arbitrarily from a large class and make the statement that John's height represented the average height of the class, you would be skeptical of the researcher's generalization. Conversely, if the researcher randomly selected 25% of the class and took their average height as the estimate of average class height, you probably would be confident that the estimate was very close. By basing a generalization on repeated observations, the researcher increased the likelihood of achieving results that are valid for the entire class population. The question of how many repeated observations (subjects) are required will be considered later.

Empirical Confirmation

An explanatory statement not only must explain the descriptive statement from which it was derived, but all other statements that can be derived from it also must be true. If any single statement that logically can be derived from an explanatory statement is found to be not true, then the explanatory statement is inadequate and must be revised or replaced by a more adequate explanation. Each time a prediction derived from an explanatory statement is found to be true, the explanation is further confirmed. Scientific laws are highly confirmed explanatory statements. However if a phenomenon inconsistent with a law is observed, the law must be revised or replaced by a more adequate law. Newton's laws of gravitation were found inadequate to explain why stars behind the sun could be observed during a total solar eclipse. It took a more adequate explanation, Einstein's theory (law) of relativity, to cover this phenomenon.

ASSUMPTIONS OF SCIENCE

Underlying the goals and principles of science are several basic assumptions about nature. Of particular importance are the assumptions of (1) determinism, (2) limited causality, and (3) contiguity (Underwood, 1966).

Determinism

Science involves the assumption that we live in a deterministic world. By determinism we mean that every event has a cause. In scientific experimentation we attempt to manipulate variable X to see what happens to variable Y to determine whether changes in X appear to cause variable Y to change. If we could not make the assumption of determinism, there could be no science.

Limited Causality

Science not only requires the assumption of determinism. It also requires the assumption that the number of things that cause another thing to occur or change is, for the most part, limited to relatively few variables. If empirical investigation of hundreds of variables were required to adequately describe the change in a single dependent variable, the scientist's tasks would be impossible. Fortunately, scientists have been able to describe changes in most dependent variables fairly accurately in terms of relatively few independent variables.

Contiguity of Events

In searching for causal relationships, the assumption is made frequently that events must be related in time and space. While this assumption appears to be correct, it is important to keep in mind that we cannot always observe the chain of events between the events that we actually see. As a result, the events we do observe may appear not to be temporally or spatially contiguous. For instance, the symptoms of many contagious diseases do not appear until a number of days after exposure to the carrier person or animal. Because of this, it often is difficult to determine who or what the carrier was. Similarly, sociological and psychological events may find their origins in events that took place months or years beforehand. Hence, tracing these events to their origins can be extremely difficult.

FIVE BASIC APPROACHES TO SCIENTIFIC RESEARCH

Generally speaking, most scientific disciplines employ five basic approaches in the study of questions in their fields: (1) correlational, (2) case history, (3) field study, (4) experimental, and (5) quasi-experimental.

CORRELATION APPROACH

Sometimes the primary interest of a researcher is not to arrive at cause–effect relationships but merely to determine whether a relationship exists and to measure the magnitude of the relationship. The researcher's ultimate aim often is to be able to predict a performance score on variable Y from knowledge of a score on variable X. While this approach can be used in a variety of situations, the most common use in the behavioral sciences is in developing and administering standardized tests for selection and placement. For this reason, the correlation approach often is referred to as the *psychometric approach.*

Establishing Validity

When the correlation approach is used in prediction, the process is referred to as *validation.* The X variable, or variable used to predict Y, is referred to simply as the *predictor.* The test is called the *predictor instrument.* The variable to be predicted from the test scores, variable Y, is the *criterion variable.* For example, the predictor variable may consist of scores on an intelligence test. The criterion variable may be academic performance in college measured by grade point average or performance on a job indicated by supervisor efficiency reports. The two common forms of validity established by use of the correlation approach are *concurrent validity* and *predictive validity.*

Concurrent Validity

Let us assume an intelligence test has been developed that we think can predict success in college. The first step in validating this experimental predictor instrument would be to administer it to a group of college students. We then would see to what extent scoring high or low on the test relates to a student's high or low grade point average. If a significant relationship between the two is found, the test has been demonstrated to have concurrent validity; it is a valid predictor of grade point average for students already in college.

Predictive Validity

While establishing the concurrent validity of our intelligence test would be very encouraging, we still would not be sure of the future performance of students not yet in college. To answer this question, the test would be administered to a group of college applicants. Then we would put the test in a drawer and wait until those selected had been in college long enough to establish a stable grade point average. Next, the test results would be analyzed to determine whether they related to the students' academic performance. If the test scores relate to grade point average, the test has predictive validity—the test results have been found predictive of future academic performance. We would then feel safe in using the test as an aid in selecting future college students.

It should be noted that we could have omitted the step of first demonstrating concurrent validity and gone directly into the predictive validation study. However, because the degree of concurrent validity possessed by our test could be determined right away, it was worth the relatively little additional effort to actually determine it. Had the test not been found to have concurrent validity, we could have saved the considerable time and effort involved in determining predictive validity. For this reason, in actual practice, concurrent validity is invariably determined prior to proceeding with a predictive validation study.

The specific statistical measures used in determining to what extent variables are related are referred to as *correlation coefficients* and will be covered in detail in Chapter 3.

Using Multiple Predictors

In the previous example, we are looking for some relationship between performance on a test and performance in college. The fact that this correlation probably would be far from perfect would indicate a certain amount of error in our prediction. It would seem that something other than the factor being measured by our test is involved in academic success. In an attempt to quantify whatever it is that our intelligence test is not measuring, we might turn to a variety of other kinds of measures to see whether they might be valid predictors of academic performance. These include interest, attitude, or skill tests, personal history factors, achievement tests, or tests designated to measure other facets of personality. If through the process of validation it was found that several of these instruments had some predictive validity, we could combine an applicant's scores on these tests with his intelligence test score and thus obtain a composite score.

We could then attempt to correlate this composite score with the grade point average criterion as a means of increasing our ability to predict academic performance. This further increase in predictive efficiency is realized by weighting the score on each predictor according to how much the predictor contributes *unique* variance to the predictor composite: determining how much one predictor measures that part of the relationship with the criterion that is not being measured by other predictor variables. As can be seen in Figure 1.1, this is a function of both the predictor's correlation with the criterion and its correlation with the other predictors.

In the figure, Test B overlaps (correlates) more with academic performance than Test A or Test C. Scores on Test B would be weighted the heaviest in deriving a composite predictor score. Test A does not overlap a great deal with academic performance. However, to the extent that it does overlap, it is measuring a part of the variance that is not being measured by Tests B and C. It would be weighted moderately heavily in our composite. Test C, while it overlaps heavily with academic performance, also overlaps (is correlated) with Test B. It is actually measuring only a small amount of whatever contributes to academic performance that has not been measured already by Test B. In fact, it adds less unique variance to our composite than Test A. Therefore, it would be weighted less than Test A in the composite score.

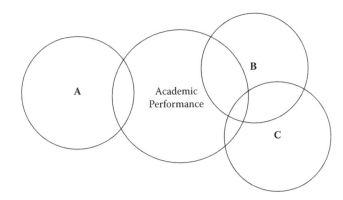

FIGURE 1.1 Prediction of academic performance with a selection of subsets (B and C are correlated, A is independent).

In more recent work on prediction, researchers not only use psychological and educational tests in their composites, but also utilize predictor indices developed as measures of exogenous variables. Examples of these variables are employment and promotion opportunities in competitive organizations, political or economic climate, and sequence of job assignments. Use of such variables has been found to increase significantly predictive efficiency in such diverse areas as personnel retention, consumer buying habits, and choices of political candidates.

Establishing Test Reliability

Thus far we have discussed validity, or the ability of a test to predict something external to itself, such as some aspect of behavior. Equally important is *internal* validity or the ability of a test to predict itself. For example, let us assume I was to take an intelligence test today. Assuming my intelligence had not changed, if I were to retake the same test in several weeks, I should get about the same score. If my test scores did not remain consistent, then they could not very well predict external criteria such as my school grade point average. For example, my high score on one administration would indicate I should obtain good grades and my low score on the second administration would predict I should obtain poor grades.

Simply stated, a test that cannot predict itself (consistently) cannot predict anything else. Thus, a test's reliability sets an upper limit on its ability to predict something external to itself, such as school grades in our example. Determining the consistency of a test is very important in behavioral research. The internal validity or consistency of a test is referred to as reliability. The reliability of a test can be established by any one of three basic methods: (1) test–retest, (2) alternate forms, and (3) split half.

Test–Retest Method

In the test–retest method, a test is administered to the same group of subjects twice. To minimize carry-over learning effects, the two administrations are usually separated by at least several weeks. The results of the first administration are then correlated with the results from the second administration to see whether the subjects tended to maintain their same relative positions within the group. If they did, the test is said to be reliable. Note that for a test to be reliable, the subjects do not have to get the same scores on both administrations; they merely need to maintain their same relative positions within a group.

Alternate Forms Method

Because of the possible problem of subjects learning the test, which might affect reliability when using the test–retest method, test constructors often will develop two presumably equivalent forms of a given test. In establishing the test's reliability, the subjects will take both forms of the test. If each subject scores about the same on both forms relative to the rest of the group, the test is considered reliable.

Split Half Method

A variation of the alternate forms method is to simply split the test in half, and compare each subject's score on one half with his or her score on the other. Thus, each half of the test is treated like an alternate form. The most common method is to score every odd-numbered test item as one half and every even-numbered item as the alternate form. When test reliability is established using the split half method, the test is said to have internal consistency.

Developing Homogeneous Subgroups

Another major use of the correlation approach is in the subgrouping of test items or people into homogeneous subsets. The basic approach is to cluster together the items or people found to correlate highly and not cluster items or people that do not correlate highly.

In the case of tests, subgrouping is sometimes done to determine whether the tests measure the same or different things. Referring to Figure 1.1, Tests B and C were found highly correlated while

Test A was unrelated to both B and C. Apparently Tests B and C measured much the same thing. Knowing this, we might want to use only one of these instruments and, along with A, use our testing time for an instrument that measures another aspect of academic performance.

The general purpose of clustering items found to be related is to identify the number and nature of underlying variables or factors. Those items that correlate highly are assumed to measure the same thing. By clustering items, researchers not only identify how many different factors they appear to be measuring, but by rationally analyzing the items in a cluster, they often can make a judgment as to what the factors are. Further, they can add together scores on the items comprising a factor to obtain a quantitative index of the amount of that factor attributable to an individual.

When the correlation approach is used to cluster items into homogeneous factors, the technique is referred to as *factor analysis*. A good example of this technique is found in the analysis of items comprising a general intelligence test. By factor analyzing the test results for a large sample of subjects, the number of specific factors or dimensions of intelligence measured can be determined and labeled. A person's score on each dimension can then be computed and utilized along with his overall intelligence score. Thus, a researcher may find that a hypothetical general intelligence test appears to measure verbal aptitude and mechanical aptitude. From a utilitarian standpoint, knowing how a person scored on each of these factors could be useful in vocational guidance or personnel selection, because some occupations appear to require more of one aptitude. From the standpoint of increasing knowledge, factor analysis is relatively simple; however, the statistical and judgmental techniques involved are sophisticated. For this reason, the study of factor analysis is normally reserved for courses at the advanced graduate level.

Clustering people into homogeneous subgroups will be discussed under the case history approach below.

CASE HISTORY APPROACH

Using the case history approach, behavioral scientists attempt to increase knowledge through intensive in-depth studies of individuals. In the process of collecting data on their subjects, they may look at several available informational sources. These may include scores on aptitude, achievement, personality, or interest tests; responses on a life history, medical, or other personal data questionnaire; written works of the individuals, records, interviews with or letters of evaluation from other people; or interviews with the subjects. Through this kind of in-depth approach, several kinds of objectives can be realized.

Solving Personal Problems

A researcher may be able to help individuals solve their personal problems. The problems may involve educational or vocational choice or be emotional in nature. The required information is gathered to formulate hypotheses about the individuals' problems and their possible causes. This in turn may lead to the collection of additional data in the process of explaining the possible causes and help individuals resolve their difficulties and adjust to their environment. Because clinical psychologists, psychiatrists, and social workers among others often use this approach in their everyday work, it often is referred to as the *clinical approach*.

Predicting and Subgrouping

The case history approach also can be used in conjunction with the correlation approach—to aid in prediction or in the development of homogeneous subgroups. In prediction, biographical factors have been found to be highly useful predictors of job success and job retention. For example, the answer to the question "how many social and fraternal organizations or clubs do you belong to?" may be related to success in a particular sales job. Rationally, this would seem possible because the more organizations to which a salesperson belongs, the more potential customers the salesperson is likely to contact on a social basis. In addition, it could be an indication that the individual is fairly outgoing and gregarious. However, many ideas that look good rationally do not hold empirically, and some that do not appear to be rationally related to a performance criterion are. Therefore we require empirical study to find out which factors are useful predictors.

By identifying and grouping people of similar backgrounds, interests, etc., we are likely to find that they will tend to behave similarly in a variety of areas such as political affiliations, values, socioeconomic aspirations, and attitudes toward a wide variety of topics. Information of this nature is extensively used in public opinion, consumer, economic, and sociological research. One of the major uses of these kinds of data in the above types of research is for drawing representative subject samples using stratified random sampling discussed earlier.

Another major use of homogeneous subgrouping of people is in experimental and correlation studies. Selecting subjects from a particular homogeneous subgroup for the experimental and control groups helps ensure that the two groups will be equivalent. In another type of study, the subjects in one group may be selected from a different homogeneous subgroup from the subjects in another group. In this case, the difference in subgroup serves as one variable and any differences of the other variable can be used to further our knowledge about ways in which the homogeneous subgroups differ. For example, let us assume that in a study the two groups were found to differ in their reactions to induced stress. That is, a difference in subgroups was found to correlate with differences in stress reaction. This information would further our knowledge about the two subgroups by revealing how they behave differently from one another under a particular kind of stressful situation.

FIELD STUDY APPROACH

The field study approach is also known as systematic or naturalistic observation and real-life research. All of these terms together provide a good description of this approach. Field study involves going out into the *field* to *systematically observe* events as they occur *naturally* in *real life*. The primary concern of a researcher using this approach is *realism*. By studying events in their natural environment, the behavioral scientist avoids the sterility and artificiality of the laboratory. As we already noted, the result an experimenter causes to occur in a laboratory may seldom if ever happen in real life or may do so under different conditions of motivation and expectancy.

Often, no attempt is made to control potentially causal variables in a field situation. Researchers may be trying merely to determine what variables appear related to the dependent variables of interest to them. This frequently is done as a means of identifying variables for further investigation in the laboratory. In addition, researchers may use the field study approach as a means of verifying whether a causal relationship identified in a laboratory occurs in real-life situations.

Sometimes experimenters are interested in using the field study approach as a means of trying to determine causal relationships. In this situation, to establish probable causality, they must control variables that may exert potential influence on their dependent variable of interest, but they do so as the control occurs naturally. For example, let us assume that Variables A, B, C, and D exist in a given situation and we wish to determine which one is primarily responsible for changes in some dependent variable such as worker performance. The basic approach would be to control for one variable at a time by observing worker performance under a variety of naturalistic situations. Thus, we would have to wait for situations in which different values of A, B, and C were present and Variable D was not present or did not change. Similarly, we would also have to find situations where the only variables changing were A, B, and D; A, C, and D; and B, C, and D. This procedure would allow us to establish eventually which variable appeared to be the primary causal one. We might find, for example, that in the above described situations where Variables A, C, or D were absent, we noted a change in the dependent measure, but when Variable B was constant or absent, no performance change occurred with changes in A, C, or D. Thus, we would be able to conclude that Variable B was the probable causal variable.

Although the aforementioned approach for determining causal relationships has the distinct and sometimes essential advantage of complete realism, it also is extremely expensive and time consuming when compared with laboratory experimentation.

An excellent example of this approach was the investigation to determine factors that distinguish high producing from low producing business and industrial units conducted by the Survey Research Center of the University of Michigan (Kahn & Katz, 1963). The center conducted field study investigations of a wide variety of organizations including an insurance company, automotive manufacturer, tractor company, electric utility, and maintenance-of-way section gangs on a railroad. Through this very extensive investigation over a period of years, four factors that appeared to distinguish high producing from low producing worker units were isolated. Three were found to be characteristics of the immediate supervisor. The supervisors of high producing units (1) tended to clearly distinguish their jobs as supervisors from the jobs of subordinates, (2) were employee oriented rather than production oriented (they treated subordinates as people rather than as just units of production), and (3) did not overwhelm with supervision (they told their employees what to do, not how to do it step by step). The fourth factor was a general feeling among the high producing workers that their unit was better than others in the organization.

Although a primary reason for using the field study approach is to obtain greater realism, this is not always achieved. The intrusion of researchers into a field situation can be a significant additional variable that grossly distorts the subjects' behavior. In addition, as discussed below in the section on experimental approach, just knowing that they are being studied can introduce significant changes to subjects' motivations. For these reasons, considerable thought must be given to planning and conducting field observations.

EXPERIMENTAL APPROACH

The experimental method is the scientist's primary approach for determining causal relationships. As noted earlier in the chapter, the basic approach of experimentation is to manipulate some independent variable while controlling or otherwise discounting the effects of other variables that may affect the dependent variable. Thus, when changes occur in the dependent variable, the cause of the change can more clearly and efficiently be attributed to the independent variable than with the other research approaches. For example, in discussion of the field study approach, we noted that usually one variable at a time is controlled. Determination of the causal variable or variables therefore is laborious— and is rarely as clear as when using the experimental approach. However, in some fields such as astronomy, we have no choice but to rely on the field study approach for determining causation.

Advantages of Experimental Approach

A major distinction between the experimental and the other approaches is manipulation and control of the variables. Only in experimentation can a scientist deliberately cause the independent variable to change and the extraneous variables not to interfere. Thus, as Woodworth and Schloshberg (1954, p. 2) noted:

a. The experimenter can be fully prepared for accurate observation. He can repeat his observation under the same conditions and enable other experimenters to duplicate them and make an independent check of his results.
b. He can vary the conditions systemically and note the variation in results.

Disadvantages of Experimental Approach

From the above discussion, it may seem that experimentation is the panacea of all our ills in research. This, as was alluded to in discussing the other approaches, is not the case.

Lack of Realism

Without doubt, the primary criticism of the experimental approach is lack of realism. In real-life situations, nothing occurs in isolation. Bringing an issue into the sterile environment of a laboratory may therefore change its nature. In reality, the validity of this criticism depends to some extent on the nature of the problem. For example, a study to distinguish what shapes of knobs can be distinguished from one another by feel alone (for subsequent use in combination in an aircraft cockpit) can easily be performed in a laboratory. On the other hand, factors contributing to effective leadership in a production plant would have to be studied in a laboratory with considerable caution. The purity of the laboratory situation would probably distort how people really behave in the complex social environment of a job.

Motivational Differences

Closely related to the problem of realism is the question of motivation. The strong motivation often manifested by laboratory subjects is not necessarily characteristic of their day-in and day-out motivations in real-life situations. Furthermore, in some kinds of studies, the reverse may be true. For example, the real-life situation may generate much more anxiety than the laboratory setting. Simulated versus real-life studies of pilot behavior in aircraft emergency situations is one example. In addition to differences in motivational level, laboratory subjects may be influenced by different motives than those that influence them in real life. Thus, college sophomores may be motivated primarily by the need to participate in an experiment as a requirement for completion of a basic psychology course. Intrinsic task motivation may be of only secondary, if any, importance.

The need to consider motivational difference between a research study and real life also is applicable to field studies. The classic studies carried out at the Hawthorne Plant of the Western Electric Company some years ago vividly demonstrated this. In an investigation of the effect of illumination on production performance, the workers under observation increased their production regardless of whether illumination level was increased or decreased. Through further investigation, the researchers concluded that being singled out for special attention had influenced the workers' behavior. This motivational factor had apparently overshadowed any effect illumination level might have had on their performance. Today, when it is believed that this kind of motivation factor may have influenced the outcome of a research study, the result may be called the Hawthorne effect.

Purposes of Experimentation

In the course of attempting to establish cause–effect relationships, experiments can be performed for a variety of subpurposes. These include (a) exploratory studies, (b) confirmatory experiments, and (c) pilot studies.

Exploratory Study

Sometimes experimenters, in investigating a new area, do not know enough about the variables they are working with to propose a reasonable hypothesis regarding the nature of the variables' relationship. In this case, they may carry out a series of studies in which they simply "play around" with the variables to see what happens. This kind of experimentation generally is referred to as an exploratory study. By manipulating variables and noting the nature of changes in the dependent variable of interest, scientists eventually learn what are likely to be causal variables and how they seem to operate.

One purpose of exploratory studies is to avoid what de Bono (1983) describes as "vertical thinking." It often appears easier, once you start digging a hole, to continue digging it deeper. But if the hole is in the wrong place, no amount of digging, regardless of quality, is going to put it in the right place. Put simply, it pays to keep an open mind during exploratory studies!

Confirmatory Experiment

Based on the results of the exploratory study, experimenters can intelligently propose an experimental hypothesis and conduct an experiment for the purpose of confirming or refuting the hypothesis. This is the kind of study we most often think of in conjunction with the experimental method and it generally is referred to as a confirmatory experiment.

Pilot Study

Prior to conducting a confirmatory experiment, experimenters often conduct an abbreviated study. This is done to ensure that the range of independent variable values is correct in terms of the range of the subject's sensory capabilities, to smooth out experimental procedures, or determine whether the test directions can be understood and convey the intended meaning to the subject population. Studies done for the purpose of serving as a dress rehearsal or dry run of a full-scale confirmatory experiment are referred to as pilot studies.

Experimentation Versus Demonstration

An experiment requires at least two values or levels of the independent variable. That is, the independent variable must change so that we can observe whether the dependent variable changes. In short, we have to be able to provide a *process* description of what is observed (i.e., we have to describe the change). For example, as we increase Variable X, Variable Y also increases. A change process description is essential for determining causation.

A *demonstration* on the other hand, does not require a change in Variable X. Only a single treatment condition is present or applied, and the value of Variable Y merely is noted and we need only provide a state description of what was observed. That is, we describe a fixed or static condition of the two variables.

Manipulation Versus Selection of Independent Values

In doing research, the values of the independent variable may be controlled in two ways. First, experimenters can deliberately manipulate the amount. For example, they could deliberately change the amount of a drug administered to note the effect of drug dosage on the subjects. Sometimes, however, deliberately manipulating the independent variable is impractical. For example, it would be impractical to try to change the intelligence of a group of subjects as a means of determining the effect of intelligence on academic performance in college. Instead, researchers select subjects already having high and low levels of intelligence and merely note differences between grade point averages of the two groups.

When the independent variable is deliberately manipulated, we can provide a process description of the events and thereby determine whether Y appears to change as X changes. In short, we can determine causation. However, when we select subjects already possessing a given level of Variable X, this is not the case. We are not actually changing the value of variable X and noting what happens when we change it. Hence, all we can safely establish is that a relationship exists between X and Y—not that X is causing Y. Remember: a correlation between variables does not imply causation. Only when we actually change the value of Variable X are we conducting a true experiment.

Quasi-Experimental Approach

Many behavioral studies combine features of both the field and experimental approaches. These studies are not true experiments because researchers do not exercise complete control over all the important variables; and they are not pure field studies because some aspect of the field situation is manipulated by the researchers. Thus, these studies are quasi-experimental in nature. The advantage is that the researchers obtain the benefits of both worlds—the greater realism of the field or natural environmental situation and the greater efficiency of deliberate manipulation of one or more selected IVs characteristic of the experimental approach.

Quasi-experimental designs can help us gain insight into causal relationships more efficiently than is possible with pure field study. Quasi-experiments also sometimes are referred to as *field experiments*. Because of the addition of the experimental approach, we can reduce the possibility of extraneous variable contamination. The two most common general types of quasi-experimental designs are the *time-series* and *nonequivalent control group* studies.

Time Series Design

In a time series study, a change in some IV is introduced into the field environment. At several successive points in time, both before and after the IV change, the DV value is measured. This procedure enables us to determine whether a DV change is merely a continuation of an existing trend and thus not a result of the IV change, or is a true result of the IV manipulation. It also enables us to determine whether the DV change is of a lasting or only temporary nature. For example, suppose it is hypothesized that showing violence on television contributes to violent crime in a given geographical area.

Let us assume that a state decides to test this hypothesis and outlaws the presentation of violence on television (IV change). The number of violent crimes in the state (DV) may be noted for each of the five years preceding and the five years following enactment of the law (time series). If the crime rate were to drop after enactment of the law, we could determine whether (a) the drop represented part of a trend toward less violent crime that already had been taking place, (b) it indicated a lasting change over the five-year period following the elimination of violence on television, or (c) the drop was a temporary change followed by a return to the previous crime rate.

Assuming that any potentially contaminating variables present did not change during the study period and no new ones were introduced, if (b) were true we could tentatively conclude that eliminating violence on television contributes to a lowering of the crime rate; if (c) were the outcome, it too might indicate that our IV change had some effect although only temporary in nature; in the event the outcome was (a) or if DV did not change, we would probably conclude that eliminating violence on television had no effect on the state's violent crime rate.

Nonequivalent Control Group Design

Let us assume a college teacher wishes to study the effectiveness of two different teaching methods on class performance. One of the teacher's two class sections attends in daytime and consists primarily of younger, full-time students; the other section is an evening class attended primarily by older, part-time students who work during the day. To carry out the study, the teacher has to use one teaching method for the day group and another for the evening class. Thus it is not possible to randomly assign students to each treatment condition to ensure equivalence of the two groups. Instead, the teacher must use intact groups that are not equivalent before the treatment is introduced.

To partially overcome this problem, the teacher will use both pre-tests and post-tests (*test–retest* method) and evaluate whether the pre-test to post-test change was greater for one class section than for the other. To the extent that one class showed a greater change, the teacher might conclude that the teaching method used for that section was the more effective of the two. However, it would be important for the teacher to first consider whether rival hypotheses might also account for the results before drawing this conclusion. For example, if the group that showed less change scored much higher than the other group on the pre-test, there would be less room for pre-test to post-test improvement on that group's part. That would explain why that group showed less improvement in comparison with the other; this is known as a *ceiling effect*. Field situations in which researchers are restricted to using intact groups are common in behavioral studies. As a result, this type of quasi-experimental design often is utilized, particularly in applied behavioral research.

SUMMARY

1. *Science* is the study of meaningful problems through use of the scientific method. Meaningful problems are those for which solutions can be gained through empirical observations.

2. The *scientific method* is a systematic process of investigation involving a series of seven sequential steps: (a) identify the problem, (b) formulate a hypothesis, (c) conduct pilot study, (d) collect data, (e) test the hypothesis, (f) generalize, and (g) replicate.

3. To increase our understanding of ourselves and our world, science has several basic goals. The empirical goal is description of what we observe; the theoretical goal is explanation of what we describe.

4. Implicit in the scientific method are a series of five principles or rules that scientists follow in pursuit of the goals of description and explanation. The principles of description are empirical verification, operational definition, controlled observation, and statistical generalization. The principle of explanation is empirical confirmation.

5. Science also involves certain assumptions about nature. These include determinism, limited causality, and contiguity of events in time and space. However, because we do not always see the chaining of events between the phenomena we actually observe, contiguity is not always apparent.

6. There are at least five basic approaches to scientific research. These are correlation, case history, field study, experimental, and quasi-experimental or field experiment.

   - The *correlation approach* is used to establish relationships for such purposes as prediction and subgrouping of homogeneous elements. It does not, however, enable us to determine whether the relationships between variables are causal.

   - The *case history approach* is used to assist individuals to understand and solve their personal problems. It can also produce insight into possible causal relationships that can be further investigated through experimentation. When combined with the correlation approach, it can be used to cluster people into homogeneous subgroups.

   - The *field study approach* involves observing events as they actually occur in their natural environment. It has the advantages of realism and reducing the number of possible causes of real-life events. Its major disadvantage is that it is very inefficient and expensive for determining causal relationships.

   - The *experimental approach* is the primary method for determining causal relationships. It has the advantage of controlling the variables and inducing stimulus conditions at will. This enables the experimenter to efficiently and clearly determine the effects of independent variables and to replicate experiments to determine validity. The major disadvantages are the possible loss of realism and the possibility of motivational differences between the laboratory and real-life situations.

   - *Quasi-experimental approach.* Many behavioral studies combine the features of both the field and experimental approaches. These are referred to as quasi-experiments or sometimes as field experiments. They have the advantages of the greater realism of field study and the greater efficiency of deliberate IV manipulation characteristic of the experimental approach. The two most common types are the time series and nonequivalent control group designs.

7. *Exploratory experiments* are conducted to learn more about relationships of variables in order to formulate hypotheses.

8. *Confirmatory experiments* are conducted to confirm or refute experimental hypotheses.

9. *Pilot studies* are dress rehearsals of confirmatory studies for the purposes of refining experimental procedures and predicting final results.

10. *Demonstrations* differ from experiments in that they do not require manipulating variables. They lead to static rather than process descriptions of observations.

11. An independent variable (IV) can be controlled in two ways: by actually manipulating or changing its value in the experiment or by selecting subjects already having high and low independent variable values. Only when the value of an IV is changed in the course of a study is the effort a true experiment.

KEYWORD DEFINITIONS

Case History Approach: Assisting an individual to understand and solve his or her personal problems. It can also produce insight into possible causal relationships that may be investigated further through experimentation.

Caterus Parabus: State, condition, or environment in which all variables are controlled except the one manipulated by the investigator (i.e., everything is the same or equal).

Concurrent Validity: If a test is demonstrated to have a significant relationship with the previous results of a different test, it is said to have demonstrated concurrent validity.

Confirmatory Experiment: Experiment conducted to confirm or refute experimental hypotheses.

Contiguity of Events: Assumption that observed events are temporally or spatially contiguous (adjacent).

Control Group: One of two subject groups in an experiment is called the control group and its members usually receive the normal or standard amount of the treatment variable.

Controlled Observation: To control or otherwise discount the effects of all other variables that may affect Y in order to make the descriptive statement that "a change in variable X produces a change in variable Y."

Correlational Approach: To establish relationships for such purposes as prediction and subgrouping of homogeneous elements.

Dependent Measure (DM): The same as dependent variable (DV), but the word *measure* may make this concept easier to learn, recall, and understand because the dependent variable is most often the measurement of interest, performance measured, or variable demonstrating a change caused by the independent variable or treatment.

Dependent Variable (DV): The variable being measured whose value or amount appears to be dependent on the value or amount of the independent variable.

Determinism: The philosophy that every event has a cause.

Empirical: Based on results of direct observation.

Empirical Confirmation: Explains the descriptive statement from which it was derived while all other statements that can be derived from it are also found to be true.

Empirical Verification: A descriptive statement can be regarded as true only if it is found to correspond with reality.

Experimental Approach: Primary method for determining causal relationships. It can control the variables and induce stimulus conditions at will, enabling a researcher to efficiently and clearly determine the effects of independent variables and replicate experiments to determine validity.

Experimental Group: Non-control group that receives the experimental level of the treatment variable.

Exploratory Experiment: Experiment conducted to learn more about relationships of variables to formulate hypotheses. Confirmatory experiments are conducted to confirm or refute experimental hypotheses.

External Validity: Ability of a test to predict something external to itself.

Field Study Approach: Observation of events as they occur in their natural environment. The advantages of the approach are realism and reducing the number of possible causes of real-life events.

Hawthorne Effect: A motivational factor that apparently overshadowed any effect of the IV on performance, thus influencing the outcome of a research study.

Independent Variable (IV): Experimental variable that is deliberately changed or made different for the distinct subject groups.

Internal Consistency: Establishing test reliability (consistency or conformity) using the split half method.

Internal Validity: Ability of a test to predict itself.

Laws of Science: Experimental hypotheses that have been confirmed so many times that they are generally taken to be true.

Limited Causality: Assumption (by science) that the number of events that cause an occurrence of change is, for the most part, limited to relatively few variables.

Meaningful Problem: Problem that can be solved through direct observation.

Nonequivalent Control Group: Common type of quasi-experimental design.

Operational Definition: Definition of all terms in a descriptive statement relative to the operations involved in manipulating or observing them (i.e., how they are to be measured).

Pilot Study: Study performed to serve as a dress rehearsal or dry run of a full-scale confirmatory experiment.

Predictive Validity: A test demonstrates a significant relationship with the future results of a different test such as a performance measure.

Predictor: Independent measure that can be used to predict performance of dependent measures.

Quasi-Experimental Study: Behavioral study combining features of both field and experimental approaches. It exhibits the greater realism of a field study and better efficiency of deliberate IV manipulation characteristic of the experimental approach.

Random Sampling: Selection method that ensures that each subject in a population has an equal and constant chance of selection.

Realism: Going into the field to systematically observe events as they occur naturally in real life.

Reliability: The consistency of a test; its ability to accurately measure the same thing each time it is conducted or the same thing for each subject who takes it.

Sampling With Replacement: In selecting a representative sample, to keep the chances of selection from a given population constant, each selected subject is returned to the population before the next random selection is made.

Science: Study of meaningful problems via scientific method.

Scientific Method: Systematic process of investigation involving sequential steps followed by a researcher to identify and answer a problem.

Statistical Generalization: Results shown to hold for an adequate sample of a given population before attempting to generalize the results of an experiment to that population.

Stratified Random Sampling: After identifying variables that may affect the outcome of a study and stratifying (classifying) the population with respect to these variables, subjects for the sample are randomly drawn from these classes in the same proportion that they occur in the population.

Validity: The ability of a test to accurately measure what it was expected to measure.

EXERCISES

1. What is science?
2. What are the seven components of the scientific method?
3. What are the primary goals of science?
4. What are the five principles of science?
5. What are three assumptions of science?
6. What are the five basic approaches to scientific research?
7. In what ways do experiments differ from demonstration?
8. What is a meaningful problem?

9. What are the primary advantages of the quasi-experimental approach?
10. What is the difference between validity and reliability?
11. List three methods by which the reliability of a test can be established.
12. Differentiate experimental studies from correlation studies.

EXERCISE ANSWERS

1. Science is the study of meaningful problems through the use of the scientific method.
2. The seven components of the scientific method are identify the problem, formulate a hypothesis, conduct a pilot study, collect data, test the hypothesis, generalize, and replicate.
3. Science has two primary goals. The first is description—the empirical goal—we describe what we observe. The second is explanation—the theoretical goal—we explain what we describe.
4. The five principles of science are empirical verification, operational definition, controlled observation, statistical generalization, and empirical conformation.
5. The three assumptions of science are determinism, limited causality, and contiguity.
6. The five basic approaches to scientific research are the correlational approach, case history approach, field study approach, experimental approach, and quasi-experimental approach.
7. Experiments differ from demonstrations in that they require the manipulation of variables. Experiments lead to process descriptions of what was observed.
8. A meaningful problem is one whose solution can be gained through empirical observations.
9. The primary advantage of the quasi-experimental approach is its ability to combine features of both field study and experimental approach. It combines the greater realism of the field or real world with the greater efficiency of deliberate manipulation of the independent variable.
10. Validity is the ability of a test to predict something external to itself: its ability to accurately measure what it is expected to measure. Reliability is the consistency of a test: its ability to accurately measure the same thing each time it is performed or the same thing for each subject who takes it.
11. Three ways that reliability of a test can be established are the test–retest method, alternate forms method, and the split half method.
12. Experimental studies allow the determination of the effect of one variable on the change of another variable. Correlational studies determine only relationships between variables, not necessarily causal relationships.

REFERENCES

Anderson, B. F. (1971). *The psychology experiment* (2nd ed.). Belmont, CA: Brooks/Cole.
Brain, W. R. (1965). Science and antiscience. *Science, 148*, 192–198.
Brooks, H. (1971). Can science survive in the modern age? *Science, 174,* 21–30.
de Bono, E. (1983, Nov.). The use of lateral thinking. *Pennsylvania State University Graduate Bulletin.* University Park, PA: Graduate Studies Printing.
Dubos, R. (1961). Scientist and public. *Science, 133*, 1207–1211.
DuBridge, L. A. (1969). Science serves society. *Science, 164*, 1137–1140.
Forscher, B. (1963). Chaos in the brickyard. *Science, 142*, 359.
Gliner, J. A., Morgan, G. A., & Leech, N. L. (2009). *Research methods in applied settings: An integrated approach to design and analysis* (2nd ed.). New York: Routledge/Taylor & Francis Group.
Jackson, S. L. (2006). *Research methods and statistics: A critical thinking approach* (2nd ed.). Belmont, CA: Thomson Wadworth.

Johnston, J. M., & Pennypacker, H. S. (2009). *Strategies and tactics of behavioral research* (3rd ed.). New York: Routledge/Taylor & Francis Group.

Kahn, R. L., & Katz, D. (1963). Leadership practices in relation to productivity and morale. In R. A. Suttermeiter (Ed.), *People and productivity*. New York: McGraw-Hill.

Kemeny, J. (1959). *A philosopher looks at science*. Princeton, NJ: Van Nostrand.

Kerlinger, F. N. (1973). *Behavioral research: A conceptual approach*. New York: Holt, Rinehart & Winston.

Kerlinger, F. N. (1986). *Foundations of behavioral research* (3rd ed.). New York: Holt, Rinehart & Winston.

Krathwohl, D. R. (1985). *Social and behavioral science research: A new framework for conceptualizing, implementing, and evaluating research studies*. San Francisco: Jossey-Bass.

Kuhn, T. (1962). *The structure of scientific revolutions*. Chicago: University of Chicago Press.

McGuigan, F. J. (1960). *Experimental psychology*. Englewood Cliffs, NJ: Prentice-Hall.

McPherson, B. D. (1978). *Avoiding chaos in the sociology of sport brickyard*. Quest Monograph No. 30, 72–79.

Popper, K. (1959). *The logic of scientific discovery*. New York: Harper & Row.

Reynolds, P. (1971). *A primer in theory construction*. Indianapolis: Bobbs-Merrill.

Seltzer, M. (1975). The quality of research is strained. *The Gerontologist, 15,* 503–506.

Simpson, G. G. (1963). Biology and the nature of science. *Science, 139*, 81–88.

Thomson, G. (1960). The two aspects of science. *Science, 132*, 996–1000.

Underwood, B. J. (1966). *Experimental psychology*. New York: Appleton-Century-Crofts.

Woodworth, R. S., & Schloshberg, J. (1954). *Experimental psychology*. New York: Holt, Rinehart & Winston.

2 Methods of Describing Data

Following data collection comes the phase of research known as data description. During this phase, data are examined or explored and then reduced or simplified to communicate the characteristics observed. This chapter begins with a distinction of samples and populations and then goes into graphic illustrations and finally into descriptive statistics of an individual variable (univariate) and two variables examined together (bivariate). The methods introduced are relatively standard and/or common for most academic disciplines. The topics covered are

Samples and Populations
Consideration of Numbers in Statistics
Graphical Methods of Description
Univariate Descriptive Statistics
Summary
Keyword Definitions
References
Exercises
Exercise Answers

KEYWORDS

Abscissa
Asymptotic
Average Deviation
Average Squared Deviation
Bar Chart
Centile (Percentile)
Central Tendency
Continuous
Descriptive Statistic
Deviation
Discrete
Frequency
Frequency Distribution
Frequency Polygon
Graph
Histogram

Interval Scale
Kurtosis
Mean
Median
Mode
Moments of a Distribution
Nominal Scale
Normal Curve
Ordinal Scale
Ordinate
Parameter
Pie Diagram (Pie Chart)
Population
Quartile Deviation
Range
Ratio Scale

Root Mean Square Deviation
Sample
Semi-Interquartile Range (Q)
Skewness
Standard Deviation
Standard Score
Statistic
Sum of Squares
Univariate
Variability
Variance
Weighted Mean
x-Axis
y-Axis
z-Score

In experimentation, the data collected consist of measurements of one kind or another. These measurements often are numerous and difficult to interpret. Therefore, the researcher's first problem is usually to reduce the data to some manageable and meaningful volume, to reduce many numbers to a few numbers, to translate from complexity to simplicity. Graphs and statistics are the most efficient and effective techniques for this purpose. People often think of graphs and statistics as making things more complicated. Nothing could be further from the truth. *The primary purpose of graphs and statistics is to simplify* so that we may better understand and coherently represent the data.

This chapter presents (1) samples and populations, (2) the consideration of numbers in statistics, (3) graphical methods of description, and (4) univariate descriptive statistics.

SAMPLES AND POPULATIONS

To clarify what a statistic is, we must understand the distinction between a sample and a population. This perhaps is best done with an illustration. Let us assume that we are interested in knowing the height of high school seniors in the United States. We have measured the height of the seniors from a single school and computed their average height as the best estimate. Further, we have measured the height of the seniors from a single school and computed their average height as the best estimate of all senior classes until all senior classes in the United States can be measured.

If we computed a grand mean (average height of all high school seniors in the United States) of an infinite number of single school samples, we would have the best estimate possible of all high school seniors in the United States. The average of all our sample averages is used to tell us something about the population value. All high school seniors make up the population from which our various samples were drawn. A value or property that describes the population or a corresponding population value is referred to as a *parameter*. Thus, a parameter is an index or description of the *population distribution* for a given characteristic. In our example, the population's mean height is a parameter.

Technically, a parameter may have two meanings: (1) a statistical context as just mentioned and (2) as any variable that influences an empirical relationship (experimental context). Thus, the schedule of reinforcement during the shaping of a behavior is a parameter of the learning paradigm (e.g., fixed concurrent vs. variable interval intermittent reinforcement schedule). Note that parametric research is designed to establish a functional relationship between one or more quantitative independent variables (IVs) and a dependent measure (DM) by testing at several levels of each IV. Testing multiple levels of a treatment is necessary to determine the "parameters" of the prediction equation relating the variables. *Parametric tests* are rare significance tests in which the null hypothesis is a statement about the parameter(s) of the population(s) from which the samples were drawn. Typically, these tests do not include distribution-free contrasts and are based on the assumption that sampling was from binomially and normally distributed populations.

In contrast, a *statistic* is an index or description of a *sample distribution* such as the mean height for one of our samples in the above example. Statistics are used to estimate parameters. As may be noted intuitively, the larger the size of the sample, the better estimate a statistic value is likely to be in representing the corresponding parameter value. For instance, the mean height of a sample from 75% of the senior classes is likely to be a more accurate estimate of the mean height of all seniors in the United States than the mean for the seniors from a single high school. One of the major aspects of research is to make inferences about population characteristics on the basis of one or more samples studied. Sampling statistics serve as vehicles for making these inferences and for seeking generality.

In the above illustration, all high school seniors in the United States made up the population universe from which our various samples were drawn. A population, as the term is used in statistics, is an arbitrarily defined group. We could just as readily have defined our population as all senior class students in a given school district if the height of that total group had been of interest. One further note about statistical populations: they need not even exist, but they must be defined. An example of a nonexistent statistical population would be the senior class of the year 2050.

CONSIDERATION OF NUMBERS IN STATISTICS

Descriptive and inferential statistics are applied to variables (i.e., a characteristic of a person, place, or item that can have different values). Values or qualitative variables (aka categorical variables) differ in kind (i.e., type or quality) rather than in amount or volume (e.g., sex, race, ethnicity). Values of quantitative variables express differing quantities of a characteristic (e.g., age, IQ, fitness).

Continuous Versus Discrete Data

Measurement is generally defined as the assignment of numbers to objects according to rules. With this in mind, we can classify data into two types: *continuous* and *discrete*. Continuous data (e.g., feet, pounds, or minutes) may be measured or expressed in varying degrees of fineness or proportion of one unit. On the other hand, some kinds of data may be expressed only as whole units. Good examples of this discontinuous or *discrete* type of data are people, dogs, and cars because their quantities increase or decrease by single units. In statistics, however, discrete data frequently are treated as continuous. Thus, we often see such statements as "the average American family has 1.6 cars."

Four General Scales of Measurement

Four scales or levels of measurement are most commonly distinguished (based on S. S. Stevens's work at Harvard in 1940s). Ranging from lowest to highest levels of complexity, these are nominal, ordinal, interval, and ratio measurement scales. In accordance with our general definition of measurement, the rules for the way in which numbers are assigned constitute the essential criteria defining the scales. The higher level scales require more restrictive rules, that is, the greater are the number of the basic postulates of numbers that apply (see Guilford, 1954, *Psychometric Methods*, p. 11, for a listing of these postulates). Of particular relevance to us, there are differences in how much can be done in the way of statistical operations at the different levels of measurement. The higher the level of scale, the more we can do with the numbers we obtain in measurement.

Nominal Scale

With a *nominal scale*, a number is used to designate membership or assignment of items, objects, characteristics, or individuals into mutually exclusive categories. The members of a class are regarded as equal or equivalent in some respect and different from the members of other categories without magnitude assessment. For example, a population may be classified into three groups according to eye color: group 1 – brown, group 2 – blue, and group 3 – other.

Classification is the lowest form of measurement and only the simplest of statistical operations may be performed. We can (1) count the number in each class and obtain frequencies; (2) determine the most populous class or *mode* of the distribution of classes (we can use this to make categorical predictions); (3) or, if some objects are classified on the basis of two aspects or principles of classification, determine the interdependence of those two aspects by computing a coefficient of contingency, such as the phi coefficient (discussed in the next chapter).

Ordinal Scale

An *ordinal scale* is one in which measurements are placed in rank order, starting with the largest or smallest one first. When going from a nominal to an ordinal scale, one adds a *greater than* concept to the *different from* concept. As long as the item ranked is ranked along a single dimension, only one ranking is possible. However, when the ranking is based on a composite of several variables, we must adopt relative weights for the variables so that one and only one ranking is possible (a basic rule of ordinal scaling).

Most behavioral variables are actually composites of several dimensions. Accordingly, in ranking behavioral variables, we are actually forcing a multidimensional quantity onto a linear scale. Sometimes the weighting is explicit, as in the use of multiple regression equations. In the case of ratings along a psychological dimension made by human observers, however, the weighting is intuitively accomplished. We have no way of knowing whether the weights are differentially applied (unless we resort to the use of multidimensional scaling or factor analysis). It should be noted that ordinal scaling does not imply equal spacing of categories along the scale. Thus, only transitive relationships may be established using this scale. For example, using an ordinal scale, it is possible to determine that Americans rank teaching as a less prestigious occupation than television broadcasting, but it is not possible to quantify the degree of difference.

In addition to the statistics usable with nominal scales, the principle of order makes possible the use of *medians, centiles, semi-interquartile ranges,* and *Spearman rank order correlations* (discussed in the Chapter 3).

Interval Scale

As the second highest scale of measurement, the unique feature of the *interval scale* over the lower levels is the use of equal (constant) units of measurement. The commonly used Fahrenheit and Centigrade thermometers are scales of this type. Although we may speak of addition of intervals, we may not have achieved the property of additivity if the zero point for an interval scale is not absolute. A zero point may be placed in any arbitrary position (the freezing point of water, not the complete absence of heat). Accordingly, as the zero point is moved, the sum of any two numbers on the scale varies. For example, we cannot say 80°F is twice as warm as 40°F. Likewise, intelligence quotient (IQ) is measured on an interval scale so we cannot say that someone with an IQ of 150 is twice as intelligent as someone with an IQ of 75. Technically, an interval scale has an arbitrary zero point.

At this level of scaling, most common statistical operations (i.e., the *mean, standard deviation, Pearson product moment correlation*) and other statistics that depend on these values such as the *t* and the *F* (inferential statistics) that will be covered in Chapter 4 are possible.

Ratio Scale

A *ratio scale* has all the features of the interval scale plus a nonarbitrary (absolute) zero point. For instance, mass is measured on a ratio scale; therefore, 200 kg is twice as heavy as 100 kg and a 0 kg measurement describes weightlessness.

Measurements in feet, pounds, gallons, and the like are on a ratio scale. Therefore, it is possible to say one length is twice or half the size of another. When measurements are on a ratio scale, meaningful comparisons can be made. With this level of measurement, all of the mathematical and statistical manipulations may be accomplished.

Scaling Behavioral Dimensions

In actual practice, most of the measurements we make in the behavioral sciences do not exceed the interval level of scaling. A physicist may describe absolute zero on a heat scale; but what does zero IQ mean? What constitutes having twice as much intelligence as a person with an IQ of 60? Inspection of certain psychological measures such as intelligence quotients leads us to wonder whether even an assumption of interval scaling is completely valid. However, Ghiselli (1964) pointed out at least three good reasons for treating measurements of behavioral dimensions as interval scales.

Behavioral Meaning

It is often argued that psychological measures cannot be viewed as interval scales because a given score interval (say, 10 units) has different behavioral meanings at different levels on the scale. For example, in the ability to learn abstract subject matter, moving from an IQ of 105 to one of 115 may be quite different from moving from 115 to 125. However, this argument is not relevant. The Fahrenheit scale is said to be an interval scale, yet the behavior of water over a temperature increase of 10 degrees differs greatly, depending upon the location on the scale over which the change occurs (compare, for example, the range around 32°F with that around 212°F). Simply arguing that similar scale intervals have different behavioral meanings at various scale points is irrelevant to the issue. The definition of scale score meanings in terms of human behavior can be established by careful research. Hence, behavioral markers must be established for any scale—nominal, ordinal, or interval.

Averaging Out

Even if immediately adjacent scale intervals differ somewhat in length—including some units of relatively greater length, some relatively shorter, and so on—it is likely that longer intervals will "average out" so that, for all practical purposes, intervals from different parts of the scale may be treated as equal. This is probably the case in most psychological measures that include a broad sampling of test items of various types placed randomly throughout a test.

Scaling Methodology

Methods (paired comparison, confidence estimation, etc.) are available for scaling stimuli on the basis of the relative ease of discriminating distances or making judgments about test items or stimuli. These methods result in true interval scales. In actual use, however, such scales yield results showing little, if any, practical difference from results obtained with measures that may not have strictly defined interval properties.

GRAPHICAL METHODS OF DESCRIPTION

UNIVARIATE FREQUENCY DISTRIBUTION

The most useful method for describing scores on one variable is the univariate frequency distribution, more commonly and simply called the *frequency distribution*. The *univariate* term (*uni* = one, *variate* = variable) refers to the distribution of scores for a single variable. The frequency distribution (*f*) shows the number of times something occurs. We will take the following steps in setting up a frequency distribution.

Determine Range

This is defined as the highest score minus the lowest score plus one. Using the hypothetical data from Table 2.1 for 40 college freshmen, the range would be 86 − 28 + 1 or 59.

Determine Number and Size

Generally, 10 to 20 intervals are used. With fewer than 10 intervals, the groupings become too coarse for good accuracy; with more than 20, the work becomes quite laborious. Dividing the range by 15 will give you a good idea of the appropriate size for your class intervals. Most often, odd-numbered class interval sizes are used because the midpoint of each interval is a whole number. The notable exception to this rule is the use of 10 which, when appropriate, is a convenient interval size to work with. From our data in Table 2.1, 59 ÷ 15 = 3.93. Thus we can conveniently use an odd numbered interval size of either 3 or 5.

Set Up Frequency Distribution

Beginning at the bottom, let the lowest interval begin with a number that is a multiple of the class interval. For the data in Table 2.1, the low score was 28. Using 5 as the interval, the lowest interval would thus begin with 25. The frequency distribution for the data in Table 2.1 is depicted in Table 2.2.

Tally Scores

This involves making a tally mark in the appropriate interval for each score from Table 2.1, as shown in Table 2.2.

TABLE 2.1
Scores on Mathematical Aptitude Test

60	56	78	62	37	54	39	62
56	42	82	38	72	62	44	54
62	56	55	57	65	68	47	52
42	48	56	55	66	42	52	48
86	28	41	50	42	47	48	53

TABLE 2.2
Frequency Distribution for Data in Table 2.1

X	Tally Marks	f
85–89	I	1
80–84	I	1
75–79	I	1
70–74	I	1
65–69	III	3
60–64	IIIII	5
55–59	IIIII II	7
50–54	IIIII I	6
45–49	IIIII	5
40–44	IIIII I	6
35–39	III	3
30–34		0
25–29	I	1
		$\Sigma f = 40$

Post Tallies

Count the number of tallies in each interval and enter this sum in a column labeled *f* (frequency).

Add *f* Column

This sum of the frequencies should agree with your total number of scores. Note from Table 2.2 that the Σf for our data was 40 and that the same number of scores appears in Table 2.1.

GRAPHING RESULTS

Frequency Polygon

Having determined the frequency distribution, the next step is to graph the results. Of the various graphic devices, the one most commonly used for this purpose is the *frequency polygon*. The popularity of this graphic device probably results from the fact that it is extremely easy to construct and interpret. In all work with graphs, two axes are used. The vertical one is called the y-axis, and the values taken along this axis are called *ordinates*. The other axis, the x-axis, is called the *abscissa*. These two axes cross at right angles at a point called the *origin* (0,0). The frequency values are always placed on the y-axis.

The scores or values on the variable of interest are along the x-axis. In plotting frequencies, the values on the variable of interest are plotted along the x-axis. In plotting frequencies, the values are placed above the corresponding midpoints of the class intervals. Once plotted, these points are connected with straight lines. Rather than leave the graph suspended in space, assume there is another interval above and below your range and each has a frequency of 0. Draw connecting lines accordingly. A frequency polygon for our data is depicted in Figure 2.1.

After constructing a frequency polygon for a relatively small group, you may note a ragged shape to the resulting graph. It is natural to wonder whether the picture might have been different had a larger sample been used. Many of these irregularities are chance fluctuations. As the number of cases is increased in any sample, these chance fluctuations tend to become smoother. For small samples, the *method of running averages* for smoothing curves may be used to compensate for the small frequencies. This involves adding two more intervals to the table, one at the top and one at the bottom, each with a frequency of zero. Then, for each interval in the original table, we add the

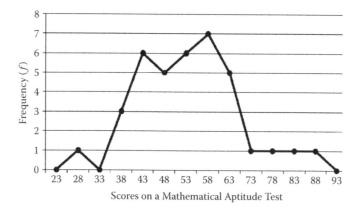

FIGURE 2.1 Frequency polygon for data in Table 2.2.

frequency for that interval to the frequency for the adjacent interval on each side and divide by 3 to get the smoothed frequency.

This process is repeated for each interval until we have the smoothed frequencies for the entire table. These smoothed frequencies are then plotted. This time, however, we should connect them by a freehand curve rather than by straight line. A smoothed frequency polygon for our data is depicted in Figure 2.2. This smoothed curve may be used as an estimate of the distribution had there been more cases.

Histogram

In addition to the frequency polygon another type of graph, the *histogram*, may be used to depict a frequency distribution. The histogram is identical in its construction to the frequency polygon up to the point of plotting the frequencies. The frequencies in the histogram are represented by vertical bars, each of which extends across the class interval. The histogram is less frequently used than the frequency polygon because it is more time consuming to construct and only one frequency distribution can clearly be placed at any set of axes. A histogram for our data is depicted in Figure 2.3.

In deciding whether to use a frequency polygon or a histogram to depict your data, another factor to consider is whether you want to illustrate categorical or continuous data. A frequency polygon

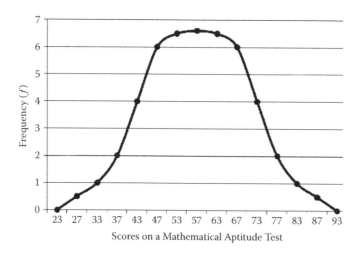

FIGURE 2.2 Smoothed frequency polygon for data in Table 2.2.

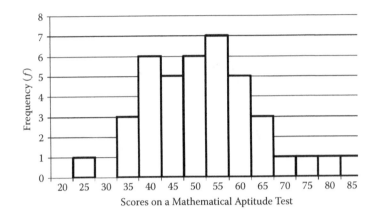

FIGURE 2.3   Histogram for data in Table 2.2.

implies continuity and should not be used for categorical data. A histogram may be used with either continuous or categorical data.

Other Types of Graphs

Several graphs are particularly useful for displaying proportions and percentages. One of the most common is the *bar chart*. Another one often encountered is the *pie diagram* or *pie chart*. Both of these types of graphs may be seen daily in newspaper and magazine articles. Table 2.3 presents the number of students attending each major college of a hypothetical university. The frequencies have been converted into the percentages of the total enrollment they represent. Figure 2.4 presents a bar graph and a pie diagram for the percentage data in Table 2.3.

Cumulative Frequency Distribution

Our data from Table 2.1 also could have been plotted as a *cumulative frequency distribution*. Here, the plot of each interval midpoint includes *all* cases up to and including that interval. The cumulative frequency distribution for our data appears in Figure 2.5.

An advantage of a cumulative frequency distribution is that it may readily be used to obtain both percentiles and centile ranks (described later). A *percentile* is defined as a specific point in a distribution where the percent of cases below it corresponds to the value of the percentile. For example, suppose we wanted the 75th percentile. Because there are 40 scores, we go up to the y-axis to the point corresponding to 75% of 40 = 30 scores. A dotted line is drawn parallel to the base until it intercepts the curve (see Figure 2.5). Dropping the dotted line vertically from the point of intercept, it touches the x-axis at about 61. Thus, $P_{75} = 61$ (or very close to it). Had we started with a particular

---

TABLE 2.3
University College Enrollments

College	Enrollment	Percentage
Engineering Sciences	2,900	29
Physical Sciences	2,400	24
Agriculture	1,500	15
Social Sciences	1,400	14
Arts and Humanities	1,000	10
Home Economics	800	8
	10,000	100%

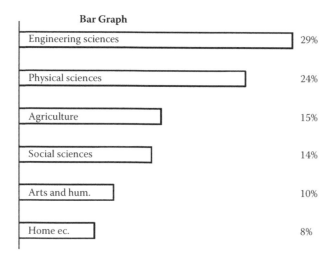

Bar Graph

Engineering sciences	29%
Physical sciences	24%
Agriculture	15%
Social sciences	14%
Arts and hum.	10%
Home ec.	8%

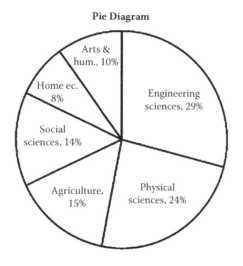

Pie Diagram

Arts & hum., 10%

Home ec. 8%

Social sciences, 14%

Agriculture, 15%

Engineering sciences, 29%

Physical sciences, 24%

FIGURE 2.4 Graphical presentations of data in Table 2.3.

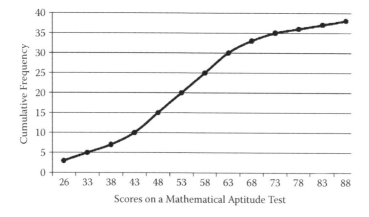

FIGURE 2.5 Cumulative frequency polygon for data in Table 2.2.

score and wanted to determine its corresponding percentile rank, we would have drawn the parts of the line in reverse order and seen where it intercepted the y-axis.

UNIVARIATE DESCRIPTIVE STATISTICS

After the frequency distribution is plotted, the data are usually reduced still further to descriptive statistics. A *descriptive statistic* is a single number used to describe a distribution. Many numbers of descriptive statistics are possible, for example, the highest score in the distribution, the product of the two midpoint scores in the distribution, or the sum of all odd-numbered scores. However, the number of descriptive statistics that provide *useful* information about the distributions they characterize is very small. These may be subdivided into *moments of the frequency distribution*: (1) measures of central tendency, (2) measures of dispersion or variability, (3) measures of skewness or shape, and (4) measures of kurtosis or peakedness (see Appendix B glossary for further distinctions and Newell & Hancock, 1984, for details).

MEASURES OF CENTRAL TENDENCY

Measures of *central tendency* are used to locate the center of the distribution. Three common measures of central tendency are the mode, median, and the mean.

Mode

The *mode* is the score that occurs most frequently in the distribution. It is the only common measure of central tendency that may be used with nominal data. In dealing with certain discrete series like the size of a family, the modal value is apt to be more typical than the other common measures of central tendency. Perhaps the greatest advantage of this measure is that it will reveal the bimodality of U-shaped distributions that would be concealed by the mean or median (see Figure 2.6). The mode does have certain disadvantages, however. First, it provides only a relatively crude estimate of a central tendency. Second, it is not an algebraic function. Finally, it is influenced by a change in the step interval (grouping error; see Downie & Heath, 1970, for further discussion of grouping error).

Median

The *median* is that value of a variable that divides the distribution into two halves that are equal with regard to the sum of the frequencies in each—the point above or below which lies an equal number of cases. The median may be computed by arranging the scores in order and then finding the value of the middle-most score if N is odd, or if N is even, the value between the two middle scores. The median may be used with ordinal or higher level data. Its chief advantages are its ease of computation and the fact that it is not influenced by extreme scores. Its disadvantages are that it is not an algebraic function and that it tends to fluctuate from sample to sample more than the mean.

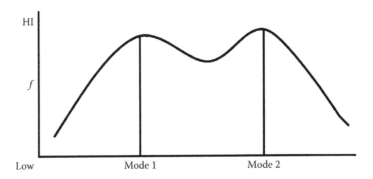

FIGURE 2.6   Bimodal frequency distribution.

When data have been grouped, such as in our frequency distribution in Table 2.2, the median can be determined directly from the grouped data. First, count up 50% of the cases from the bottom of the distribution. In our example in Table 2.2, we would count up to 20 cases, because 40 represents the total cases. The upper limit of the interval 50–54 has 21 cases below it (one more than the needed 20). We therefore must interpolate to find the median. Of the six cases in the interval 50–54, we only need five. We assume that these six cases are spread equally through the interval. Because we need five of them, we must go 5/6 of the distance up through the interval. We therefore add 5/6 of the five (interval size) to the upper limit of the interval 45–49 or 49.5.

$$\text{Median} = 49.5 + (5/6 \text{ of } 5) = 49.5 + 25/6 = 49.5 + 4.2 = 53.7$$

The 49.5 was used as the upper limit of the interval 45–49 rather than 49 because it is the point halfway between 49 and 50 along the underlying continuum. We can check our computation of the median by starting at the top of the frequency distribution and counting down 50% of the cases. For the data in Table 2.2 we subtract 1/6 of 5 from the *lower* limit of the interval 55–59 or 54.5.

$$\text{Median} = 54.5 - 1/6(5) = 54.5 - 5/6 = 54.5 - 0.8 = 53.7$$

Mean

The *mean* is the sum of all the scores in a distribution divided by the number of scores:

$$\overline{X} = \frac{(\Sigma X)}{n}$$

Mean is another term for arithmetic average. A major advantage of the mean over the other measures of central tendency is that every score contributes to it. Because the mean is an algebraic function, it is the most useful of the measures of central tendency. It too, however, has certain disadvantages. First, it may be used only with interval or higher level data. Second, it may be unduly influenced by extreme scores.

Averaging Means

At times you may have scores for two or more samples where the means for the individual samples have already been computed. To compute the grand mean for all samples, you do not have to add all the scores and divide by the total *n*. Rather, you may simply compute the *weighted mean*. This is obtained by multiplying each sample mean by its corresponding *n* or sample size, adding the products for each sample together, and then dividing by the total *n*. Table 2.4 illustrates this method for three samples.

Note that the weighted mean was 56.25 and simply averaging the three means would have yielded a result of 60. Only when sample sizes are equal can you merely average the sample means to obtain the grand mean.

TABLE 2.4
Computing Weighted Mean for Three Samples

	X	n	X*n
Sample A	50	40	2000
Sample B	60	30	1800
Sample C	70	10	700
	$\Sigma n = 80$		$\Sigma Xn = 4500$

$$\overline{X}_T = \frac{4500}{80} = 56.25$$

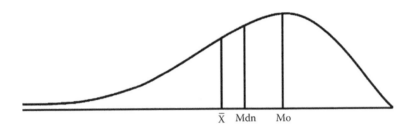

$\overline{X}$ Mdn   Mo

FIGURE 2.7   Relative positions of different measures of central tendency in negatively skewed frequency distribution.

When to Use Different Measures of Central Tendency

Of the three measures of central tendency, the mean is the most frequently encountered. However, there are situations even when we have interval data in which the use of the mean is not justified. With extremely high or low scores, the median is preferred to the mean. When a distribution is badly skewed, the mean is pulled toward the tail and is always higher (or lower) than the median and mode (see Figure 2.7). On the other hand, the mean is used whenever more computations are to be made. As we shall see, many statistics are determined from the mean. The three measures of central tendency are the same only when the distribution is symmetrical.

The frequency distribution in Figure 2.7 is referred to as a negatively skewed distribution because the tail is toward the left or negative end, and the mean is the furthest to the left. In a positively skewed distribution the tail would be toward the right (high or positive) end of the distribution. Of the three measures of central tendency, the mean would be the farthest to the right.

As a memory tool to help you remember the direction of skewness, think of which way the tail points on a number line: When the tail stretches to the left (as is the case above), the distribution is negatively skewed and when the tail stretches to the right the distribution is positively skewed.

Centiles and Quartiles

While these statistics are not averages, they are very similar in their computation to the median and therefore are discussed here. A *centile* (also called a *percentile*) is a specific distribution value that corresponds to the percent of cases below it. For example, the 78th percentile is the point in a distribution that identifies 78% of the cases below it. All centile points may be computed in the same fashion as the median. Several centile points have special names; the 50th centile is called the median. The centile point C25 is known as the first quartile (Q_1). Similarly, C75 is known as the third quartile (Q_3). In addition to these, are nine *decile* points that divide the distribution into ten equal parts. C10 is equivalent to the first decile, C20 to the second, etc.

Measures of Dispersion, Variability, or Spread

Averages alone tell us very little about a distribution. They locate its center but tell us nothing about how the scores or measurements are arranged in relation to the center. Invariably, whenever an average is needed, there is also a need to measure the scattering, dispersion, fluctuation, spread, or variability of the individuals within a group—the extent to which they are grouped closely about the central tendency or spread widely from it. The more common measures of variability are the range, semi-interquartile range, average deviation, variance, and standard deviation.

Range

The *range* has already been defined as the high score minus the low score plus one. We will spend little time on it here other than to note that of all measures of variability, the range is the most unreliable because it varies more from sample to sample than any of the others. Normally it is justifiably used only when we want a quick measure of variability and do not have time to compute one of the other measures. In dealing with a population rather than a sample, however, the range becomes more useful.

Semi-Interquartile Range

The *quartile deviation* or *semi-interquartile range* (Q) is simply half of the distance between the 25th and 75th centiles.

$$Q = \frac{(Q_3 - Q_1)}{2}$$

In a normal distribution, if we take the median and add and subtract one quartile deviation on each side of it, we cut off approximately 50% of the cases. Eight Qs will approximately cover the entire range. Because the quartile deviation is associated with the median, it follows that *whenever the median is used as the measure of central tendency, the quartile deviation should be used as the appropriate measure of variability.* A major advantage of Q over the range is that it is a more stable measure of variation.

With grouped data, the procedure for determining Q_1 and Q_3 is the same as that for computing the median discussed earlier. Thus, for our grouped aptitude test scores in Table 2.2, Q_1 would be the score of the 10th person from the bottom. Because he or she is the last person in the 40–44 interval, no interpolation is necessary. We simply use the upper limit of that interval or 44.5 as Q_1. For Q_3 interpolation is required and is computed as follows:

$$Q_3 = 64.5 - \frac{3}{5}(5) = 64.5 - 15/5 = 64.5 - 3.0 = 61.5$$

For our data then Q is as follows:

$$Q = \frac{(61.5 - 44.5)}{2} = \frac{(17.0)}{2} = 8.5$$

Interpretation of our data in terms of Q may be seen in Figure 2.8.

Average Deviation

The *average deviation* is simply the average of the absolute deviation scores. The first step in computing an average deviation is to subtract the mean from each of the scores to produce a set of

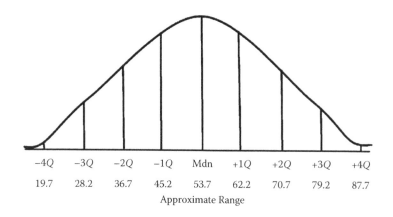

	−4Q	−3Q	−2Q	−1Q	Mdn	+1Q	+2Q	+3Q	+4Q
	19.7	28.2	36.7	45.2	53.7	62.2	70.7	79.2	87.7

Approximate Range

FIGURE 2.8 Interpretation of data in Table 2.1 in terms of Q.

deviation scores. A good computational check at this point is to see whether the deviation scores add to zero—the sum of the signed deviation scores about the mean is always equal to zero. The second step is to drop the signs of the deviation scores. The final step is to take the average of the absolute deviation scores. The average deviation, or mean deviation as it is alternately known, is no longer widely used in statistical work, having been replaced by the standard deviation. We considered it here only to make the material about standard deviation easier to understand.

Variance

Variance is the average squared deviation. Variance equals the standard deviation squared and is represented by the lower case Greek letter sigma squared, σ^2. Variances are additive, but standard deviations are not. Therefore it is generally more appropriate to apply statistics to the variance rather than the standard deviation in describing changes in distribution variability across conditions.

Standard Deviation

The *standard deviation* is the square root of the average squared deviation or root mean square deviation. Of all the measures of variability, it is the one most often used, primarily because it is needed in so many other statistical operations. When referring to the *sample* standard deviation, the symbols σ, σ_{n-1}, S, or SD are used. When referring to the standard deviation of a *population*, the σ or σ_n (lower case sigma) is used. A computational example is provided in Table 2.5.

To find the sample or unbiased standard deviation, we use the following formula and substitute in our data as shown.

$$\text{Sample } SD = s = \sigma_{n-1} = \sqrt{\frac{\sum x^2}{n-1}} = \sqrt{\frac{\sum X^2 - \frac{(\sum X)^2}{n}}{n-1}}$$

$$= \sqrt{\frac{144}{9}}$$

$$= \sqrt{16}$$

$$= 4$$

TABLE 2.5
Computing Standard Deviation From Ungrouped Data

X	x	x^2
25	7	49
23	5	25
20	2	4
19	1	1
18	0	0
18	0	0
16	−2	4
15	−3	9
14	−4	16
12	−6	36
$\sum \bar{X} = 180$	$\sum x = 0$	$\sum x^2 = 144$
$\bar{X} = 18$		(Sum of squares)

Note: $x = X - \bar{X}_x$ or distance from raw score to mean.

Computation of Standard Deviation Using Raw Score Formula

If a calculator is available, it is much easier to use the raw score formula to find the sum of the deviations squared Σx^2. This sum is referred to as the *sum of squares*. The raw score formula for obtaining the sum of squares is as follows:

$$\Sigma x^2 = \Sigma X^2 - \frac{(\Sigma X)^2}{n}$$

Thus,

$$\sigma_{n-1} = \sqrt{\frac{\Sigma X^2 - \frac{(\Sigma X)^2}{n}}{n-1}}$$

Note that the sample or unbiased estimate of the standard deviation is labeled as σ_{n-1} and contains $n-1$ in the denominator of the equation. As shown in Table 2.6, we merely square each of our scores and then sum both the X and the X^2 columns. Using the raw score formula, we then substitute as follows:

$$\Sigma x^2 = \Sigma X^2 - \frac{(\Sigma X)^2}{n}$$

$$= 3384 - \frac{(180)^2}{10}$$

$$= 3384 - \frac{32400}{10}$$

$$= 3384 - 3240 = 144$$

$$\sigma_{n-1} = \sqrt{\frac{144}{9}}$$

$$\sigma_{n-1} = 4$$

as before.

TABLE 2.6
Computing Standard Deviation
With Raw Score Formula

X	X^2
25	625
23	529
20	400
19	361
18	324
18	324
16	256
15	225
14	196
12	144
$\Sigma X = 180$	$\Sigma X^2 = 3384$

If we were dealing with the entire population instead of just a sample or subset, we would calculate the population standard deviation, also called the biased *estimate of the standard deviation*. The only difference between this and the aforementioned sample or unbiased estimate of the standard deviation is that n instead of $n - 1$ is used in the denomination. (Technically the total population should be depicted with a capital N to distinguish it from a lower case n for samples; but because many statistics books simply use a lower case n for both, it is used here in this common form. Elsewhere, the capital N refers specifically to the entire population.) Using the same data,

$$\text{Population SD} = \sigma = \sigma_n = \sqrt{\frac{\sum X^2 - \frac{(\sum X)^2}{n}}{n}}$$

$$= \sqrt{\frac{144}{10}}$$

$$= 3.795$$

Relation of Standard Deviation to Variance

The standard deviation also may be defined as the square root of the variance; the variance is defined as the average squared deviation. Notice again the distinction between variability in data representing the entire population (N) composed compared to the subset sample (n) that uses $n - 1$ in the denominator.

$$\text{Population Variance} = \frac{\sum x^2}{n} = \frac{\sum X^2 - \frac{(\sum X)^2}{n}}{n} = \sigma_n^2 = \sigma^2$$

$$\text{Also} = \frac{\sum x^2}{N} = \frac{\sum x^2 - \left(\frac{\sum x}{N}\right)^2}{N} = \sigma_N^2 = \sigma^2$$

$$\text{Sample Variance} = \frac{\sum x^2}{n-1} = \frac{\sum X^2 - \frac{(\sum X)^2}{n}}{n-1} = \sigma_{n-1}^2 = s^2$$

The variance is a measure of variation but is a square rather than a linear measure. Note that variances are additive and standard deviations are not. This means you can perform statistical operations on variances but not on standard deviations.

Averaging Standard Deviations

We noted earlier that standard deviations are not additive. Therefore, when standard deviations are to be averaged, you should use the following formula:

$$\sigma_T = \sqrt{\left(\frac{n_A\left(\overline{X}_A^2 + \sigma_A^2\right) + n_B\left(\overline{X}_B^2 + \sigma_B^2\right)}{(n_A + n_B)}\right) - \overline{X}_T^2}$$

where

σ_T = standard deviation of the combined group

σ_A, σ_B = standard deviation of two groups

n_A, n_B = number of individuals in each of the two groups

$\overline{X}_A, \overline{X}_b$ = means of the two groups

$\overline{X}_T$ = mean for the combined group

Interpretation of Standard Deviation

Normal Curve

The standard deviation is associated with the normal curve. Accordingly, a brief discussion of what is meant by *normal curve* is in order. During the 18th century, gamblers were interested in the chances (probability) of beating various gambling games, and because they were pragmatic, they called upon mathematicians for help. De Moivre (1733) was the first to develop the mathematical equation of the normal curve. During the early 19th century, the concepts of probability and normal distribution were further developed by Gauss and Laplace. The normal curve also is commonly known as the bell-shaped curve, Gaussian curve, De Moivre's curve, or the curve of error.

The *normal curve* is shaped like a bell and its maximum height (ordinate) is the mean. About this point the curve is symmetrical with all other ordinates (Ys) shorter. The normal curve is also *asymptotic*; its tails theoretically never quite touch the base line but extend infinitely in either direction.

The great descriptive value of the normal curve lies in two factors. First, the normal curve may be precisely described mathematically. Second, the distribution of many psychological and physical traits in large populations are very close to the normal curve in shape. We therefore may assume that they are normally distributed. This, in turn, enables us to determine the probability that a given value of a trait (event) will occur. In terms of probability, the events lying closest to the mean are more likely to occur than those events at either extreme away from the middle. A table of normal probability gives us in percentages the fractional parts of the total area under the normal curve found between the mean and the ordinates (Ys) erected at various distances from the mean (see Table B in Appendix A).

From a distance one standard deviation (*SD*) on either side of the mean (i.e., from –1 *SD* to +1 *SD*), the area between these points will always include 68.28% of the total area. In terms of probability, we can state that the chances are two out of three of a score in any sample falling within the area of one standard deviation on either side of the mean. A second standard deviation measured beyond the first one cuts off 13.59% of the area. The area included by two standard deviation units on both sides of the mean thus accounts for more than 95% of the area or cases. A third sigma on each side of the mean will include 99.74% of the total. Only 0.26% of the cases are beyond three standard deviation units from the mean (i.e., 0.13% to either side). Figure 2.9 depicts the interpretation of our data in Table 2.6 in standard deviation units. Note that the interpretation is in terms of the population to which our sample is assumed to be representative, that is, in terms of a very large group in which the trait measured (mathematical aptitude) is assumed to be normally distributed.

Although of great descriptive value to us, the primary importance of the normal curve is not the fact that trait scores are assumed to be normally distributed. Rather, it is the fact that the distributions of various statistics are known or assumed to be normal. This is extremely important in sampling (rather than descriptive) statistics. This will become apparent later when sampling statistics are covered. The equation for plotting a normal curve is:

$$Y = \frac{N}{\sigma\sqrt{2\pi}} e - (x - m)^2 \Big/ 2\sigma^2$$

where

Y = height of curve for a particular value of X

π = constant 3.1416

e = base of nonlinear logarithms = 2.7183

N = number of cases, the total area of the curve

σ = standard deviation of the distribution

m = population mean

Relationship of Range to Standard Deviation

We noted earlier that approximately six standard deviations cover the range. This is true only when the number of cases is large (500 or more). As *n* decreases, the number of standard deviation units needed to include all of the cases decreases. For example, with a sample of 100 cases only 5 standard deviations will include the range. Similarly, with *n* = 50, 4.5 standard deviations cover the range; with an *n* of 30, 4.1 standard deviations are required; when *n* = 10, the number of standard deviations needed reduces to 2.1. However, based upon our sample standard deviation, we can estimate the range for our population of interest (assuming the population is large) by adding and subtracting three standard deviations from the mean as illustrated in Figure 2.9.

Standard Score

Definition of Standard Scores

A *standard score* or *z-score* may be defined as a raw score expressed in standard deviation units. This is accomplished by dividing a score's deviation from the mean by its standard deviation:

$$z = \frac{X - \overline{X}_x}{\sigma_{n-1}}$$

$$= \frac{x}{\sigma_{n-1}}$$

When a distribution of scores has been converted to *z*-scores, the mean becomes 0 and the standard deviation is 1. As we noted earlier, three standard deviation units above and below the mean of our sample will include practically all the cases in our population. Thus, the highest *z*-score we are likely to encounter is +3.00 and the lowest is –3.00. In other words, we may interpret *z*-scores just as we do standard deviation.

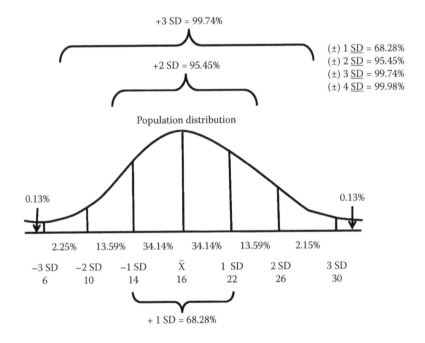

FIGURE 2.9   Interpretation of data in Table 2.6 in terms of standard deviations.

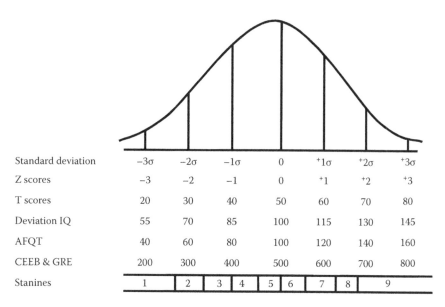

Standard deviation	-3σ	-2σ	-1σ	0	$^+1\sigma$	$^+2\sigma$	$^+3\sigma$
Z scores	−3	−2	−1	0	$^+1$	$^+2$	$^+3$
T scores	20	30	40	50	60	70	80
Deviation IQ	55	70	85	100	115	130	145
AFQT	40	60	80	100	120	140	160
CEEB & GRE	200	300	400	500	600	700	800
Stanines	1	2	3 4	5 6	7	8	9

FIGURE 2.10 Commonly used standard scores.

For example, a student with a z-score of +2.00 is two standard deviations above the mean. From this we would estimate that he scored higher than all but a little over 2% of the population from which his class was representative (the population may consists of all students who take the same course under the same conditions). This may be seen by turning to Table B in Appendix A. In the far left column are z-scores ranging from 0.00 to 3.70. Whenever we take a point on a curve and erect a perpendicular, we divide the curve into two areas, a larger one and a smaller one. Thus, in the next two columns in Table B the portions of the normal curve lying in the larger area and in the smaller area for each z-score are listed.

Because the z-score of our student was +2.0 or two standard deviations above the mean, it follows that the corresponding "area in larger portion" of 0.9772 represents the proportion of cases scoring the same or less than our student's score. Similarly, the corresponding "area in smaller portion" of 0.0228 or a little over 2% represents our estimate of the students who performed or were expected to perform better than our student. Table 2.7 lists the scores from Table 2.6 along with the corresponding z-scores and proportions above or below them.

TABLE 2.7
z-Scores and Corresponding Proportions for Test Scores in Table 2.6

X	X	z-Score	Area Below	Area Above
25	7	1.75	0.96	0.04
23	5	1.25	0.89	0.11
20	2	0.50	0.69	0.30
19	1	0.25	0.60	0.40
18	0	0.00	0.50	0.50
16	−2	−0.50	0.30	0.69
15	−3	−0.75	0.23	0.77
14	−4	−1.00	0.16	0.84
12	−6	−1.50	0.07	0.93
$\bar{X}_x = 18$	$\sigma_{n-1} = 4$			

Use of Standard Score

Because z-scores are equal units of measurement, they can be manipulated mathematically. When we use z-scores in averages and correlations, we get the same results that we would achieve by using the corresponding raw scores. Just as we cannot readily interpret the standard deviation when distributions are noticeably skewed, we also cannot readily interpret z-scores with highly skewed distributions.

The primary advantage of using z-scores over other standard scores is that they allow us to combine or compare an individual's scores on two variables that are expressed in different units and/or have different standard deviations, but have approximately the same shape distributions. For example, let us assume we wish to combine the results of two different tests for a class of 50 to determine a composite score for each student. Let us further assume the teacher wishes to have each test count for half of a student's total grade. Table 2.8 depicts what might happen for a given student both with and without first converting to standard scores.

If the two raw scores were simply averaged, our student would find that his very fine showing relative to the rest of the class on Test A would count very little because the average for both tests is 80 and the average standard deviation is 7.2. His average raw score of 86 for both tests would correspond to a z of only 0.83—a far distance from the average z of 1.8 obtained by first converting the scores to z-scores before combining them. In short, the tests were not really weighted equally when we merely added the raw scores together. Rather, they were weighted by their standard deviations. Test B thus carried a lot more weight.

When the tests to be averaged have the same standard deviations but different means, it is not essential to convert to z-scores before combining the results. Instead, a constant can be added (or subtracted) to the scores on one test to make the means equal. For example, if the mean of Test A is 75 and of Test B is 80, we merely have to add 5 points to each score on Test A prior to combining the results. However, because tests rarely have the same standard deviation or mean, it is only occasionally practical not to convert to z-scores prior to combining test results.

Linear Transformation of z-Scores

As may be noted from our discussion, z-scores are expressed in decimals and half of them are negative in any given distribution. Because of these factors, z-scores are rather cumbersome to work with. To avoid these drawbacks, linear transformations often are made by increasing the size of the mean and standard deviation. The basic equation for a transformation (T) is standard score = z (new standard deviation) + new mean:

$$T = z(\sigma) + \overline{X}$$

The most commonly used transformation uses a mean of 50 and a standard deviation of 10.

$$\text{Standard score} = z\,(10) + 50$$

TABLE 2.8
Means, Standard Deviations, and Scores for a Single Student on Two Tests

	Class		Student	
	$\overline{X}$	Σ	X	z-Score
Test A	80	2	86	3.0
Test B	80	10	86	0.6
Both tests $\overline{X} = 80$		Average $\sigma_{n-1} = 7.2$	Average raw score = 86	Average z = 1.8

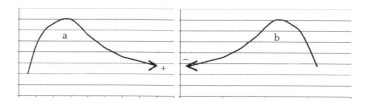

FIGURE 2.11   Positively (a) and negatively (b) skewed frequency distributions.

This particular transformation is called a *T*-score. Another common transformation often used with intelligence tests is deviation IQ. It has a mean of 100 and a standard deviation of 15. The Armed Forces Qualifying Test (AFQT) uses a mean of 100 and a standard deviation of 20. The College Entrance Examination Board (CEEB) and the Graduate Record Examination (GRE) use a mean of 500 and a standard deviation of 100. The last of the more common linear transformations are called *stanines* or *standard nines* and utilize a mean of 5, a standard deviation of 2, and a range from 1 to 9. Each stanine score is actually used to cover a range of raw scores. For example, a stanine of 5 is used for all *z*-scores from +0.25 to +0.75. While this is a rather crude way of converting raw scores, it has the advantage of allowing each variable recorded in stanine form to be entered into a single column on a computer spreadsheet. This is particularly efficient when large masses of data must be entered.

Measures of Distribution Skewness

The shape of a frequency distribution can be described in terms of directional skewness. Distributions that are biased, shifted, or sketched to the right are said to be *positively skewed* (see Figure 2.11a) and those shifted to the left are *negatively skewed* (see Figure 2.11b).

Measures of Distribution Kurtosis

Kurtosis is the fourth of four moments of a frequency distribution and it describes distribution shape in terms of peakedness. Figure 2.12 illustrates frequency distributions that are leptokurtic, normokurtic, and platykurtic.

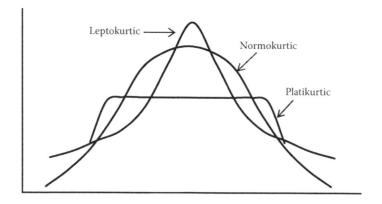

FIGURE 2.12   Frequency distributions that are leptokurtic, normokurtic, and platikurtic.

SUMMARY

1. A population is an arbitrarily defined group in which one is interested. A sample is a subgroup of the population. Numbers used to describe populations are called *parameters*. Numbers used to describe samples or to make inferences about populations from samples are called *statistics*.

2. Data that may be expressed only as whole units are called *discrete*. Data that may be described in varying degrees of fineness are called *continuous*. Statistics often treat discrete data as though they were continuous.

3. The four general levels of measurement from lowest to highest are *nominal, ordinal, interval,* and *ratio*. The higher the level, the more stringent are the underlying assumptions. Many statistics assume an interval level of measurement even though in actual practice our measures only approach rather than actually meet the assumptions of interval scaling.

4. Common graphical methods for describing a set of data or *frequency distribution* are the *frequency polygon, histogram,* and *cumulative frequency distribution*. Proportions and percentages are often illustrated using a *bar graph* or a *pie diagram*.

5. A *descriptive statistic* is a single number used to describe a distribution. The two common types of descriptive statistics are *measures of central tendency* and *measures of variability*, or *dispersion*.

6. The three basic measures of central tendency are (a) the *mode* or most frequently occurring score, (b) the *median* or score that divides the distribution in half, and (c) the *mean* or arithmetic average.

7. The mean is normally used with interval or higher level data, the median with ordinal data, and the mode with nominal data. However, with interval data, the median is a better measure of central tendency if the distribution is highly skewed, and the mode may be more descriptive if the shape of the distribution has more than one hump (peak).

8. A *centile* is a point in a distribution where the percent of cases below it corresponds to the value of the centile. Several centile points have special names such as the *median* for the 50th centile, the *first quartile* for the 25th, and the *third quartile* for the 75th centile. Nine *deciles* divide the distribution into ten equal parts.

9. The basic measures of variability are the *range*, the *quartile deviation* or *semi-interquartile range (Q)*, and the *standard deviation* (σ). The range is the least reliable indicator of variability. Q is the measure of variability used with the median, and the standard deviation is used with the mean. Approximately 8 Qs or 6 σs will cover the range of a large, normally distributed population. However, the smaller the sample, the fewer the number of Qs or σs required to cover the sample range.

10. The *variance* also is a measure of variability, but it is square rather than linear. Variances are additive, whereas standard deviations, the square roots of variances, are not.

11. The *normal curve* or curve of probability is of value because (a) it can be precisely described mathematically, (b) the distribution of many traits in large populations are close to the normal curve in shape, and (c) the distributions of many statistics are known or assumed to be normal in shape.

12. A *standard score* or *z*-score is a raw score expressed in standard deviation units. A *z*-score has a mean of 0 and a standard deviation of 1. Some common linear transformations of standard scores are *T*, stanine, deviation IQ, AFQT, CEEB, and GRE scores. The primary value of standard scores is in comparison of two or more variables expressed in different units or having different standard deviations. This is possible because standard scores convert all distributions to a common mean and standard deviation.

KEYWORD DEFINITIONS

Abscissa: Value taken along the x-axis.

Asymptotic: The tails of a normal curve theoretically never quite touch the base line and extend infinitely in either direction.

Average Deviation: Average of absolute deviation scores; also called mean deviation.

Bar Chart: Format of illustrating data in which the values of the independent variable (X) are indicated by rectangular columns (bars), where the height of each column represents the amount of Y (dependent variable) associated with each value of X. Bar charts illustrate proportions or percentages.

Centile: Point in a distribution in which the percent of cases below it corresponds to the value of the centile.

Central Tendency: Statistic that describes the tendency of numerical data to fall about a middle value between extremes of a set of measures. Statistics include the arithmetic mean, median, and mode.

Continuous: Data that can be described in varying degrees of fineness.

Descriptive Statistic: Single number used to describe a distribution.

Discrete: Data that can be expressed only as whole units.

Frequency: Number of scores in an interval.

Frequency Distribution: Most useful method for describing scores on one variable; also known as univariate frequency distribution.

Frequency Polygon: Graph of frequency distribution.

Graph: Illustration used to simplify data (e.g., bar chart and pie diagram).

Histogram: Type of graph used to depict frequency distribution.

Interval Scale: Second highest scale of measurement.

Kurtosis: Degree to which a frequency or probability distribution differs from a normal shape (mesokurtosis) by being more peaked (leptokurtosis) or flatter (platykurtosis).

Mean: Arithmetic average.

Median: Score that divides the distribution redundant half; 50th centile.

Mode: Most frequently occurring score; most populous of the class distributions.

Moments of a Distribution: Statistics that describe a frequency distribution in terms of its average value, relative scatter of observations, symmetry, etc. Generally the four moments are central tendency, variability, skewness, and kurtosis.

Nominal Curve: Curve or probability that is of value because (a) it may be precisely described mathematically, (b) the distribution of many traits in large populations are close to the normal curve in shape, and (c) the distributions of many statistics are known or assumed to be normal in shape.

Nominal Scale: Lowest of four general scales of measurement where items are placed in mutually exclusive categories.

Normal (Bell-Shaped) Curve: Frequency distribution that takes the symmetrical appearance of a bell.

Ordinal Scale: Second lowest scale of measurement consisting of ranking of different items.

Ordinate: Value taken along vertical y-axis.

Parameter: Number describing a population; an index or description of the population distribution for a given characteristic.

Pie Diagram (Pie Chart): Format for illustrating percentages or proportions as sizes of "slices" of a full circle (or pie).

Population: Arbitrarily defined group of interest.

Quartile Deviation: Half the distance between the 25th and 75th centiles; also called semi-interquartile range.

Range: Least reliable indicator of variability.

Ratio Scale: Highest scale of measurement featuring equal units and absolute zero.

Root Mean Square Deviation: Square root of average squared deviation.

Sample: Subgroup or subset of a population.

Semi-Interquartile Range (Q): Measure of variability used with a median (standard deviation is
used with a mean).

Skewness: Degree to which a frequency distribution shifts to the right (positively) or left
(negatively).

Standard Deviation: Square root of average squared deviation; square root of variance; measure
of variability used with a mean.

Standard Score: Also known as z-score; raw score expressed in standard deviation units; has a
mean of 0 and a standard deviation of 1.

Statistic: Index or description of a sample distribution.

Sum of Squares: Sum of deviations (from mean) squared.

Univariate: Involving one variable; refers to distribution of scores for a single variable.

Variability: Spread or dispersion of scores in a distribution. Statistics include standard deviation,
range, interquartile range, etc.

Variance: Measure of variability; a square rather than linear measure. Variances are additive;
standard deviations, the square roots of variances, are not.

Weighted Mean: Result of multiplying each sample mean by its corresponding n or sample size,
adding the products for all samples together, and dividing by the total n.

x-Axis: Abscissa or horizontal axis.

y-Axis: Ordinate or vertical axis.

z-Score: Standard score; raw score expressed in standard deviation units with a mean of 0 and a
standard deviation of 1.

EXERCISES

1. Given the job aptitude scores for 40 employees shown in the following table:
   a. Construct a frequency distribution for the data.
   b. Prepare:
      (1) A frequency polygon
      (2) A smoothed frequency polygon
      (3) A histogram
      (4) A cumulative frequency distribution
   c. Determine the (1) mean, (2) median, and (3) mode.
   d. Determine the (1) range, (2) Q, and (3) σ_{n-1}.

63	58	62	64	62	58	62	57	54	40
64	72	68	52	57	52	76	75	82	47
72	88	66	80	48	92	68	66	72	66
75	67	52	58	76	96	38	51	60	84

2. Based on the following breakdown of employees by department, construct (a) a bar graph and (b) a pie diagram for the data.

Department	N
Production	42
Administration	28
Sales	15
Engineering	12
Custodial	3

3. Based on the following statistics for three samples, compute the weighted mean.

Sample	Mean	n
A	80	20
B	77	34
C	70	26

4. Given the following sets of exam data,
   a. Determine each person's composite class standing, assuming that math and English are to be weighted equally.
   b. Convert the scores on each exam to the distribution having a mean of 50 and a standard deviation of 10 (i.e., to T-scores).

Exam Scores		
Student	Math	English
1	96	84
2	92	87
3	88	92
4	84	85
5	81	90
6	79	80
7	74	78
8	68	81
9	66	76
10	64	74

5. The following scores were obtained on a mathematics entrance examination by a freshman class:

64	62	94	72	78	49	72	85	75	70	84
75	57	85	60	82	92	77	94	72	75	91
88	91	68	55	81	79	67	90	64	66	74
89	93	73	63	78	72	90	73	81	76	76
73	73	71	45	68	83	87	76	85	85	57
67	72	62	93	71	82	82	56	83	78	49
68	78	65	82	70	69	61	78	71	98	83
77	79	55	84	81	64	76	69	87	72	68

a. Construct a frequency distribution for the data.
b. Construct a frequency distribution of the grouped scores having 14 intervals.
c. Express the grouped frequency distribution of b above as (1) a relative frequency, (2) a cumulative frequency, and (3) a cumulative percentage distribution.
d. Using the cumulative frequency derived in c above, determine (1) P_{75} and (2) P_{40}.

6. Determine the weighted arithmetic mean for the net profit-to-sales ratio of the three subdivisions of the XYZ Corporation.

Division	Net Profit to Sales Percentage (X)	Sales (w)	Net Profits (wX)
A	8	$15,000,000	$1,200,000
B	12	$25,000,000	$3,000,000
C	5	$10,000,000	$500,000
		$50,000,000	$4,700,000

EXERCISE ANSWERS

1.a. Interval frequencies for 40 employees.

X	f
95–99	1
90–94	1
85–89	1
80–84	3
75–79	4
70–74	3
65–69	6
60–64	7
55–59	5
50–54	5
45–49	2
40–44	1
35–39	1

Note: This assumes an interval size of 5, but you may also use 3 based on text recommendations.

1.b.1. Frequency polygon.

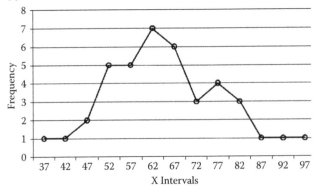

1.b.2. Smoothed frequency polygon.

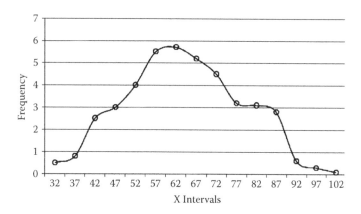

1.b.3. Histogram (vertical bar chart).

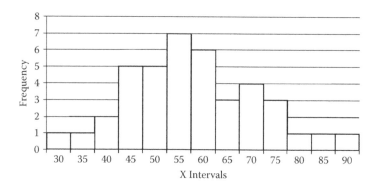

1.b.4. Cumulative frequency distribution.

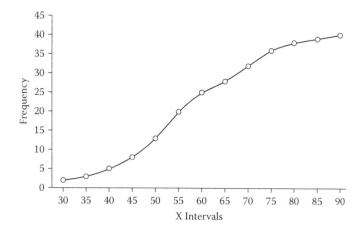

1.c. (1) $\bar{X} = 65$; (2) median = $59.5 + (6/7)(5) = 63.79$; (3) mode = interval 60–64.
1.d. (1) $96 - 38 + 1 = 59$; (2) $Q_1 = 59.5 - (4/5)(5) = 55.5$, $Q_3 = 74.5$, $Q = 9.5$; (3) $\sigma_n = 13.07$; $\sigma_{n-1} = 13.24$.

2.a. Bar Graph

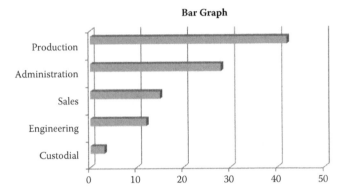

2.b. Pie diagram.

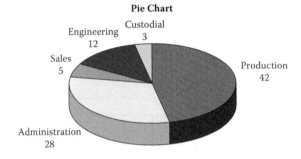

3. Computation of weighted mean.

	$\bar{X}$	N	$\bar{X}n$
A	80	20	1600
B	77	34	2618
C	70	26	1820
	$\Sigma n = 80$		$\Sigma \bar{X}n = 6038$

$$\bar{X}_r = \frac{6038}{80} = 75.48$$

4.a. Composite class standings.

Student	Math	English	Z_1	Z_2	$\bar{z}$
1	96	84	1.51	0.22	0.86
2	92	87	1.15	0.72	0.94
3	88	92	0.79	1.56	1.18
4	84	85	0.43	0.39	0.41
5	81	9	0.16	1.23	0.70
6	79	80	−0.02	−0.45	−0.24
7	74	78	−0.47	−0.79	−0.63
8	68	81	−1.01	−0.29	−0.65
9	66	76	−1.19	−1.13	−1.16
10	64	74	−1.37	−1.46	−1.42
	$\bar{X}_1 = 79.2$	$\bar{X}_2 = 82.7$			
	$\sigma_{1_n} = 10.52$	$\sigma_{2_n} = 5.64$			
	$\sigma_{1_{n-1}} 11.09$	$\sigma_{2_{n-1}} = 5.95$			

4.b. Conversion from z-scores to T-scores. $T = Z(10) + 50$.

Student	T_1	T_2
1	65.1	52.2
2	61.5	57.2
3	57.9	65.6
4	54.3	53.9
5	51.6	62.3
6	49.8	45.5
7	45.3	42.1
8	39.9	47.1
9	38.1	38.7
10	36.3	35.4

5.a. Frequency distribution of raw scores.

Score	f	Score	f	Score	f	Score	f
98	1	84	2	70	2	56	1
97	0	83	3	69	2	55	2
96	0	82	4	68	4	54	0
95	0	81	3	67	2	53	0
94	2	80	0	66	1	52	0
93	2	79	2	65	1	51	0
92	1	78	5	64	3	50	0
91	2	77	2	63	1	49	2
90	2	76	4	62	2	48	0
89	1	75	3	61	1	47	0
88	1	74	1	60	1	46	0
87	2	73	4	59	0	45	1
86	0	72	6	58	0		
85	4	71	3	57	2		

5.b. Frequency distribution with 14 intervals.

Class Interval	Real Limits	f
96–99	95.5–99.5	1
92–95	91.5–95.5	5
88–91	87.5–91.5	6
84–87	83.5–87.5	8
80–83	79.5–83.5	10
76–79	75.5–79.5	13
72–75	71.5–75.5	14
68–71	67.5–71.5	11
64–67	63.5–67.5	7
60–63	59.5–63.5	5
56–59	55.5–59.5	3
52–55	51.5–55.5	2
48–51	47.5–51.5	2
44–47	43.5–47.5	1
		88

5.c. Grouped frequency distribution expressed in relative frequency, cumulative frequency, and cumulative percentage distributions.

Class Interval	f	Relative f	Cumulative f	Cumulative %
96-99	1	0.01	88	100.00
92-95	5	0.06	87	98.86
88-91	6	0.07	82	93.18
84-87	8	0.09	76	86.36
80-83	10	0.11	68	77.27
76-79	13	0.15	58	65.91
72-75	14	0.16	45	51.14
68-71	11	0.12	31	35.23
64-67	7	0.08	20	22.73
60-63	5	0.06	13	14.77
56-59	3	0.03	8	9.09
52-55	2	0.02	5	5.68
48-51	2	0.02	3	3.41
44-47	1	0.01	1	1.14
	88	1.00		

5.d. $P_{75} = 0.75 \times 88$ scores = 66; go up 66 scores to interval 80–83, then interpolate to get 82.70. P_{40} = 0.40 × 88 = 35.2; so 72.70.

6. $$\text{Weighted Mean} = \overline{X}_{weighted} = \frac{\sum wX}{\sum w} = \frac{4,700,000}{50,000,000} = 9.4\%$$

REFERENCES

Downie, N. M., & Heath, R. W. (1970). *Basic statistical methods* (3rd ed.). New York: Harper & Row.

Ghiselli, E. E. (1965). *Theory of psychological measurement*. New York: McGraw-Hill.

Guilford, J. P. (1954). *Psychometric methods* (2nd ed.). New York: McGraw-Hill.

Hays, W. L. (1963). *Statistics*. New York: Holt, Rinehart & Winston.

Newell, K. M., & Hancock, P. A. (1984). Forgotten moments: A note on skewness and kurtosis as influential factors in inferences extrapolated from response distributions. *Journal of Motor Behavior, 16*(3), 320–335.

Sullivan, M., III. (2011). *Fundamentals of statistics* (3rd ed.). New York: Prentice-Hall.

3 Bivariate Descriptive Statistics

After learning to create descriptive statistics on a single variable (univariate), it is only natural to progress to describing two variables (bivariate) at the same time. It is particularly valuable to understand bivariate distributions, the relationship or correlation of two variables, as a means of predicting one variable when the other is known (prediction from regression). Thus, this chapter addresses these concepts and more in their simplest form—via an assumption that the data are normally distributed and the relationship of the two variables is linear. Topics covered are

Bivariate Frequency Distributions
Correlation: The Pearson r
Other Correlation Coefficients
Prediction and Concept of Regression
Summary
Keyword Definitions
References
Exercises
Exercise Answers

KEYWORDS

Abscissa
Biserial r
Bivariate
Bivariate Frequency Distribution
Coefficient of Determination
Coefficient of Nondetermination
Correlation Coefficient
Correlation Ratio (ETA)
Degrees of Freedom (*df*)
Direction
Forced Dichotomy
Homoscedasticity
Inter-Judge Reliability

Kendall's Coefficient of Concordance
Linearity
Lurking Variable
Machine Formula
Magnitude
Monotonic Relationship
Negative Relationship
Negatively Accelerated Curve
Ordinate
Pearson Product Moment
    Correlation (r)
Phi Coefficient
Phi Correlation

Point Biserial r_{pb}
Positive Relationship
Positively Accelerated Curve
Regression
Scattergram
Spearman Rank Order
    Correlation Coefficient
Standard Error of Estimate
Underestimate
Univariate

BIVARIATE FREQUENCY DISTRIBUTIONS

Thus far, we have been concerned with graphical and statistical techniques for describing the distribution of scores for a single variable and explored *univariate* graphical and statistical techniques. In this chapter, we will discuss graphical and statistical techniques for describing bivariate frequency distributions. As the name implies, a *bivariate* frequency distribution describes simultaneously the distribution of scores on two variables. In behavioral research, one is often the independent variable (IV) and the other is the dependent variable, in this chapter called the dependent measure (DM).

The basic question to be answered in looking at a bivariate distribution is "what are the nature and magnitude of the relationship between two sets of scores or variables?" The degree of relationship between two sets of scores may vary from 1.0 or perfect correlation to 0.0 or no correlation. In addition to varying in *magnitude,* correlations may also vary in *direction,* that is, they may be *positive* or *negative*. If they are positive, the higher a score is on one variable, the higher the score is likely to be on the second variable. If the correlation is negative, the higher a score is on one variable, the *lower* the score is likely to be on the second variable. For example, for X and Y variables, a perfect positive relationship is seen in the set of numbers: 1–1, 2–2, 3–3, ..., 9–9. A perfect negative relationship is revealed in the set 1–9, 2–8, 3–7, ..., 9–1.

GRAPHING RELATIONSHIP BETWEEN TWO VARIABLES

To illustrate a bivariate frequency distribution, let us assume that the students who took our mathematics aptitude test in the example in Chapter 2 just completed a course in college algebra. Let us further assume that we want to know whether our mathematics aptitude test is predictive of student performance in college algebra. In other words, did the students who scored high on our aptitude test also get the high grades in college algebra and vice versa? Table 3.1 shows the scores for each student for both variables.

We designated mathematics aptitude as variable X and college algebra as variable Y. Figure 3.1 graphs the bivariate frequency distribution. In keeping with the standard graphing procedure, the X variable has been placed along the horizontal axis or *abscissa* and the Y variable along the vertical

TABLE 3.1
Mathematics Aptitude Test and College Algebra Course Scores for 40 Subjects

Subject	X Math Apt	Y Algebra	Subject	X Math Apt	Y Algebra
1	28	55	21	54	70
2	37	61	22	55	75
3	38	67	23	55	82
4	39	64	24	56	77
5	41	65	25	56	72
6	42	68	26	56	80
7	42	62	27	56	83
8	42	67	28	57	79
9	42	72	29	60	76
10	42	74	30	62	74
11	44	76	31	62	80
12	47	72	32	62	88
13	47	78	33	62	93
14	48	82	34	65	95
15	48	71	35	66	88
16	48	73	36	68	86
17	50	76	37	72	94
18	52	74	38	78	92
19	53	80	39	82	96
20	54	87	40	86	97

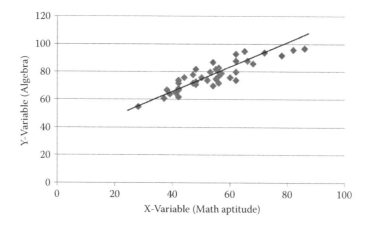

FIGURE 3.1   Scattergram of data in Table 3.1.

axis or *ordinate*. This kind of graph is called a scatter plot, scatter diagram, or simply *scattergram*. Each student's set (pair) of scores (X and Y values) is represented by a single point on the scattergram. Inspection of a scattergram provides an intuitive appreciation of the degree of relationship between the X and Y variables.

A line has been drawn through the scatter points such that the summed magnitude of the deviations of scatter points from it is as minimal as is possible (actually, the summed magnitude of the *squared* deviations is minimized). The line therefore is referred to as the *line of best fit* for the data, that is, it fits or describes the scattergram pattern better than any other single line. If all the scatter points were directly on this line, we would have a perfect correlation ($r = 1.00$)—no scatter about our line of best fit. Thus, knowing a student's score on one variable would enable us to predict exactly his or her score on the other variable (see Figure 3.2a). With a moderately high correlation, such as in our example in Figure 3.1, the scattergram pattern is elliptical about the line of best fit; many of the scores deviate to some extent from it. Thus, knowing a person's score in variable X only enables us to approximately predict her score on variable Y. For example, in Figure 3.1 we can see that if a student has an aptitude score of 62, her algebra grade may lie between 74 and 93. When you have a zero correlation, knowing a score on variable X gives you no idea of the score expected on variable Y. In that case, the scattergram will be circular as depicted in Figure 3.2b. In summary, as the shape of a scattergram approaches a straight line, the correlation progressively increases from 0 to 1.00.

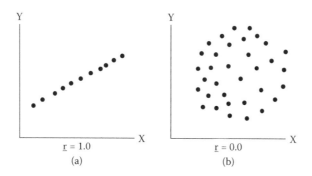

FIGURE 3.2   Scattergrams of (a) perfect and (b) zero correlations.

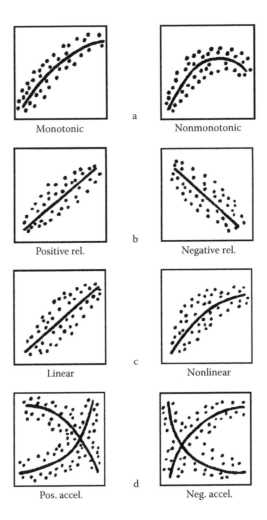

FIGURE 3.3 Shapes of bivariate distributions.

SHAPES OF BIVARIATE FREQUENCY DISTRIBUTIONS

Bivariate frequency distributions may be (a) monotonic or nonmonotonic, (b) positive or negative, and (c) linear or curvilinear. Curvilinear distributions may be (d) positively or negatively accelerated. The various shapes of bivariate distributions are illustrated in Figure 3.3.

In *monotonic relationships,* increases in one variable are always accompanied by increases in the other (positive relationship) or are always accompanied by decreases in the other (negative relationship). In *nonmonotonic relationships,* increases in one variable are accompanied by both increases and decreases in the other.

In *positive relationships* the line of best fit will run from the lower left to the upper right corner of a scattergram. In *negative relationships,* the line of best fit will run from the upper left to the lower right corner. A *positively accelerated* curve is one whose increases on variable X are accompanied by progressively larger increases or decreases on variable Y. A *negatively accelerated* curve is one whose increases on variable X are accompanied by progressively smaller increases or decreases on variable Y.

A behavioral researcher is likely to encounter any of these distributions. However, it is particularly important to note that the distribution shape may merely indicate some *portion* of one of the above distributions. For example, if a researcher were to investigate the full range of values for the independent variable (IV) X, he might obtain a nonmonotonic relationship. However, if he were merely to investigate the lower (or upper) range of IV values, his scatterplot would have a monotonic distribution.

Two researchers investigating the effect of the same IV on a given dependent measure (DM) Y might get monotonic relationships but in opposite directions simply because one used a high and the other a low range of IV values. Similarly, a researcher investigating the full range of IV values might get a curvilinear relationship. However, if he were to investigate only a portion of the range, he might get a linear or near-linear scattergram.

CORRELATION: THE PEARSON *r*

NATURE OF CORRELATION COEFFICIENTS

The magnitude and direction of relationships between two variables are described with bivariate descriptive statistics known as *correlation coefficients*. The magnitude of the relationship is expressed by the numerical value of the correlation coefficient. As we noted earlier, this value may range from 1.00 for a perfect relationship to 0.0 for no relationship. The direction of the relationship (positive or negative) is expressed by placing a plus (+) or minus (–) sign in front of the numerical value of the correlation coefficient.

Note that the direction of relationship in no way determines the accuracy with which we can predict a score on one variable from knowledge of the score on the other variable. Accuracy of prediction is strictly a function of the magnitude or degree of relationship as expressed by the numerical value of the correlation coefficient. Thus a correlation of +0.80 is just a predictive as one of –0.80.

However, remember from our discussion about the correlational approach in Chapter 1 that it is possible to have a strong correlation even if one event or value does not actually cause the other. Cases of association without causality are often results from a third, underlying variable (Z) that influences both X and Y variables in some systematic (correlated) way.

As we already have noted, a scattergram can indicate the magnitude and direction of a relationship and also determine the degrees of linearity and *homoscedasticity* that exist in the data. Perhaps nowhere else is the saying that "a picture is worth a thousand words" more applicable than when analyzing and interpreting experimental data. Correlation coefficients describe the picture of the scatter in terms of both magnitude and direction.

The ability to describe the relationship of two variables is important in research. Although association does not imply causality, it usually is a prerequisite. Regardless of whether X causes Y (or vice versa), the correlation coefficient makes a statement about the probability or likelihood that two items or events will occur together.

A variety of correlation coefficients are widely used in research. The particular coefficient one computes depends largely on the nature of the data and the assumptions that may be made about the bivariate distribution. Specifically, the following must be considered to analyze the nature of data:

Are they nominal, ordinal, or interval?
Is one or are both variables a dichotomy?
If a dichotomy, is it true or a forced dichotomy?
What is the nature of the distribution?
Is it linear or nonlinear?
Does the distribution possess good homoscedasticity?

Pearson Product–Moment Correlation (*r*)

The Pearson *r* is by far the most widely used correlation coefficient. It was named for its originator, Karl Pearson, whose mathematical development of the concept of correlation established him as

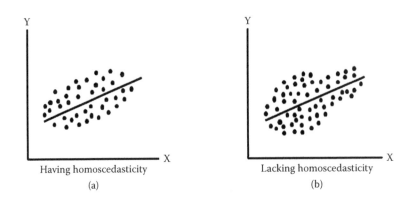

FIGURE 3.4   Examples of scattergrams (a) having and (b) lacking homoscedasticity.

one of the great fathers of the behavioral sciences. The other correlation coefficients that we shall consider later are essentially adaptations of the Pearson r.

Assumptions of Pearson r

The assumptions underlying the Pearson product–moment correlation coefficient are as follows:

The means of the arrays fall along a straight line. In other words, the line of best fit is straight rather than curvilinear. To the extent that the distribution does not meet this assumption, the computed r will be an *underestimate* of the true relationship between the variables.

The standard deviations (or variances) of the rows and columns tend to be equal. In Figure 3.4, scattergram a has good homoscedasticity; the shape is an ellipse. Scattergram b, on the other hand, lacks homoscedasticity. All the columns at the right have smaller variances than those at the left; the relationship in the lower portion of the distribution is weaker than in the upper portion. To the extent that the assumption of homoscedasticity is violated, the Pearson r will be an *underestimate* of the true relationship.

A Pearson r cannot be computed with nominal or ordinal data. Both variables must contain interval or ratio data and both variables involve values that are continuous. (Warning: If any assumption or measurement is violated, use of the Pearson product moment correlation is inappropriate. The good news is that easy alternative correlations can be applied based on each violation.)

Computation of Pearson *r*

Standard Score Formula

The Pearson r may be defined as the mean of the z-score products of two paired variables (i.e., $r = \overline{Z_x Z_y}$). The sum of the sum of the squares of the z-scores in a given distribution (i.e., Σx_x^2 or Σz_y^2) is equal to n (when n rather than $n - 1$ has been used in computing the standard deviations). When correlation is perfect, the sum of the z-score products will be the same as the sum of the squared z-scores for either variable (i.e., $\Sigma z_x z_y = \Sigma z_x^2 = \Sigma z_y^2$ when $r = 1.0$); thus, $\Sigma z_x z_y$ will equal n and the correlation coefficient will equal unity. Table 4.2 presents the X and Y variable scores for ten subjects, the corresponding deviation scores, the z-scores, and the cross products of the z-scores. Substituting $\Sigma z_x z_y$ into the z-score formula we obtain:

$$r = \frac{\Sigma z_x z_y}{n}$$

$$r = \frac{7.672}{10}$$

$$r = 0.767 = 0.77$$

TABLE 3.2
Computation of Pearson *r* From *z*-Scores

X	Y	x	y	z_x	z_y	$z_x z_y$
20	14	6.7	5.1	1.61	1.40	2.254
18	11	4.7	2.1	1.13	0.58	.655
16	15	2.7	6.1	0.65	1.67	1.086
15	10	1.7	1.1	0.41	0.30	0.123
14	8	0.7	−0.9	0.17	−0.25	−0.042
14	9	0.7	0.1	0.17	0.03	0.005
12	4	−1.3	−4.9	−0.31	−1.34	0.415
10	5	−3.3	−3.9	−0.80	−1.07	0.856
8	9	−5.3	0.1	−1.28	0.03	−0.038
6	4	− 7.3	−4.9	−1.76	− 1.34	2.358

$\bar{X}_x = 13.3$ $\bar{X}_y = 8.9$ $\qquad\qquad\qquad\qquad$ $\Sigma z_x z_y = 7.672$

$\sigma_x = 4.15$ $\sigma_y = 3.65$

Note: The standard deviation in this case uses N in the denominator because the entire population (not just a sample [n]) is represented.

Raw Score Formula

Researchers frequently work with raw score data rather than data converted into *z*-score form. Therefore, the following formula is the one most often used.

$$r = \frac{n\Sigma XY - (\Sigma X)(\Sigma Y)}{\sqrt{[n\Sigma X^2 - (\Sigma X)^2][n\Sigma Y^2 - (\Sigma Y)^2]}}$$

This also is referred to as the *machine formula* because it was most readily used when researchers worked with calculators.

Table 3.3 presents the raw scores of the ten subjects that were used in deriving the *z*-scores in Table 3.2. Substituting the summed values from Table 3.3 into the raw score formula, we have

$$r = \frac{n\Sigma XY - (\Sigma X)(\Sigma Y)}{\sqrt{[n\Sigma X^2 - (\Sigma X)^2][n\Sigma Y^2 - (\Sigma Y)^2]}}$$

$$r = \frac{10(1300) - (133)(89)}{\sqrt{(10(1941) - (133)^2)(10(925) - (89)^2)}}$$

$$r = \frac{13000 - 11837}{\sqrt{(19410 - 17689)(9250 - 7921)}}$$

$$r = \frac{1163}{\sqrt{(1721)(1329)}}$$

$$r = \frac{1163}{\sqrt{2287209}}$$

$$r = \frac{1163}{1512.4}$$

$$r = 0.769 = 0.77$$

TABLE 3.3
Computation of Pearson *r* From Raw Scores

X	Y	X²	Y²	XY
20	14	400	196	280
18	11	324	121	198
16	15	256	225	240
15	10	225	100	150
14	8	196	64	112
14	9	196	81	126
12	4	144	16	48
10	5	100	25	50
8	9	64	81	72
6	4	36	16	24
$\Sigma X = 133$	$\Sigma Y = 89$	$\Sigma X^2 = 1941$	$\Sigma Y^2 = 925$	$\Sigma XY = 1300$

If the machine formula is used as a manual formula, much time may be saved when working with large numbers if the data are coded. This generally is accomplished by subtracting a quantity equal to the smallest number in each column (X and Y columns) from each of the numbers in that column. For instance, in our example in Table 3.3, we could have subtracted 6 from each score in the X column and 4 from each score in the Y column before proceeding to compute our squared and cross product values. Table I in Appendix A may be used to save computational time.

Raw Score Deviation Formula

If you already have computed your deviation scores, the following formulas may be used:

$$r = \frac{\Sigma xy}{n\sigma_x \sigma_y} \quad \text{or} \quad r = \frac{\Sigma xy}{\sqrt{(\Sigma x^2)(\Sigma y^2)}}$$

Scattergram Method

Although not illustrated here, it also is possible to compute the Pearson *r* directly from a scattergram of scores.

Effect of Range on Value or Coefficient

The size of the correlation coefficient is directly related to the range in the two variables being correlated. When we have a restricted range of scores on one or both variables, the size of the *r* will be lowered. It should be noted here that the range of scores and not the size of the sample affects the size of the correlation.

At times we must deal with restricted ranges. For example, let us consider selection work in the Air Force. To become a pilot, an individual must first pass a series of rigorous aptitude and physical tests. If they pass, the select group of aviation cadets then undergoes flight training during which some "wash out" or voluntarily drop out. If we attempt to use only the final, highly select group that successfully completes flight training on which to validate selection tests, we would find all correlation coefficients very low. Formulas for correcting correlation coefficients for restriction in range effects have been developed and may be found in various texts on tests and measurement.

Interpretation of Correlation Coefficients

Students often ask, "What constitutes an acceptably high correlation coefficient?" The answer is that there is no absolute answer. To a large extent, the relative answer depends on three factors:

(a) the nature of the group studied, (b) the nature of the variables studied, and (c) the planned use of the coefficient.

Nature of Group

If we were attempting to correlate grade point average in first grade with college GPA for a group of subjects, a correlation of 0.60 would be considered very good. On the other hand, the same correlation between college junior and senior year grade point averages for a group of subjects would be somewhat low. In the first case, many other variables known to influence college grades may have affected students during the intervening period between first grade and college. Thus, we would expect the correlation to be moderate. In the second example we would not expect factors affecting student grades as seniors to be much different from those that affected them as juniors and would thus expect a high correlation.

Nature of Variables Studied

If we were to correlate scores on an achievement motivation test with job performance as measured by supervisor ratings, a correlation of 0.35 would be considered very acceptable. On the other hand, if we correlated scores on an experimental general intelligence test with college grade point average, a correlation of 0.35 would be disappointingly low. In the first example, many factors other than achievement motivation may affect job performance. Some of these other factors are ability, experience, specific work conditions, and other motives such as survival, safety, and social needs. Therefore, we would expect achievement motivation to account for only part of an individual's performance level and the correlation between a measure of achievement motivation and a measure of performance should reflect this.

In the second illustration, because intelligence tests purportedly measure an individual's ability to learn and also reflect past learning achievements, we would expect the correlation to be at least moderately high. However, because other factors such as current motivation also influence academic achievement, we would not expect an extremely high correlation.

Use of Coefficient

If we correlated general intelligence test scores with job performance (a validation study), a correlation of 0.50 would be considered very high. Conversely, if we administered the same general intelligence test to a group of subjects on two occasions a month apart (a reliability study), a correlation of 0.50 would be disappointingly low. As noted above, many factors affect job performance. Thus, a correlation between job performance and a single predictor or factor such as general intelligence would not be expected to be high. In the second case, we evaluated the ability of the test to predict itself. Accordingly, we would not expect other factors to differentially affect performance on two separate administrations of a test. Of course, if the testing conditions were very different for the two administrations or if a much longer time elapsed, thereby allowing various factors (such as considerable additional learning) to change the subjects' test scores, then a lower correlation would be expected.

Equivalence of Two Sets of Scores

The fact that two sets of scores yield a numerically high correlation does not imply that the scores are equivalent. For example, let us assume that an achievement test is administered to a group of students on two separate occasions. The resulting high correlation between the two administration results does not mean that the students have not learned anything more about the subject. It merely means that the students who were high achievers on the first administration are still the high achievers and vice versa. In other words, the students maintained their relative positions in the class.

Correlation Coefficient: Not a Proportion or Percent

The method of expressing correlation coefficients as decimal fractions in no way implies that they are to be interpreted as proportions or percents. A correlation coefficient represents merely an index of the relationship between two sets of data.

Correlation Versus Causality

As mentioned in Chapter 1, the correlation of two variables does not necessarily mean that one causes the other. For example, changes in some other variable or variables may simultaneously cause both variables under examination to change. Thus a high correlation does not necessarily indicate a causal relationship.

Consumption of ice cream in Wisconsin increases in the summer months when more drownings occur so ice cream consumption and drownings are positively correlated. However, eating ice cream does *not* increase a risk of drowning. The relationship arises because people swim less in the winter when they also eat less ice cream.

Interpretation of r^2 (Coefficient of Determination)

The square of a correlation coefficient is an extremely useful statistic. It can be demonstrated that r^2 is the ratio of two variances. The denominator of the ratio is the total variance of one of the variables (Y). The numerator of the ratio is that part of this total Y variance that can be accounted for by the second variable (X). r^2 then, is that part of the variance of variable Y that can be accounted for by variable X. For example, if the correlation of general intelligence test scores with college grade point averages is 0.50 for a group of students, r^2 would be 0.25. Thus, we could say that 25% of the grade point average variance was accounted for by our measure of general intelligence. r^2 often is referred to as the *coefficient of determination* because it represents the proportion of the variance of one variable that can be determined by the other variable. Similarly, $1 - r^2$ may be called the *coefficient of nondetermination,* because it represents the portion of the variance that cannot be determined (predicted) by the other variable

$$1 = r^2 + (1 - r^2)$$

or all variance = shared variance + unique variance in Y.

OTHER CORRELATION COEFFICIENTS

POINT BISERIAL r_{pb}

The point biserial is simply a special case of the Pearson product moment correlation. It is used when one of the variables is continuous and the other is believed to be a true dichotomy (i.e., having two discrete categories such as true and false). One of the most common uses of this statistic is in test item analysis. It is used to calculate the correlation between the scores on a single test item (right or wrong) with scores on the total test. The calculated r_{pb} serves as an index of how well the given test item separates the better students from the poorer ones. Items that are found not to discriminate well (have low r_{pb} values) can thus be identified and discarded in revising the test for subsequent use.

Computation of Point Biserial

In Table 3.4, the interval groupings for total test scores on an exam are shown in Column 1. In the next two columns, the number of students answering a given test item correctly or incorrectly is shown for each interval grouping of total test scores. For example, if the first test paper reviewed had a total score of 83 and the particular item we were examining was answered incorrectly, we would mark a tally in the *incorrectly* column for the interval from 80 to 84. This procedure would be repeated for that specific item for each test paper. The most common formula for the point biserial is as follows:

$$r_{pb} = \frac{\overline{X}_p - \overline{X}_t}{\sigma_t} \sqrt{\frac{p}{q}}$$

TABLE 3.4
Computation of r_{pb} for Given Test Item

	Number of Students Answering Item		
Total Test Score	Correctly	Incorrectly	F_t
95–99	3	0	3
90–94	2	1	3
85–89	3	1	4
80–84	5	2	7
75–79	5	3	8
70–74	5	4	9
65–69	3	3	6
60–64	1	2	3
55–59	2	2	4
50–54	0	2	2
45–49	0	0	0
40–44	0	1	1

where

$\overline{X}_p$ =the mean scores of those answering the item correctly

$\overline{X}_t$ =the mean of all the test scores

σ =the standard deviation of the test

p =the proportion of the total group answering the item correctly

$q = 1 - p$

Although the computations are not shown here, the mean for the proportion of students with correct answers in Table 3.4 is 78.2, the $\overline{X}_t = 74.3$ and $\sigma_t = 12.5$. Computing p and q, we have $p = 0.58$ and $q = 0.42$. Substituting these values into our equation we obtain

$$r_{pb} = \frac{78.2 - 74.3}{12.5} \sqrt{\frac{(0.58)}{(0.42)}}$$

$$= \frac{(3.9)}{(12.5)} \sqrt{1.38}$$

$$= 0.312(1.17)$$

$$= 0.365 = 0.36$$

Because this coefficient is a product moment correlation, we interpret it as we interpreted the Pearson r. Our test item appears to do a fair job of discriminating better and poorer students.

Assumptions Underlying Point Biserial

The r_{pb} has two major underlying assumptions.

Uniform distribution within dichotomies — We noted earlier that one of the two variables is assumed to be a true dichotomy. This means that the distribution within each of the dichotomous categories is uniform, i.e., the r_{pb} considered all students who answered the specific test item correctly were equally correct. Although some authorities insist that the dichotomous variable must meet this requirement to be defensible, many statisticians believe r_{pb} to be fairly robust (insensitive) with respect to this assumption. In practice, therefore, the point biserial also is used occasionally with a forced dichotomy, defined as a continuous variable arbitrarily divided into upper and lower proportions such as pass and fail on an examination.

Fairly equal dichotomous proportions — Authorities warn that if the proportions falling in the two dichotomies are extreme (for example, 0.90 and 0.10), the r_{pb} obtained will be unreliable. This means that among samples drawn from the same population, the magnitude of r_{pb} will differ. The more evenly divided the proportions, the more reliable the point biserial becomes.

BISERIAL r

When one of the two variables is continuous and the other is actually continuous but has been forced into a dichotomy, the most frequently used correlation coefficient is the biserial r. One of the most common situations encountered of this type in research is where passing or failing results or being successful or unsuccessful may be considered on a continuum ranging from extremely high to extremely low. Thus, the high criterion group is comprised of all those scoring above an arbitrarily chosen point on the continuum and the low criterion group is comprised of all those scoring below that point. In research we sometimes encounter situations in which the only available criterion data about some group of interest are in this form.

For example, let us assume we have scores from an engineering aptitude test for a group of electrical engineers and wish to know how well their scores predict (correlate with) their actual on-the-job performance. If the only performance measure we had was a rating by a supervisor as to which engineers were considered outstanding and which were considered less able, we could use the biserial r as our correlation coefficient.

Computation of Biserial r_b

The biserial r_b usually is calculated using the following formula:

$$r_b = \frac{\overline{X}_p - \overline{X}_t}{\sigma_t}\left(\frac{p}{y}\right)$$

where
X = variable recorded as continuous
$\overline{X}_p$ = mean of high criterion group
$\overline{X}_t$ = mean of total group
σ_t = standard deviation of total group
p = proportion of subjects in high criterion group
y = ordinate of normal curve that cuts off to the right of its associated abscissa (area equal to p)

All the symbols used here are the same as for the point biserial except y. The y ordinate may be obtained for any proportion in a normal probability table (Table B in Appendix A). For the example given earlier, let us assume that 60% of the electrical engineers were judged as high performers. Let us also assume that the mean engineering aptitude score for the high on-the-job performers was 130 and the mean aptitude score for the total group was 125. The standard deviation for the aptitude test was found to be 15. The biserial r between aptitude and job performance is

$$r_b = \frac{130 - 125}{15}\left(\frac{0.60}{0.3857}\right)$$

$$= 0.333(1.56)$$

$$= 0.519 = 0.52$$

An alternate formula is as follows:

$$r_b = \frac{\overline{X}_p - \overline{X}_q}{\sigma_t}\left(\frac{pq}{y}\right)$$

where $q = 1 - p$ or the proportion of the subjects in the low criterion group, $\overline{X}_q$ = mean of the low criterion group, and all other symbols are as in the previous formula.

Assumptions Underlying Biserial r

Normal Distribution of Dichotomized Variable

The biserial r is based on the assumption that the dichotomized variable is normally distributed. Because this assumption rarely is established, mathematically oriented statisticians tend to shy away from its use. When this assumption of normality is grossly violated, it is possible to calculate a biserial r greater than unity.

Fairly Equal Proportions in Dichotomy

As with the point biserial r, the more extreme the proportions, the less reliable is the statistic.

Interpretation of Biserial r

The biserial r is an estimation of the product moment correlation coefficient that would have been derived if both variables had been treated as continuous. It has no direct relationship to the Pearson r. However, its values range from -1 to $+1$. Although the standard error of the biserial r is known to be greater than that of the Pearson r, the sampling distribution of the biserial r is not well understood. The biserial r can be and is used in virtually every situation where the point biserial r *is* used.

Spearman Rank Order Correlation Coefficient (Rho)

The rho is an adaptation of the Pearson r for use with ordinal data. This coefficient is most effectively used when (a) the original data are in the form of ranks, (b) the score distribution is less meaningful than the ranks (as, for example, when the scores are extremely heterogeneous), and (c) the number of cases is below 30 and the computation is to be done manually. This correlation may be used to describe the relationship of academic grades to degree completion.

Calculation of Spearman Rho

The two sets of raw score data for the subjects from Table 3.3 appear in the first two columns of Table 3.5. You may recall that the Pearson r obtained for these data was 0.77.

In computing the rho from raw scores we first convert the scores to ranks. When the scores are tied, we assign an average rank to both subjects. Thus, for example, individuals 7 and 10 on the Y variable each received 9.5 as the average of Y ranks 9 and 10. Having converted the data to ranks, the next step is to compute a difference or D column. This is done simply by subtracting the X rank from the Y rank or vice versa for each individual. (Note that if we had initially started with ranked data, computation of this difference between ranks would have been our first step.) Next, we square each difference score and then sum the D^2 column. From this point, calculation of rho is relatively simple.

$$\rho = 1 - \frac{6\sum D^2}{N(N^2 - 1)}$$

TABLE 3.5
Computation of Spearman Rho From Raw Scores

Individual	X	Y	X-Rank	Y-Rank	D	D²
1	20	14	1	2	1.0	1.00
2	18	11	2	3	1.0	1.00
3	16	15	3	1	−2.0	4.00
4	15	10	4	4	.0	.00
5	14	8	5.5	7	1.5	2.25
6	14	9	5.5	5.5	.0	.00
7	12	4	7	9.5	2.5	6.25
8	10	5	8	8	.0	.00
9	8	9	9	5.5	−3.5	12.25
10	6	4	10	9.5	−0.5	.25
					$\Sigma D^2 = 27.00$	

where N = the number of pairs; ρ = Spearman rho; 6 = a constant. Substituting, we have

$$\rho = 1 - \frac{6(27)}{10(100-1)}$$

$$= 1 - \frac{162}{990}$$

$$= 1 - 0.1636$$

$$= 0.8364$$

Note that this correlation coefficient differs somewhat from the Pearson r value calculated from the raw scores. This is the case even though the rho may be interpreted as a Pearson r computed from ranks (i.e., it is a product moment correlation coefficient) and provides a good approximation of the Pearson r computed from interval level data. Failure of the two to more accurately agree here probably stems from the partial violation of the underlying assumption explained below.

Assumption Underlying Spearman Rho

A basic assumption of the rho is that the two underlying variables are continuous and that no two individuals have the same quantity of one of the traits. If the preciseness of measurement is crude enough to result in a large number of tie scores when ranking the data, the resulting rho will be grossly affected. Under such conditions, the use of rho is questionable unless one of the available formulas is applied to correct these ties. In actual practice, however, a case where such action is warranted is rare.

Use of Spearman Rho

When n is small, rho is much easier to compute than the Pearson r. Because it also provides a good estimate of the Pearson r, it is particularly appropriate for classroom use. However, in research where maximum precision is desired, the Pearson r is preferred if the data meet its underlying assumptions.

KENDALL'S COEFFICIENT OF CONCORDANCE (*W*)

Kendall's *W* is used to determine the relationship among three or more sets of ranks. Recall that rho is used to quantify the relationship of two sets of ranks. Like *W*, rho is derived from the Pearson r. In

TABLE 3.6
Calculation of *W* for Four Sets of Ranks

Employees	Managers' Rankings				Sum of Ranks	D	D²
	1	2	3	4			
1	2	3	1	4	10	12	144
2	3	1	3	1	8	14	196
3	1	2	4	2	9	13	169
4	4	5	2	3	14	8	64
5	6	4	5	7	22	0	0
6	7	6	8	5	26	−4	16
7	8	7	6	6	27	−5	25
8	5	9	7	9	30	−8	64
9	9	8	10	8	35	−13	169
10	10	10	9	10	39	−17	289
					$\Sigma = 220$		$\Sigma D^2 = 1136$

behavioral research, it is not uncommon to have judges or experts evaluate the performance behaviors of subjects when more suitable DM measures are not available. For example, let us assume that four managers have been asked to rank the job performances of ten employees. To have confidence in the ability of the managers to assess the performance of these employees on the same basis, we would expect that the managers will rank the employees fairly similarly; in other words, the correlation or relationship among the four sets of ranks would be high. Table 3.6 presents the employee rankings by the four managers.

Computation of *W*

To calculate *W*, we first sum the managers' rankings for each employee. We next compute the average sum of ranks (220/10 = 22). *D* is obtained by determining the difference of each of the sum of ranks from the average sum of ranks. Each *D* score is squared and the squares are added. We now are ready to compute *W* using the following formula:

$$W = \frac{12\Sigma D^2}{m^2(n)(n^2 - 1)}$$

where 12 = a constant; *m* = the number of sets of ranks; and *n* = the number of objects ranked. Substituting, we have

$$W = \frac{12(1136)}{(4)^2(10)(100 - 1)}$$

$$= \frac{13632}{(16)(10)(99)}$$

$$= \frac{13632}{15840}$$

$$= 0.861 = 0.86$$

The high correlation indicates high *inter-judge reliability* (i.e., the managers rated the employees in a similar or reliable fashion). Apparently they used the same subjective criteria and intuitively weight the criteria consistently in ranking the employees.

PHI COEFFICIENT (Φ)

When the data on both dimensions are in the form of dichotomies, the most commonly used correlation coefficient is the phi. The phi may be regarded as an extension of the Pearson and the point biserial r to the situation in which both variables are dichotomized.

Computation of Phi

Let us assume that the scores of 100 subjects have been dichotomized into upper and lower halves of two variables (X and Y). The resultant two-by-two bivariate frequency table may look like the one in Table 3.7. Note that the cells and the marginal values have been assigned letters. The formula for computing the phi coefficient is

$$\phi = \frac{(ad - bc)}{\sqrt{(k)(l)(m)(n)}}$$

Substituting from Table 3.7, we have

$$\phi = \frac{(30)(40) - (20)(10)}{\sqrt{(50)(50)(40)(60)}}$$
$$= \frac{1200 - 200}{\sqrt{6,000,000}}$$
$$= \frac{1000}{2449}$$
$$= 0.408$$
$$= 0.41$$

Assumptions Underlying Phi

The assumptions that apply to the single dichotomy of the point biserial apply to both dichotomies of the phi coefficient. These are (1) uniform distribution within dichotomies and (2) fairly equal dichotomous proportions. Note that unlike the Pearson r, the phi makes no assumptions about the shape of the distribution of scores. Unity correlation is not possible unless the subjects are divided into equal parts on both dimensions.

Special Uses of Phi Coefficient

Because the phi coefficient is robust with respect to the assumptions of both *linearity* and *homoscedasticity*, dividing interval level research data into upper and lower halves and computing the phi coefficient can serve as a quick check on the extent to which the data meet these two assumptions. To the extent that the phi coefficient is larger than the Pearson r, the data does not meet one or both

TABLE 3.7
Dichotomized Scores for 100 Subjects

		X		
		Hi	**Low**	
Y	Hi	30_a	20_b	$_k50$
	Low	10_c	40_d	$_l50$
		40_m	60_n	100_n

of these assumptions and the Pearson r is an underestimate of the true relationship. Because phi is relatively simple to compute, researchers may initially divide their data into upper and lower halves and compute the phi to quickly estimate the variable relationship.

CORRELATION RATIO (ETA)

With decidedly nonlinear relationships, use of the Pearson r and most of the other common correlation coefficients is inappropriate. Under these circumstances, the correlation ratio or eta coefficient can be computed. Computation of eta requires that one of the variables be continuous (interval level of measurement) but the other may be nominal data. From our discussion of scattergrams, you may recall that a scatterplot will be circular when no correlation exists (see Figure 3.2b). Under this circumstance, the mean value of Y for each X column or value is the same as for every other X value and identical to the overall Y mean.

The standard deviation of the Y means (mean value of Y for each X value or column) is therefore 0. In contrast, with perfect correlation, all values fall along a line that may be linear or curvilinear. Knowing the value of X allows you to determine precisely the value of Y. Thus the standard deviation of the Y means is as great as the overall standard deviation for the Y values. Therefore, the relationship between two sets of scores may be defined as

$$\text{relationship} = \frac{\sigma'_y}{\sigma_y}$$

where

$\sigma'_y =$ standard deviation of Y means for each X
$\sigma_y =$ overall standard deviation for Y

This correlation ratio will range in value from 0 (where the means of Y rows or values are all the same and thus have a 0 standard deviation) to 1.0 (when the standard deviation of the Y means is exceeds 0 and is exactly equal to the overall Y standard deviation).

Calculation of Correlation Ratio

Table 3.8 depicts the scores of 25 subjects on 2 different exams. The values of X are depicted in Column 1. Column 2 shows the various Y scores corresponding to each X value. For example, two

TABLE 3.8
Computation of Eta

X	Y	M'_y	$M'_y - M_y$	$(M'_y - M_y)^2$	n	$n(M'_y - M_y)^2$
20	14	14	6	36	1	36
18	10, 12	11	3	9	2	18
15	10, 14	12	4	16	2	32
14	7, 9, 11	9	1	1	3	3
13	6, 9, 12, 13	10	2	4	4	16
12	6, 8, 8, 10	8	0	0	4	0
11	4, 6, 11	7	−1	1	3	3
10	4, 6	5	−3	9	2	18
8	1, 3, 5	3	−5	25	3	75
5	1	1	−7	49	1	49
$\bar{X}_y = 8.0$		$M_y = 8$			$\Sigma_n = 25$	$\Sigma = 250$
$\sigma_y = 3.77$						

subjects scored 18 on test X, one scored 10 on variable Y, and the other scored 12. The third column (M'_y) contains the Y means for each X value. The number of scores on which each M'_y value is based is shown in the sixth column (n).

To compute the correlation ratio, the deviations of the Y means from the overall Y mean are computed. These are shown in Column 4 ($M'_y - M_y$). These values are then squared (Column 5). Next, the squared values are weighted according to the number of scores (Column 7) and these values are summed, yielding a value of 250. Dividing this value by n and calculating the square root yields a standard deviation of the Y means:

$$\sigma'_y = \sqrt{\frac{\sum n(M'_y - M_y)^2}{n}} = \sqrt{\frac{250}{25}} = \sqrt{10}$$
$$= 3.16$$

In our example, the overall standard deviation of Y was found to be 3.77. Substituting into our formula for eta we have

$$\eta = \frac{\sigma'_y}{\sigma_y} = \frac{3.16}{3.77} = .82$$

A common use of eta (η) is to compare it with the Pearson r when a scatterplot is not prepared for the data. To the extent that eta exceeds r, the assumptions of linearity and/or homoscedasticity are not met and the Pearson r is an underestimate of the true relationship. In our example, the Pearson r was 0.80. Thus, our data met the underlying assumptions of linearity and homoscedasticity quite well.

PREDICTION AND CONCEPT OF REGRESSION

In the first chapter, we noted that the primary use of correlation in scientific research was for prediction. Inherent in prediction is the phenomenon of regression.

CONCEPT OF REGRESSION

From your past experience you probably are aware of the strong tendency for very tall parents to have tall offspring, short parents to have short children, etc. What you may not know, however, is that the children of tall parents are on average not quite as tall as their parents. Similarly, offspring of short parents will not be quite as short as their parents. This phenomenon is known as regression toward the mean. *Regression* is characteristic of prediction whenever correlation is less than perfect. This chapter presents an introduction to simple linear regression.

With perfect correlation, prediction of Y from knowledge of X will be precise. When correlation is less than perfect, the prediction of Y from X is based partially on knowledge and partially on ignorance. From our discussion of the normal curve, you know that there is a greater probability that cases will occur at and around the mean rather than toward the extremes of a distribution. Thus, to the extent that our predictions are based on ignorance (lack of correlation), we can expect that the predicted (Y) scores will be relatively closer to the mean than the X or predictor scores are close to their means (i.e., tighter distribution.)

COMPUTATION OF REGRESSION LINES

To better understand regression, we must begin by learning more about lines of best fit or regression lines. We are concerned only with linear regression. Multiple and nonlinear regression are addressed elsewhere.

X	Y
0	2
1	5
2	8
3	11
4	14
5	17

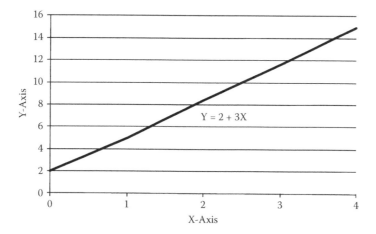

FIGURE 3.5   Straight line plot for Y = 2 + 3X.

Equation for Straight Line

The mathematical formula for any straight line is $Y = a + bX$ where Y represents the corresponding ordinate for any particular x value, a is the point of intercept with the Y axis, and b is the ratio of the change in Y with respect to the change in X. Consider the following data. For each change of one in X there is a change of 3 in Y. The straight line equation would thus be $Y = 2 + 3X$. We can see this graphically by plotting our data in Figure 3.5. Note that the line intercepts the Y axis at 2, and that for every change of 1 on X there is a change of 3 on Y.

Computation of Linear Regression Line

The equation for the straight line in linear regression differs from the equation above only in that the Y we seek is a predicted value of Y. This is designated as $\hat{Y}$ and thus $\hat{Y} = a + bX$. The actual values of Y may be expected to differ somewhat from $\hat{Y}$ unless we have perfect correlation.

To determine the regression line $(\hat{Y})$ we must first solve for the values of a and b such that the $Y - \hat{Y}$ values are minimized. We want to get as close to the line as possible. This is known as the method of least squares because it minimizes the squared deviations about the resulting regression line. In other words, the errors of prediction are minimized. Usually least squares is written $\Sigma(Y - Y)^2$. To solve for b_{yx} we may use either of the following formulas.

$$b_{yx} = \frac{\Sigma XY - \frac{(\Sigma X)(\Sigma Y)}{N}}{\Sigma X^2 - \frac{(\Sigma X)^2}{N}} = \frac{\Sigma xy}{\Sigma x^2}$$

The a coefficient is obtained by:

$$a_{yx} = \overline{X}_y - \overline{X}_x(b_{yx})$$

To illustrate this, let us solve to find the regression line for predicting Y from X using the data in Table 3.1. Based on the table, the following information was computed:

$$\overline{X}_x = 53.85$$

$$\overline{X}_y = 77.53$$

$$\Sigma x^2 = 6149$$

$$\Sigma y^2 = 4154$$

$$\Sigma xy = 4458$$

Substituting into our formulas for b and a we have

$$b_{yx} = \frac{\Sigma xy}{\Sigma x^2}$$

$$= \frac{4458}{6149}$$

$$= 0.725$$

$$a_{yx} = \overline{X}_y - \overline{X}_x(b_{yx})$$

$$= 77.53 - 53.85(.725)$$

$$= 77.53 - 39.04$$

$$= 38.49$$

If we enter the a and b values in the equation for the linear regression line, we have

$$\hat{Y} = a + b(X)$$

$$= 38.5 + 0.725(X)$$

Because any two points determine a straight line, we may select several values of X and solve for Y. The resulting regression line is depicted in Figure 3.6. Note also in the figure the plotting of the regression line for predicting X from Y. Computation of $\hat{X}$ for the plot on Figure 3.6 is as follows:

$$b = \frac{\Sigma xy}{\Sigma y^2}$$

$$= \frac{4458}{4154}$$

$$= 1.073$$

$$a_{xy} = \overline{X}_x - \overline{X}_y(b_{xy})$$

$$= 53.85 - 77.53(1.073)$$

$$= 53.85 - 83.19$$

$$= -29.38$$

$$\hat{X} = a_{xy} + b_{xy}(Y)$$

$$= -29.34 + 1.073(Y)$$

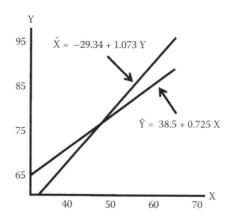

FIGURE 3.6 Plot of two regression lines.

If correlation is perfect, the two regression lines will be the same. If we translate our results to z-scores, their slope would be 45°. As the correlation decreases from unity, the slopes of the two lines would change about their means. The Y line would become progressively flatter until it became horizontal when $r = 0$. The X line would become progressively steeper until it became vertical at $r = 0$. When there is no correlation, the best prediction of Y from any X value is $\bar{X}_y$; similarly, the best prediction of X for any Y value is $\bar{X}_x$.

Relation of b_{yx} and b_{xy} to r

Computing r^2 may be useful for checking calculations of the regression slopes. For the data in Table 3.1, $r = 0.88$; thus $r^2 = 0.775$. Substituting our results from the regression calculations, we can check to see how close $(b_{yx})(b_{xy})$ is to r^2.

$$(b_{yx})(b_{xy}) = r^2$$

$$(0.725)(1.073) = 0.775$$

$$0.778 = 0.775$$

Because of this relationship, the following formulas may also be used for the regression lines:

$$\hat{Y} = \left(r \left(\frac{\sigma_y}{\sigma_x} \right) \right)(X - \bar{X}_x) + \bar{X}_y$$

$$\hat{X} = \left(r \left(\frac{\sigma_x}{\sigma_y} \right) \right)(Y - \bar{X}_y) + \bar{X}_x$$

STANDARD ERROR OF ESTIMATE

We already noted that perfect correlation allows us to precisely predict Y from knowledge of X. In other words, all our Y values will fall on the regression line, thus the difference $\hat{Y} - Y = 0$. Conversely, when we have zero correlation, our regression line becomes $\bar{X}_y$. Under this condition the difference $\hat{Y} - Y = \sigma_y$. We are saying that there is an error in our predictions and in standardized form it is equal to the standard deviation of Y. Thus, we can say that 68% or 2/3 of our subjects will fall within plus or minus one standard deviation of the Y mean, assuming a normal distribution of the Y variable.

As the magnitude of our correlation increases from zero toward unity, this error in prediction will become progressively smaller than the standard deviation of Y. In standardized form, this error of prediction is called the standard error of estimate (SE_{est}). The SE_{est} is defined as the standard deviation of the errors of prediction. Mathematically we can express the SE_{est} as

$$SE_{est} = \sqrt{\frac{\sum(Y - \hat{Y})^2}{N - 2}}$$

Computation of SE_{est}

A mathematically equivalent formula that usually is considerably easier to compute than the one above is

$$SE_{est} = \sigma_y \sqrt{1 - r^2}$$

where

$$\sigma_y = \sqrt{\frac{\sum Y^2}{N - 2}}$$

Note that in computing the standard deviation we must use $N - 2$ *degrees of freedom* in our denominator. This is necessary to ensure that our SE_{est} is an unbiased estimate of the population parameter. Substituting in our data from Table 3.1, we have

$$SE_{est} = \sqrt{\frac{4154}{38}} \sqrt{1 - .78}$$

$$= \sqrt{109.316} \sqrt{0.22}$$

$$= (10.46)(0.469)$$

$$= 4.9$$

Interpretation of SE_{est}

Using the standard error of estimate, we can make inferences about the population from which our sample was drawn. Thus, we can say that approximately 2/3 of our population will fall within ± 1 SE_{est} of our regression line and that approximately 6 SE will cover the population range. This kind of information is extremely useful for predicting performance (Y, DM, or criterion variable) from predictor scores (X, IV, or predictor variable). The information allows us to know at any desired level of probability what performance range will include a subject with a given X score.

SUMMARY

1. *Bivariate frequency distributions* describe simultaneously the distribution of scores on two variables. In research, often one of these variables is the IV and the other the DM.
2. The graph of a bivariate frequency distribution is called a *scattergram*. The X variable is placed on the horizontal axis or *abscissa*. The Y variable is placed on the vertical axis or *ordinate*. The more the shape of the scattergram approaches a line, the stronger the relationship; the more the shape approaches a circle, the weaker the relationship.

3. Bivariate frequency distributions may be (a) monotonic or nonmonotonic, (b) positive or negative, and (c) linear or curvilinear, or (d) positively or negatively accelerated.

4. The magnitude and direction of the relationships between two variables are described with bivariate descriptive statistics known as correlation coefficients. The magnitude is expressed by the numerical value of the coefficient; the direction is indicated by the sign of the coefficient.

5. A variety of correlation coefficients are used in research. The particular coefficient utilized depends largely on (a) the nature of the data and (b) the assumptions that may be made about the bivariate distribution.

6. The *Pearson product–moment correlation (r)* is by far the most widely used measure of linear correlation and the most precise index when the underlying assumptions are met. The Pearson r requires (a) interval level of measurement, (b) linearity, and (c) homoscedasticity (equal variance among rows and columns).

7. There is no absolute answer to "What constitutes an acceptable high correlation coefficient?" To a large extent, the relative answer depends on the (a) nature of the group studied, (b) nature of the variables studied, and (c) use to which the coefficient is put.

8. r^2 is called the *coefficient of determination*. It is a measure of that part of the Y variance that can be accounted for (determined) by the second variable (X). Similarly $1 - r^2$ is referred to as the *coefficient of nondetermination*.

9. The *point biserial (r_{pb})* is a special case of the Pearson r. It is used when one of the variables is continuous and the other is believed to be a true dichotomy. r_{pb} is robust to its underlying assumption of uniform distribution within dichotomies and is often used when the second variable is a forced rather than a true dichotomy. However, r_{pb} is sensitive to gross violation of the assumption of fairly equal dichotomous proportions.

10. The *biserial r (r_b)* is used when one variable is continuous and the other is a forced dichotomy. The biserial r is robust with respect to its assumption of a normal distribution of the dichotomized variable and therefore may be used with a forced dichotomy. Like the point biserial, the biserial r is sensitive to very unequal proportions in a dichotomized variable. It has no direct relationship with the Pearson r; it is an estimation of the product moment correlation.

11. The *Spearman rank order correlation coefficient* (rho) is an adaptation of the Pearson r for use with ordinal data. It is the most desirable linear index of relationship to compute if (a) the original data are in the form of ranks, (b) the score distribution is less meaningful than ranks (with extremely heterogeneous scores), or (c) the number of cases is fewer than 30 and the computation will be done manually.

12. *Kendall's coefficient of concordance (W)* is used to determine the relationship between three or more sets of ranks. It therefore requires ordinal or higher level data.

13. The *phi coefficient* is an extension of the Pearson r to the case where data on both dimensions are in the forms of dichotomies. It has the advantage of being robust with respect to the assumptions of linearity and homoscedasticity. The phi is unreliable if the dichotomous proportions are extreme. Unity correlation is only possible with equally structured dichotomies.

14. The *correlation ratio* (eta) is used to determine the maximum possible relationship regardless of the shape of the bivariate distribution. Computation of eta requires that at least one variable be continuous; the other may represent nominal data.

15. Regression is a characteristic of prediction whenever correlation is less than perfect. The smaller the correlation, the relatively closer the predicted ($\hat{Y}$) scores will be to their mean than the predictor (X) scores. When $r = 0.0$, there is total regression and the predicted ($\hat{Y}$) score from any predictor (X) is the mean ($\bar{X}_y$).

16. The *standard error of estimate (SE_{est})* is the error of prediction expressed in standard deviation form. When $r = 0.0$, SE_{est} is at its maximum and equal to σ_y. As r increases, SE_{est} decreases until when $r = 1.0$ the $SE_{est} = 0.0$. SE_{est} may be used to determine at any desired level of probability what performance range will include a corresponding (Y) for a given (X) score.

KEYWORD DEFINITIONS

Abscissa: X variable placed on a horizontal axis.

Biserial r (r_b): Used when one variable is continuous and the other is a forced dichotomy.

Bivariate: Frequency distribution describing simultaneous distribution of scores on two variables.

Bivariate Frequency Distribution: Simultaneous distribution of scores on two variables.

Coefficient of Determination (r^2): Measure of part of Y variance that can be accounted for (determined) by a second variable (X).

Coefficient of Nondetermination ($1 - r^2$): Portion of variance that cannot be determined (predicted) by the other variable.

Correlation Coefficient: Magnitude and direction of relationship between two variables are described with bivariate descriptive statistics.

Correlation Ratio (Eta): Determines maximum possible relationship regardless of the shape of bivariate distribution.

Degrees of Freedom: Number of values (or scores) that are free to vary.

Direction: Correlations may vary in direction; they may be positive or negative. For positive correlations, the higher a score is on one variable, the higher the score is likely to be on the second variable. Conversely, for negative correlations, the higher a score is on one variable, the lower the score is on the second variable.

Forced Dichotomy: Continuous variable arbitrarily divided into upper and lower proportions (e.g., pass and fail on an examination).

Homoscedasticity: Assumption that the standard deviations of Y scores along the regression line should be fairly equal; otherwise the standard error of the estimate is not a valid index of accuracy.

Inter-Judge Reliability: Consistency in scoring by two or more independent judgments of the same performance.

Kendall's Coefficient of Concordance (W): Relationship of three or more sets of ranks requiring ordinal or higher level data; often is used to determine inter-judge reliability.

Linearity: Relationship between two variables. As the value of one variable increases, the value of the other increases by a constant amount. Linearity is graphically depicted as a dot cluster that approximates a straight line.

Lurking Variable: Nuisance or intervening variable that confounds a study and makes straightforward interpretation of the results difficult.

Machine Formula: Formula most readily used when solving equations with a calculator.

Magnitude: Amount or quantity of a variable.

Monotonic Relationship: An increase in one variable is accompanied by a systematic increase or decrease in the other.

Negative Relationship: Relationship in which the line of best fit extends from the upper left to the lower right corner of a scattergram.

Negatively Accelerated Curve: Curve in which increases of variable X are accompanied by progressively smaller increases or decreases on variable Y.

Ordinate: Y variable placed on vertical axis.

Pearson Product Moment Correlation (*r*): Most widely used measure of linear relationship between two variables; the most precise index to use when the underlying assumptions can be met; requires interval or higher level data.

Phi Coefficient: Coefficient of contingency.

Phi Correlation: Extension of Pearson *r* to where data on both dimensions are in the forms of dichotomies.

Point Biserial (r_{pb}): Special case of the Pearson *r* used when one variable is continuous and the other is believed to be a true dichotomy.

Positive Relationship: Relationship in which the line of best fit extends from the lower left to the upper right corner of a scattergram.

Positively Accelerated Curve: Curve in which increases on variable X are accompanied by progressively larger increases or decreases on variable Y.

Regression: Characteristic of prediction whenever correlation is less than perfect.

Scattergram: Graph of bivariate frequency distribution.

Spearman Rank Order Correlation Coefficient (Rho): Adaptation of Pearson *r* for use with ordinal data.

Standard Error of Estimate (SE_{est}): Error of prediction expressed in standard deviation form.

Underestimate: Calculated value that is lower than the true (or actual) value.

Univariate: Frequency distribution describing the distribution of scores on one variable.

EXERCISES

1. Job aptitude test scores (X) and job performance ratings (Y) for a group of skilled employees are shown in the table below.
   a. Calculate the Pearson *r*.
   b. Convert the data to ranks and calculate the Spearman rho.
   c. Dichotomize both sets of scores into upper and lower halves and calculate phi.
   d. Based on a comparison of the phi and the Pearson *r* and inspection of the scattergram, how well are the underlying assumptions of the Pearson *r* met by the data?

Subject	Job Aptitude Test (X)	Job Performance Rating (Y)	Subject	Job Aptitude Test (X)	Job Performance Test (Y)
1	15	9	11	10	4
2	14	8	12	9	8
3	13	7	13	9	2
4	12	10	14	8	3
5	12	7	15	6	2
6	11	8	16	6	1
7	11	6	17	5	5
8	10	10	18	4	5
9	10	7	19	3	6
10	10	6	20	1	4

2. Based on the following rankings of potential for graduate work for ten undergraduates by three of their instructors, determine the degree of inter-judge reliability (Kendall's coefficient of concordance).

Undergraduate Students	Instructors		
	1	2	3
1	1	4	2
2	2	3	3
3	3	1	1
4	4	2	4
5	5	5	6
6	6	7	5
7	7	6	7
8	8	8	9
9	9	10	10
10	10	9	8

3. Assume that you are conducting an analysis of the individual items on an experimental management aptitude test. The data for one of the items are as shown below. Calculate the (a) point biserial r and (b) biserial r for these data. For (b) assume p = 0.56 and $\bar{X}_p = 64$; other variables are the same.

Total Test Score	Number of Managers Answering Item	
	Correctly	Incorrectly
75–79	3	1
70–74	2	0
65–69	3	2
60–64	5	3
55–59	4	5
50–54	4	3
45–49	3	3
40–44	1	3
35–39	1	3
30–34	1	2
25–29	0	2
20–24	0	1
$\bar{X}_p = 61$	$\bar{X}_t = 55$	$\sigma_t = 15$

4. Given the following data:

$$\bar{X}_x = 78$$

$$\bar{X}_y = 52$$

$$\Sigma x^2 = 4712$$

$$\Sigma y^2 = 6804$$

$$\Sigma xy = 4863$$

a. Calculate the X and Y regression equations.
b. Determine r^2 and r from these data.
c. Calculate the SE_{est} for N = 40.

5. The owner of a coffee shop wants to determine whether the shop waitresses would perform better if they took two 10-minute coffee breaks during the day. Their productivity would be measured by averaging daily receipts on two Monday evenings. Ten waitresses were randomly selected for the test. During the first week, five waitresses took coffee breaks and the other five had no breaks. During the second week, the order was reversed so that each waitress worked a week with breaks and a week without breaks during the experiment. Daily receipt totals are shown in the table below.

Subject	Coffee Break	No Coffee Break
1	8.56	9.79
2	10.99	8.70
3	8.90	8.63
4	9.78	8.52
5	10.86	10.36
6	10.04	12.44
7	10.47	9.23
8	7.91	10.35
9	8.28	8.76
10	11.33	8.34
Daily Receipts	97.12	95.12

a. Calculate the Pearson r.
b. Based on the correlation coefficient determined, what conclusions can be drawn about productivity and coffee breaks?

6. The marketing research department of RXL Enterprises, a major seller of products for home and garden, is attempting to estimate the relationship of sales to an independent variable factor to be utilized in future planning for marketing and advertising expenditures. The second variable is customer awareness—a measure of the proportion of customers who have heard of each product within six months after its introduction. Use the data from the table below.
a. Calculate the regression equations for advertising expenditures.
b. Calculate the regression equations for customer awareness.
c. Calculate the standard error of the estimate for variable.
d. Calculate the r^2 and r for each variable.

Product	Sales ($ million)	Customer Awareness (%)	Advertising Expenditures ($ million)
XES-40	43	25	1.0
EZ-Grass	112	70	2.5
Snail Roundup	11	5	0.8
Banana Grow	17	15	0.4
Spiffy Bowl	24	20	0.7
Window Shine	21	15	0.5
Spot Away	82	50	1.8
Dust No-More	55	40	1.2
Roach-B-Gone	105	75	2.5
Shine-M-Up	80	60	1.6
Grease Slasher	79	50	1.5
Miracle Mop	46	45	1.2
Handy Wash	30	30	1.0
Oven Bright	65	60	1.5

EXERCISE ANSWERS

1.a. Pearson r calculation using, the raw score formula.

$$\sum X = 179$$
$$\sum XY = 1165$$
$$\sum Y = 118$$
$$\sum X^2 = 1869$$
$$n = 20$$
$$\sum Y^2 = 828$$

$$r = \frac{(20)(1165) - (179)(118)}{\sqrt{((20)(1869) - 179^2)((20)(828) - 118^2)}}$$

$$= \frac{23300 - 21122}{\sqrt{(5339)(2636)}}$$

$$= \frac{2178}{3751.48}$$

$$r = .580$$

1.b. Rho calculation on ranks.

X	Y	R_x	R_y	D	D^2
15	9	1	3	2	4
14	8	2	5	3	9
13	7	3	8	5	25
12	10	4.5	1.5	-3	9
12	7	4.5	8	3.5	12.25
11	8	6.5	5	-1.5	2.25
11	6	6.5	11	4.5	20.25
10	10	9.5	1.5	-8	64
10	7	9.5	8	-1.5	2.25
10	6	9.5	11	1.5	2.25
10	4	9.5	15.5	6	36
9	8	12.5	5	-7.5	56.25
9	2	12.5	18.5	6	36
8	3	14	17	3	56.25
6	2	15.5	18.5	3	9
6	1	15.5	20	4.5	9
5	5	17	13.5	-3.5	20.25
4	5	18	13.5	-4.5	12.25
3	6	19	11	-8	64
1	4	20	15.5	-4.5	20.25
					$\sum D^2 = 433.5$

$$\rho = 1 - \frac{(6)(433.5)}{(20)(20^2 - 1)}$$

$$= 1 - \frac{2601}{7980}$$

$$= 1 - 0.326$$

$$= 0.674$$

1.c. Phi calculation.

$$\text{hi}_x = 9\text{--}15 \qquad \text{hi}_y = 6\text{--}10$$

$$\text{low}_x = 1\text{--}8 \qquad \text{low}_y = 1\text{--}5$$

		X		
		Hi	Low	
Y	Hi	11	1	12
	Low	2	6	8
		13	7	20

$$\phi = \frac{(11)(6)-(2)(1)}{\sqrt{(12)(8)(13)(7)}}$$

$$= \frac{64}{93.47}$$

$$= 0.685$$

1.d. Assumptions of linearity and homoscedasticity. Because phi is larger than the Pearson r and based on the shape of the scattergram, the assumptions of linearity and homoscedasticity are not met; thus, the Pearson r is an underestimate of the true relationship.

2. Inter-judge reliability.

Student	Ranks			Sum of Ranks	D	D²
1	1	4	2	7	9.5	90.25
2	2	3	3	8	8.5	72.25
3	3	1	1	5	11.5	132.25
4	4	2	4	10	6.5	42.25
5	5	5	6	16	0.5	0.25
6	6	7	5	18	1.5	2.25
7	7	6	7	20	3.5	12.25
8	8	8	9	25	8.5	72.25
9	9	10	10	29	12.5	156.25
10	10	9	8	27	10.5	110.25
				$\Sigma = 165$		$\Sigma D^2 = 690.5$

Average sum of ranks = 16.5.
Applying Kendall's coefficient of concordance,

$$W = \frac{12(690.5)}{(3^2)(10)(10^2-1)}$$

$$= \frac{8286}{8910}$$

$$= 0.93$$

3.a. Point biserial r calculation.

$$p = 0.49 \qquad q = 0.51$$

$$r_{pb} = \frac{(61-55)}{15}\sqrt{\frac{0.49}{0.51}}$$

$$= (0.4)(0.98)$$

$$= 0.392$$

3.b. Biserial r calculation.

$$y = 0.3945$$

$$r_b = \frac{(64-55)}{15}\left(\frac{0.56}{0.3945}\right)$$

$$= (0.6)(1.42)$$

$$= 0.852$$

4.a. X and Y regressions.

$$b_{xy} = \frac{4863}{6804} = 0.7147$$

$$a_{xy} = 78-(52)(0.715) = 40.82$$

$$\hat{X} = 40.82+0.715(Y)$$

$$b_{yx} = \frac{4863}{4712} = 1.032$$

$$a_{yx} = 52-(78)(1.032) = -28.50$$

$$\hat{Y} = -28.50+1.032(X)$$

4.b. r^2 and r calculations.

$$r^2 = (0.715)(1.032) = 0.738$$

$$r = 0.859$$

4.c. SE_{est} calculation.

$$SE_{est} = \sqrt{\frac{6804}{40-2}}\sqrt{1-0.738}$$

$$= (13.38)(0.512)$$

$$= 6.85$$

5.a. Pearson r calculation.

X	Y	XY	X^2	Y^2
8.56	9.79	83.80	73.27	95.84
10.99	8.70	95.61	120.78	75.69
8.90	8.63	76.81	79.21	74.48
9.78	8.52	83.33	95.65	72.59
10.86	10.36	112.51	117.94	107.33
10.04	12.44	124.90	100.80	154.75
10.47	9.23	96.64	109.62	85.19
7.91	10.35	81.87	62.57	107.12
8.28	8.76	72.53	68.56	76.74
11.33	7.34	94.49	128.37	69.56
$\Sigma X = 97.12$	$\Sigma Y = 95.12$	$\Sigma XY = 922.49$	$\Sigma X^2 = 956.77$	$\Sigma Y^2 = 919.29$

$$r = \frac{10(922.49) - 97.12(95.12)}{\sqrt{10(956.77) - (97.12)^2}\ \sqrt{10(919.29) - (95.12)^2}}$$

$$= -0.094$$

5.b. The value determined for the Pearson r is very close to 0.0 which should indicate that X and Y are possibly unrelated. In other words, we may assume that X and Y have no relationship. It appears likely that productivity is based on something other than coffee breaks. (Remember: Correlations cannot determine causality so we would be wrong to conclude that breaks do or do not affect productivity using this statistic. We can only discuss associations or relationships.)

6.a. Regression equations (advertising expenditures).

Y = Sales (in millions of dollars).

X = Advertising expenditures (in millions of dollars).

$$b = \frac{1271.2 - (14)(1.3)(55)}{29.22 - (14)(1.3)^2} = \frac{270.2}{5.56} = 48.597$$

$$a = 55 - (13)(48.597) = -8.176$$

$$Y = -8.176 + 48.597X$$

6.b. Regression equations (customer awareness).

Y = Sales (in millions of dollars).

X = Customer awareness (%)

$$b = \frac{39815 - (14)(40)(55)}{28750 - (14)(40)^2} = \frac{9015}{6350} = 1.42$$

This represents:

$$\frac{\sum xy - n(\bar{X}_x)(\bar{X}_n)}{\sum x^2 - n\bar{X}_x^2}$$

$$a = 55 - (40)(1.42) = -1.8$$

$$Y = -1.8 + 1.42X$$

6.c. SE_{est} calculation.

$$SE_{est} = \frac{\sqrt{56476 - (-8.176)(770) - (48.597)(1271.2)}}{12}$$

$$= 9.106 \text{ (millions of dollars)}$$

$$SE_{est} = \frac{\sqrt{56476 - (-1.8)(770) - (1.42)(39815)}}{12}$$

$$= 10.51 \text{ (millions of dollars)}$$

6.d. r^2 and r calculations.

$$r = \frac{(14)(1271.2) - (18.2)(770)}{\sqrt{((14)(29.22) - (18.2)^2)((14)(56476) - (770)^2)}}$$

$$= 0.96$$

$$r^2 = (r)^2 = 0.92$$

$$r = \frac{(14)(39815) - (560)(770)}{\sqrt{((14)(28750) - (560)^2)((14)(56476) - (770)^2)}}$$

$$= 0.95$$

$$r^2 = (r)^2 = 0.90$$

REFERENCES

Kerlinger, F. N. (1986). *Foundations of behavioral research* (3rd ed.). New York: Holt, Rinehart & Winston.
Larose, D. T. (2010). *Discovering statistics*. New York: Freeman & Company.

4 Simple Experimental Designs

The previous chapters covered graphing, summarizing, and simplifying descriptions of data. This chapter is concerned with the interpretation of data such that hypotheses may be tested and conclusions drawn. While descriptive statistics may identify differences in groups of data, we are now interested in whether those differences are significant and meaningful. Welcome to inferential statistics. This chapter covers these topics:

Introduction to Inferential Statistics
Statistical Hypothesis Testing
Two Randomized Groups Design: *t*-Test for Independent Samples
Two Matched Groups and Repeated Measures Designs: *t*-Test for Correlated Data
Nonparametric Analysis
Testing for Significance of Correlation
Summary
Keyword Definitions
References
Exercises
Exercise Answers

KEYWORDS

Alpha (α)
Alternative Hypothesis (H_1)
Area in Smaller Proportion
Asymmetric Transfer Effect
Beta (β)
Carryover Effect
Central Limit Theorem (CLT)
Chi-Square Test (χ^2)
Contingency Table
Counterbalancing
Critical Ratio
Degrees of Freedom *df*
Experimental Hypothesis
Homogeneity of Variance
Independent Data
Learning Effect
Mann–Whitney *U*

Matched Groups Design
Matching Variable
Meaningful Difference
Nonparametric
Normality
Null Hypothesis (H_0)
One-Sample Chi-Square
One-Tailed Hypothesis
One-Tailed Test
Overestimate
Parametric
Pool/Pooling
Power
Power Efficiency
Practice Effect
Probability
Randomization

Region of Rejection
Relative Power
Repeated Measure
Sequence Effect
Serendipitous Finding
Significant Difference
Spurious Result/Findings
Standard Error of Differences
   Between Means ($\sigma_{D_{\bar{x}}}$)
Standard Error of Mean (SEM) ($\sigma_{\bar{x}}$)
Transfer Effect
t-Test
Type I Error ($p = \alpha$)
Type II Error ($p = \beta$)
Wilcoxon Matched-Pairs
   Signed-Ranks Test
Within-Subject Measure
z-Test

INTRODUCTION TO INFERENTIAL STATISTICS

We noted in Chapter 1 that the great value of the normal curve in statistical hypothesis testing is that sampling distributions of many statistics are normal in shape. Let us see why this is the case.

SAMPLING DISTRIBUTION OF MEANS

Example

Suppose we were to draw many samples of a given size from a large population and compute their means. Let us further assume that we determined the deviation of each sample mean from the population mean (μ) and then computed the standard deviation. The standard deviation for a distribution of sample means has a special name; it is known as the *standard error of the mean*, indicated by the $\sigma_{\bar{x}}$ symbol, SEM, or S E. The word *error* is used in place of *deviation* to emphasize the fact that any variation of sample means from μ is due to sampling error.

By dividing each of our deviation scores by the $\sigma_{\bar{x}}$ we can convert them to z-scores. We now are in a position to determine the likelihood that any given sample mean occurred by chance alone. All we need to do is find the corresponding z-score in the normal probability table and determine the area in smaller proportion. This area represents the probability of obtaining a sample mean this far from the population mean (μ) or further. Thus, if the z-score in the normal probability table for a given sample mean was 1.64, the probability that it occurred by chance alone would be 0.05 or five chances in a hundred. Stated another way, if we were to draw 100 random samples, 5 could be expected to yield means that deviated from the true population mean by as much as this one.

Central Limit Theorem

It can be demonstrated that even with a skewed distribution of a given trait in a population, the distribution of sample means for that trait will be normal for 100 or more samples. This characteristic of sample means is called the *central limit theorem*. Except for extremely skewed population distributions, samples as small as 30 are adequate to assure a normal distribution of sample means. With normal population distributions, the result is a normal distribution of sample means with even smaller sized samples. Because of the central limit theorem, we can legitimately compute the standard error of the mean in our example above.

Relationship of Sample Size to $\sigma_{\bar{x}}$

As sample size increases, the size of the standard error of the mean decreases. Let us use an extreme example to illustrate this. Suppose that from a group of 1000 students we were to draw samples of 10 and determine the average heights of the students in each sample. It is entirely conceivable that at least some of the samples would include almost all short or all tall people. Thus, the means for these samples would deviate considerably from the average height of all 1000 students. If, on the other hand, we were to draw random samples of 500 students and compute their average height, the likelihood that we would draw predominantly tall or short students is extremely small. Thus, the means based on samples of 500 would vary little from one another as compared to the mean of a sampling distribution of means based on samples of 10.

The standard error of the mean is, in fact, equal to the standard deviation divided by the square root of the sample size.

$$\sigma_{\bar{x}} = SEM = \frac{\sigma}{\sqrt{n}}$$

Thus, when the sample size is 1, $\sigma_{\bar{x}} = \sigma$. For any larger sample size, the standard error of the mean will be systematically smaller than the standard deviation. As the sample size approaches the population size, the standard error of the mean approaches 0. Stated another way, by increasing the sample size, the standard error of the mean can be reduced to any desired level. However, the reduction is not *pro rata*. The sample size must be quadrupled to reduce the standard error by one half.

Computing Standard Error of Mean $\sigma_{\bar{x}}$

To determine the standard error of the mean, we divided σ by the square root of the sample size. However, because $\sigma_{\bar{x}}$ is an unbiased estimate of σ, we can compute an unbiased estimate of $\sigma_{\bar{x}}$ directly from the standard deviation for a single random sample.

$$SE \quad \text{or} \quad SEM \quad \text{or} \quad \sigma_{\bar{x}} = \frac{\sigma X^2}{\sqrt{n}} \quad \text{or} \quad \sqrt{\left(\frac{\sigma^2}{n}\right)} \quad \text{or} \quad \sqrt{\left(\frac{\Sigma X^2}{n(n-1)}\right)} \quad \text{or}$$

$$\sqrt{\left(\frac{\Sigma X^2 - \dfrac{(\Sigma X)^2}{n}}{n(n-1)}\right)}$$

For example, if for a given random sample of 25 persons the standard deviation was 10, we would determine the standard error of the mean as follows:

$$\sigma_{\bar{x}} = \frac{10}{\sqrt{25}} = \frac{10}{5}$$

$$= 2$$

SAMPLING DISTRIBUTION OF DIFFERENCE BETWEEN TWO MEANS $\sigma_{D_{\bar{x}}}$

Example

Instead of drawing sample means and determining their sampling distribution, we could randomly draw pairs of samples. For each pair of samples, we could then determine the difference between their means and construct a frequency polygon of the difference scores. As with the sampling distributions of means, we also could compute the standard deviation for the distribution of differences between sample means (standard error of the difference between the means $\sigma_{D_{\bar{x}}}$), convert our difference scores to z-scores, and interpret them using the normal probability table. In other words, using the z table, we can determine the probability of any size difference between the means occurring by chance alone.

$$z = \frac{\bar{X}_1 - \bar{X}_2}{\sigma_{D_{\bar{x}}}}$$

This ratio of the difference between the means to the standard error of the difference is called the *critical ratio* and as we shall see shortly serves as the basis for all statistical hypothesis testing.

Computing $\sigma_{D_{\bar{x}}}$

Just as we could compute an unbiased estimate of the standard error of the mean from a single random sample, we can compute an unbiased estimate of the standard error of the differences between the means from two independently drawn random samples.

$$S_{D_{\bar{x}}} \approx \sigma_{D_{\bar{x}}} = \sqrt{\frac{\sigma^2_{x_1}}{n} + \frac{\sigma^2_{x_2}}{n}}$$

The technique is to *pool* the two variances σ^2. The equation makes use of the information in both samples to yield an unbiased estimate of the common population variance.

STATISTICAL HYPOTHESIS TESTING

Having demonstrated that it is possible for us to determine the probability of a given difference between two sample means occurring by chance, let us now see how this information may be used in statistical testing of experimental hypotheses.

Example

Let us assume we have designed an experiment to determine the effects of illumination level on the reading performance of students as measured by a test of their knowledge of material read. The students were randomly assigned to one of two groups. One group was exposed to a high level and the second group to a low level of illumination during the reading period. Our hypothesis was that the students reading under the high illumination condition would perform significantly better on a test of the material read than the group of students who read under a low level of illumination. Let us further assume we obtained the statistical information shown in Table 4.1 from our data.

To test our experimental hypothesis statistically, we must demonstrate a probable significant and meaningful difference in performance of the two groups due to lighting. Thus we propose that a hypothesis that the difference between the means for the two samples is merely a chance occurrence cannot be supported. We will then seek to statistically refute this second or null hypothesis. In doing so, we must also decide how much of a risk of being wrong we are willing to take if we reject the hypothesis of no difference or *null hypothesis* (H_0) and conclude that there was a real difference—a difference that would have appeared if we analyzed the entire population.

In the behavioral sciences, researchers typically do not risk being wrong in rejecting the null hypothesis more than 5 times in 100. They often establish the maximum risk at 1 in 100 and occasionally even 1 in 1000. Where researchers establish their levels of risk often depends on the seriousness of the consequences are if they are wrong in rejecting the null hypothesis. The level of risk is referred to as *alpha* and is symbolized by the Greek letter α.

Let us assume that in our experiment we are willing to risk being wrong in rejecting the null hypothesis 5 times in 100 ($\alpha = 0.05$). We begin by stating our null hypothesis designated H_0.

$$H_0 : \mu_1 - \mu_2 \leq 0$$

where
μ_1 = mean of population receiving high illumination treatment condition
μ_2 = mean of population receiving low illumination treatment condition

This hypothesis states that the difference between the average performance of the population receiving high illumination treatment and the population receiving low illumination is equal to or less

---

TABLE 4.1
Summary of Results of Illumination Experiment

	Group 1: High Illumination	Group 2: Low Illumination
$\bar{X}$	24.0	20.0
σ	6.0	6.5
n	100.0	100.0

---

than zero. (Because we already know that any real difference must arise because the high illumination helps performance relative to the effect of low illumination, any negative difference between our means must be due to chance.) We next state the alternative hypothesis H_1.

$$H_1 : \mu_1 - \mu_2 > 0$$

This hypothesis states that the average performance of the high illumination population is better than that of the low illumination population. It constitutes our *experimental hypothesis.* If we can reject the null hypothesis at our chosen probability level α, we then may assert (accept) the alternative or experimental hypothesis.

The next step is to determine from Table B in Appendix A the z-score required for α to be equal to or less than our chosen value of 0.05. Entering the z-table for an area smaller than 0.05, we find that the corresponding z-score is 1.64. Therefore, to reject the null hypothesis, our observed z-score or critical ratio must be equal to or larger than 1.64. For any smaller observed z value, the α or area in the smaller proportion would be greater than 0.05.

Returning to our data in Table 4.1, let us compute our observed z-score or critical ratio.

$$z = \frac{\bar{X}_1 - \bar{X}_2}{\sigma_{D_{\bar{x}}}} = \frac{\bar{X}_1 - \bar{X}_2}{\sqrt{\sigma\frac{2}{X_1} + \sigma\frac{2}{X_2}}}$$

$$= \frac{24 - 20}{\sqrt{\left(\frac{6.0}{\sqrt{100}}\right)^2 + \left(\frac{6.5}{\sqrt{100}}\right)^2}} = \frac{4}{\sqrt{(.60)^2 + (.65)^2}}$$

$$= \frac{4}{\sqrt{.36 + .422}} = \frac{4}{\sqrt{.782}} = \frac{4}{.884}$$

$$= 4.52$$

Because our observed value is greater than our tabled value of 1.64, we may reject the null hypothesis at the 0.05 point. This allows us to assert (affirm or accept) the alternative hypothesis (H_1) that the difference between our two sample means was a real or *significant difference*: as illumination level increased, performance increased. Note that causality is not included in this interpretation.

ONE-TAILED VERSUS TWO-TAILED HYPOTHESES

Our example assumed we knew enough about the effects of illumination at the intensity levels employed to state that if any effect occurred, the higher illumination had to improve performance. This is known as a *one-tailed hypothesis.* It is so named because the only way it is possible for us to be wrong in rejecting the null hypothesis is to say the *higher* illumination resulted in improved performance when it did not. Thus, our statistical test is a *one-tailed test* because the α or *region of rejection* is concentrated at one end of the normal distribution (see Figure 4.1). If we had not been confident that any effect of illumination on performance could only be in one direction, we would have conducted a two-tailed test starting by stating our statistical hypothesis as follows.

$$H_0 : \mu_1 - \mu_2 = 0$$

$$H_0 : \mu_1 - \mu_2 \neq 0$$

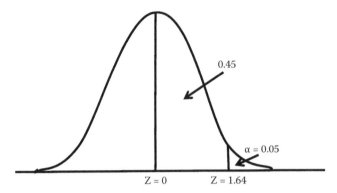

FIGURE 4.1   Area in the smaller portion (tail) of a normal distribution and *z*-score value for alpha of 0.05.

In other words, our null hypotheses would state that there was no difference between the two populations. Our alternative hypothesis would state that a difference occurred but it would not specify
a direction; theoretically, we could have found the performance of the low illumination group to
have been significantly *better* than the high illumination group. In this case, our region of rejection
should have been split between both ends of the normal distribution as depicted in Figure 4.2.

In a two-tailed test, we must use the tabled value of *z* corresponding to our area in a smaller proportion of 0.025. The corresponding *z*-score is 1.96. Now however, an observed *z* equal to or greater
than a positive or a negative 1.96 would lead to rejection of the null hypothesis. In the one-tailed test,
only a positive (or only a negative) *z*-score could result in rejection.

Researchers frequently do not know enough about their variables to be certain that the outcome of an experiment will be only in one direction. Because of this uncertainty, most behavioral
research utilizes two-tailed tests.

If we had been unable to reject H_0, the conclusion would not necessarily be that illumination has
no effect on performance. Failure to reject the null hypothesis could also indicate a difference but
that the samples were too small to detect it. Another reason for finding no difference could be that
the experimental levels of the IV were too close to significantly affect the DM. Finally, failure to
properly control for one or more contaminating or extraneous variables is another possible reason
for not being able to reject H_0. For example, if during the period when the experimental group was
performing the task, workmen started banging on pipes, the distraction could have inhibited the
performance on the DM. All these alternative explanations represent faulty research and will be
discussed in greater detail in Chapter 6.

Remember: we *reject* or *fail to reject* the null hypothesis. We *affirm, assert, or accept* the alternative hypothesis. All we can do when our statistical test is nonsignificant is *fail to reject* the null

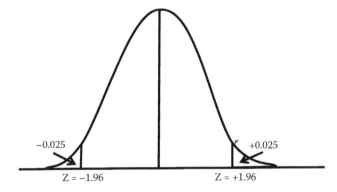

FIGURE 4.2   Region of rejection for two-tailed test at the 0.05 level of significance.

hypothesis. Just as statistics can never *prove* anything, we cannot demonstrate that people or events are similar; we can only show that they are different. If we test for the effects of a drug, we can say that we found drug or failed to find drug effects. Only a foolish statistician or naive student would say that a drug had no effects. A more correct statement would be, "If the drug causes effects, our study failed to find such evidence."

TYPE I AND TYPE II ERRORS

In statistical hypothesis testing, it is possible to make two different types of errors. The first is rejection of the null hypothesis when in fact it should not be rejected. This is called a *Type I error* and the probability of this type of error is equal to alpha (α). The second type of error is to fail to reject (assert, affirm, or fail to reject) the null hypothesis when in fact it should have been rejected. This is referred to as a *Type II error*. When researchers want to assert the alternative hypothesis (H_1), they will be conservative and guard against Type I error (guard against saying there is a difference when there is not) by setting the level of α numerically very small.

When researchers hope to be able to fail to reject the null hypothesis, they will be conservative and guard against Type II error (guard against saying there is no difference when there is). Sometimes, researchers also will guard against Type II error in exploratory studies. They want to ensure that they do not miss a *possible* IV effect and thus fail to follow up with additional research. Researchers guard against Type II error by setting the level of alpha numerically larger (e.g., at 0.10 or 0.20) than when the primary concern is Type I error.

Although as the level of α increases the probability of a Type II error decreases, it does so at a nonlinear and unspecifiable rate. In other words, Type II error is not $1 - \alpha$. The probability of a Type II error is said to be equal to beta (β) or $1 - power$. We cannot specify β because it depends on how far away from the true value of μ_1 and μ_2 the sample means happen to fall. The precise population parameters are seldom known in the usual research situation. However, because precision or power increases with increasing sample size, one way to better control for both Type I and Type II error is to increase n.

POWER OF STATISTICAL TESTING

The ability to reject the null hypothesis when it should be rejected is called *power*. Power plus the probability of Type II error (β) is equal to 1. The larger the sample size and/or the numerically larger the level of significance chosen (α), the more powerful is the statistical test.

Statistical tests differ in their power. For example, *t*-tests and *F*-tests normally are more powerful than other statistical tests of the null hypothesis. The relative ability of other statistical tests to reject the null hypothesis when compared to the *t*- and *F*-tests is called the *relative power* or *power efficiency*. Relative power is expressed as the ratio of the sample size required to reject a null hypothesis with either the *t*- or *F*-tests to the sample size required with the statistic under consideration. For example, the relative power of the Mann–Whitney *U*-test is about 0.95. This means that if 95 subjects were required to reject a given null hypothesis with the *t*-test, 100 subjects would be needed using the *U*-test. Both the *t*-test and the Mann–Whitney *U*-test will be discussed in this chapter. The *F*-test will be discussed in Chapter 5.

TWO RANDOMIZED GROUPS DESIGNS: *t*-TEST FOR INDEPENDENT SAMPLES

TWO RANDOMIZED GROUPS (BETWEEN GROUPS) DESIGN

Our hypothetical experiment to study the effects of illumination on performance is a good example of the two randomized groups design. You may recall that we began by randomly subdividing our randomly selected sample into two groups. One group was subjected to the high illumination treatment condition and the other group to the low illumination condition. We then observed their reading performance by administering a test of reading comprehension (DM) and tested the results for significance using the z-test.

We assumed that by randomizing the assignment of subjects to the two experimental groups, differences between the groups on a wide spectrum of extraneous variables would *randomize out*. For example, by chance alone there are likely to be as many highly motivated subjects in one group as in the other. Therefore, we assume that the groups *on the average* will be equal in motivation. The same would be true of such variables as intelligence, reading ability, attention span, and other subject-related variables that might conceivably affect the outcome of the experiment. Therefore, any differences between the groups on the DM may be attributed to the differences in illumination level (IV). This assumption of equivalency of the groups is more likely to be true of large samples, but may hold fairly well even for small ones.

t-TEST FOR INDEPENDENT DATA

Earlier in this chapter, we tested our null hypothesis using the z-test. This was entirely appropriate with a very large sample. However, it has been empirically demonstrated that the distributions of small samples may differ markedly from the normal distribution. The first person to discover and record this fact was a young chemist named William S. Gossett. Shortly after the beginning of the 20th century, while working for a brewery in Dublin, Ireland, Gossett became concerned with problems of quality control. He soon discovered that the traditional normal curve was an inadequate model for statistically analyzing small samples.

He therefore secured the body measurements of over 3000 British criminals, recorded them on cards, and proceeded to draw large numbers of samples of a given size. Through this empirical procedure, Gossett was able to establish different sampling distributions for different sizes of samples. He later worked out the mathematical expression for these sampling distributions with the aid of a university professor and published the results in 1908 under the pen name "Student." Some 20 years later, these findings were included in the first textbook on modern statistics written by Sir Ronald A. Fisher and finally received their much deserved attention. The development of the Student's t-distribution marked the beginning of modern statistical inference.

Concept of Degrees of Freedom

The t-distribution changes as a function of its *degrees of freedom* (*df*). The concept of degrees of freedom is closely related to n and refers to the number of scores in a distribution that are free to vary. The number of degrees of freedom is equal to the number of observations minus the number of algebraically independent linear restrictions placed upon them. Usually, these restrictions are the underlying statistics that must be determined to calculate the particular statistic used to test the hypothesis. For example, the sample $\overline{X}$ is used to estimate $S_{\overline{X}}$ or $\sigma_{\overline{X}}$. In general, one degree of freedom is lost for every independent mean about which the scores vary. For instance, if our hypothetical experiment included 10 subjects in each of 2 independent groups, the degrees of freedom would be $n_1 - 1 + n_2 - 1 = 9 + 9$ or 18. Another way to think of degrees of freedom is to consider a mean of a given size in which 9 of 10 scores may have any value—they are free to vary. However, for a mean to have a particular value, the 10th score must be a particular value and thus is not free to vary.

Use of t-Test in Statistical Hypothesis Testing

Table C in Appendix A gives the distribution of t for many different degrees of freedom and for the levels of significance most commonly used in statistical hypothesis testing. As the degrees of freedom increase from 1 toward infinity (∞), the value of t required to reject the null hypothesis at a given level of α systematically decreases. It should be noted that when $df = \infty$, $t = z$. For this reason, it is fairly common practice to use the t-distribution in testing statistical hypotheses rather than the z-distribution regardless of sample size.

Computation of the t-test is identical to that for the z-test — The only difference is that we look up the tabled value for our *critical ratio* in the t-table (Table C in Appendix A) rather than in Table B. For example, you may recall that in our hypothetical experiment our observed critical ratio was 3.20. Had our total number of subjects been only 20, we would have entered the t table (Table C) at the .05 level with $n_1 - 1 + n_2 - 1$ df, or 18 df. The corresponding t value is 2.101. Thus, we could have rejected the null hypothesis. Had our observed critical ratio been 1.98, we could not have rejected the null hypothesis, but could have if df had been equal to or greater than 120 (see Table C for 120 df, $\alpha = 0.05$).

LIMITATIONS OF RANDOMIZED GROUPS DESIGN

We noted earlier that when using the between-groups design we assume differences between the groups on a wide spectrum of variables will randomize out. This assumption tends to hold fairly well for large samples. This assumption becomes more tenuous as sample size decreases, that is, the likelihood that the two groups will be less than equivalent with respect to some important extraneous variable increases.

In experiments to be conducted over a period of time, it is always possible to lose a number of subjects for a wide variety of personal or circumstantial reasons not controllable by a researcher. Because of this violation of the random assignment of subjects to groups, the likelihood that the groups will no longer be equivalent increases greatly. In addition, regardless of the experimental design used, loss of subjects may cause the total sample to no longer be representative of the population from which it was drawn. Hence, generalizing results to a population becomes improper and cannot be done with confidence.

A third limitation of the randomized groups design arose from the way researchers employed it in many experimental situations. Researchers often cannot really assign subjects to groups using a true randomized procedure, but instead must utilize already defined and intact groups such as a class section in a school. In some cases, previous assignment to one or the other of the intact groups may have been achieved by randomizing. Too often, however, this is not the case. For example, let us assume that we wish to investigate the effect of some IV on some aspect of performance (DM) using the employees in accounting departments in two different divisions of a manufacturing firm. We can further assume that the accounting department from Division A is our experimental group and the department in Division B is our control group. Because of differences in geographic location, organizational environments, and local populations from which the employees were hired, to name a few potentially relevant dimensions, it is unlikely that the two groups are equivalent. We discussed this problem in Chapter 1 and noted that one approach to dealing with these kinds of conditions is resorting to quasi-experimental (nonequivalent control group) design that requires repeated measures (pre- and post-treatment) of both groups.

Whenever any of the above three conditions exist (small sample size, likelihood of loss of subjects during the experiment, or the use of intact groups not developed on a randomized basis), serious consideration should be given to using repeated measures or matched groups design as appropriate.

TWO MATCHED GROUPS AND REPEATED MEASURES DESIGNS: t-TEST FOR CORRELATED DATA

In using the two randomized groups design, we assumed that chance assignment of subjects would produce two essentially equal groups. As noted previously, this assumption tends to hold fairly well for large samples, but as sample size decreases, the assumption becomes more tenuous. To increase the precision of our control over extraneous variables, two alternative simple experimental designs may be used. These are the *two matched groups* and *repeated measures* designs.

Two Matched Groups Design

Sometimes the experimenter is aware of an extraneous variable that is known to correlate with the dependent measure. For example, in our experiment on the effect of illumination on performance, let us assume that intelligence quotient (IQ) is known to correlate with scores on our performance test.

If our samples were quite small, it is very possible that randomly assigning subjects to the two treatment conditions would result in groups that were unequal in IQ. As a result, we would expect to obtain a difference between the groups on our performance measure even if the difference in illumination levels had no effect. To avoid this possibility we could first administer an IQ test (or otherwise obtain the subjects' IQs from existing records). Next, we could place our subjects in rank order according to IQ scores. Then, beginning at the top (or bottom) of the list, divide the list into pairs of subjects. The subjects within each pair would then be randomly assigned to the two treatment conditions. This procedure would ensure that the two groups were equal with respect to IQ (see Table 4.2).

The dimension on which our subjects were paired is known as the *matching variable*. By using it we decreased the likelihood that the groups would be unequal by balancing the groups on at least one important extraneous variable. To control for other potentially contaminating extraneous variables, we continued by randomly assigning the subjects within each pair to the control or experimental group. This step also is particularly important to prevent experimenter bias from affecting the assignments. For example, an experimenter could unwittingly assign more highly motivated subjects to a particular group.

t-Test for Correlated Data

As with the two random groups design, the statistical test most frequently used with the two matched groups design is the *t*. However, because we matched our subjects we no longer have two completely independent groups. In both groups, performance of the high IQ subjects may be expected to be higher than the performance of the low IQ subjects (i.e., the performance scores of the two groups are related). As a result, the procedures for computing the *t*-test and degrees of freedom will be different from the computations for noncorrelated data.

Computation of *t* for Correlated Data

Let us assume that in our hypothetical illumination experiment discussed earlier, we obtained the data depicted in the second and third columns in Table 4.3. The first step in computing *t* is

---

TABLE 4.2
Development of Two Matched Groups on Basis of Intelligence Scores

High Group Subject	Illumination IQ	Low Group Subject	Illumination IQ
1	135	2	130
4	125	3	128
6	123	5	125
7	120	8	118
10	115	9	115
11	114	12	112
13	112	14	111
16	110	15	108
18	100	17	105
20	92	19	97
	$\Sigma = 1147$		$\Sigma = 1149$

TABLE 4.3
Data from Illumination Experiment: Matched Groups Design

Pair	High Illumination	Low Illumination	D	D^2
1	26	21	5	25
2	22	10	12	4
3	18	19	−1	1
4	18	12	6	36
5	16	10	6	36
6	14	16	−2	4
7	15	10	5	25
8	13	10	3	9
9	11	7	4	16
10	10	6	4	16
			$\Sigma D = 32$	$\Sigma D^2 = 172$

determining the difference between the performance scores for each matched pair of subjects (D in Column 4). We next square our difference scores (Column 5, D^2), and then add the D and D^2 columns. To compute t, we use the following formula:

$$t = \frac{\bar{X}_1 - \bar{X}_2}{S_{D_{\bar{X}}}} = \frac{\bar{D}}{\sqrt{\dfrac{\Sigma d^2}{n(n-1)}}} = \frac{\dfrac{\Sigma D}{n}}{\sqrt{\dfrac{\Sigma D^2}{n(n-1)}}}$$

where
   $\bar{D}$ = mean difference between pairs
   n = number of matched pairs

Next we solve for Σd^2

$$\Sigma d^2 = \Sigma D^2 - \frac{(\Sigma D)^2}{n}$$

$$= 172 - \frac{(32)^2}{10}$$

$$= 172 - \frac{1024}{10}$$

$$= 172 - 102.4 = 69.6$$

Substituting the formula for t, we have

$$t = \frac{\dfrac{32}{10}}{\sqrt{\dfrac{69.6}{10(9)}}} = \frac{3.2}{\sqrt{\dfrac{69.6}{90}}} = \frac{3.2}{\sqrt{.77}} = \frac{3.2}{.879} = 3.64$$

To test our null hypothesis, we would proceed as follows:

1. $H_0 : \mu_1 - \mu_2 = 0$; $H_0 : \mu_1 - \mu_2 \neq 0$.
2. t-test for correlated data, $n = 10$ pairs: $df = n - 1 = 9$.

Note that our degrees of freedom were n pairs -1 rather than $n_1 - 1 + n_2 - 1$. This is because the two sets of scores are not independent. Remember from our discussion of degrees of freedom that generally one degree of freedom is lost for each *independent* mean about which the data are summed. Here the independent mean is $\bar{D}$.

1. $\alpha = 0.01$ level, two-tailed test: Tabled value of $t = 3.25$.
2. t observed (3.64) greater than tabled value of t (3.25); reject H_0 and assert H_1.

Correlation and Two Matched Groups Design

When the data between two sets of scores are correlated, the formula for $\sigma_{D_{\bar{x}}}$ may be written

$$\sigma_{D_{\bar{x}}} = \sqrt{\sigma_{\bar{X}_1}^2 + \sigma_{\bar{X}_2}^2 - 2(r_{12})(\sigma_{\bar{X}_1})(\sigma_{\bar{X}_2})}$$

where r_{12} denotes the correlation between the dependent variable scores of the two groups. This is the general formula for the standard error of the mean. However, in the two randomized groups design the term $-2 (r_{12})(\sigma_{\bar{X}_1} - \sigma_{\bar{X}_2})$ drops out because we may assume $r = 0.0$ (i.e., the two groups are uncorrelated or independent).

The larger the value of r_{12}, the larger the value subtracted from $\sigma_{D_{\bar{x}}}$. Therefore, the denominator of our critical ratio will be smaller and the derived t larger than would be the case using the randomized groups design (where $r = 0$). Rationally this makes sense. By equating our groups on a known contaminating extraneous variable, we have reduced the variability in our experiment due to extraneous variables (i.e., we have reduced experimental error $\sigma_{D_{\bar{x}}}$).

This increase in precision is the major advantage of the matched groups design. However, there are tradeoffs. As you may recall, the degrees of freedom for correlated data are only one half the number for the two randomized groups design ($df = n$ pairs $- 1$ rather than $n_1 - 1 + n_2 - 1$). The smaller the df, the larger the value of t that must be obtained to reject (H_0). Therefore, in comparison with the two randomized groups design, the advantage of the two matched groups design is a gain in precision that in turn will result in deriving a larger critical ratio (observed t). Its major disadvantage is that because you lose one half of the df, the t value that must be obtained to reject (H_0) also increases. Figure 4.3 depicts this relationship between n and r_{12}. If the correlation between the matching variable and dependent variable (r_{12}) exceeds that shown in Figure 4.3 for the number of subject pairs (n), the matched groups design should be used rather than the randomized groups design. If the r_{12} value is less than that depicted, the two randomized groups design is preferred.

Another disadvantage of the matched groups design is that often a lapse of time occurs between (a) the testing or measuring of potential subjects on the matching variable, (b) matching, ranking, and establishment of the groups, and (c) the actual conduct of the experiment. The longer the wait, the poorer the matching is likely to be, simply because of at least some loss of reliability of test scores over time. Also related to the lapse of time between matching and conduct of the experiment is the disadvantage of possible loss of subjects. Because the loss of one subject of a matched pair means you also cannot use the matched subject, the likelihood of losing subjects to the point of no longer having a sample representative of the population is much greater than with the random groups design. For this reason a researcher my choose not to use a matched groups design whenever loss of subjects seems likely.

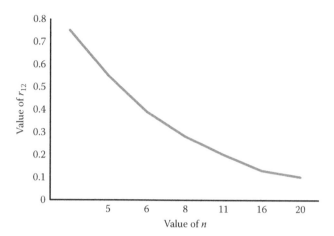

FIGURE 4.3   Relationship between n and r_{12}. Enter with a value of n and read the your expected value of r_{12} that intersects the curve at that point. If your expected value of r_{12} exceeds the value obtained from the curve, a matched groups design is preferred.

A third limitation of particular importance is that the more precise the matching and/or the more variables on which the subjects are matched, the greater the number of subjects that will be unusable—they will not match up with anyone else in the sample. As a result, the remaining sample will not be representative of the total population of interest. With precise matching, often the extreme scores cannot be used. Thus the range of subjects is restricted and the results of the experiment may be generalized only to a restricted portion of the total population. When more than one or two variables are used for matching, large numbers of subjects are likely to be lost for lack of a partner who matches up on all the variables. Because of this factor, matching on the basis of a number of variables rarely is practical.

Repeated Measures (Within Subjects) Design

In the two matched groups design, the purpose of matching was to treat each pair of subjects as one subject with respect to at least one of the variables influencing the DM. Because this increases the precision of an experiment, it follows that using the same subject under both (or all) experimental conditions should increase the precision still further. This is quite true. The individual differences that are sources of random error, are controlled by using the same subject repeatedly. This design is called *repeated measures,* or *within subjects,* because we are obtaining measures (DM scores) on each subject for more than one treatment condition.

Advantages and Uses of Repeated Measures Designs

Because the variability is reduced in a repeated measures design, the experiment is more sensitive to treatment effects. Repeated measures also is a very efficient design; the need for several groups of subjects may be reduced to merely several subjects.

The repeated measures design is useful if you expect large individual differences, for example, to study the effect of a drug on pain tolerance because pain tolerance (DM value) varies widely among individuals. This technique is also used to examine changes in behavior over time, such as the effects of learning on task performance. Where the between-groups design is used to avoid *carryover effects* from one treatment to another, the within-subjects design may be used specifically to investigate those effects.

Disadvantages of Repeated Measures Designs

One problem with a repeated measures experiment is that the potential for undesirable *carryover effects*. For example, giving individual subjects a sequence of different drug dosages may produce unforeseeable carryover effects. No matter what order of dosages is used, tolerance or desensitization may occur as a confounding variable because it is not controlled. The randomized groups design is the best for controlling for both known and unknown confounding variables.

A second problem is with time-related or order effects; later measures are affected by earlier measures, regardless of the order. These are the effects of fatigue (or practice in tests not designed to study learning). Test scores may improve on a second test, even without feedback on the first test, due to increased familiarity or comfort. Order effects differ from carryover effects in that they are transitory and may be controlled by counterbalancing. Carryover effects may be permanent or semi-permanent, as in drug experiments.

Differential order or sequence effects occur when a score on a given task is affected by the specific task or tasks preceding it. For example, suppose we were studying the effects of positive versus negative reinforcement on performance. Our positive reinforcer (Treatment A) may be candy; our negative reinforcer (Treatment B) may be a fairly strong electric shock. The subjects receiving the shock first may be so anxious by the time they receive Treatment A that their performance under positive reinforcement is seriously degraded. On the other hand, prior performance under positive reinforcement would not be expected to adversely affect performance under Treatment B. Sequence effects cannot be controlled by counterbalancing.

Experiences over time may cause learning effects that improve or degrade performance on subsequent testing. Learning effects differ from order or sequence effects by demonstrating a relatively permanent change in behavior. Such changes may be called transfer effects to indicate changes transferred to the next condition. Transfer effects may interact with order or sequence effects in a directional way. Asymmetric transfer effects describe the case in which there is a greater influence on behavior of encountering Condition A before B than the reverse. Remember: the goal in designing experiments is to control or wash out (balance or neutralize) any such extraneous or nuisance effects so that outside variables do not cloud explanations or confuse interpretation of findings.

Counterbalancing in Repeated Measures Designs

The purpose of *counterbalancing* is to mitigate the effects of the order of treatments on a subject. We could randomize the order of treatments by randomly assigning the subjects in the randomized groups design. But because one advantage of repeated measures is using fewer subjects for the experiment, this technique is usually not suitable; rarely will true randomization occur.

Another infrequently used technique is to perform all possible orders of treatments. This is easy with only two treatments, but becomes prohibitively complex with more than two treatments (4 conditions require 24 orders, 5 conditions require 120 orders). For more than two treatments, incomplete counterbalancing is often used and will be discussed in Chapter 6. For two-treatment designs, counterbalancing is achieved by assigning one-half of the subjects to Treatment A first and to Treatment B second, and the other half of the subjects to B first and to A second, as illustrated in Table 4.4.

TABLE 4.4
Counterbalancing to Control for Extraneous Variables

	Experimental Session	
	First Treatment	**Second Treatment**
1/2 subjects	High Illumination	Low Illumination
1/2 subjects	Low Illumination	High Illumination

Using *t*-Test With Repeated Measures Design

With the repeated measures design, because each subject serves as his or her own control, the data for the two treatment conditions obviously are correlated. Therefore, as with the two matched groups design, the observed *t* value is computed using the formula for correlated data.

$$t = \frac{\bar{D}}{\sqrt{\dfrac{\Sigma d^2}{n(n-1)}}}$$

Table 4.5 depicts the scores for the 20 subjects in the illumination experiment assuming use of a repeated measures design in which

$$\Sigma d^2 = \Sigma D^2 - \frac{(\Sigma D)^2}{n}$$

$$= 276 - \frac{(62)^2}{20}$$

$$= 276 - \frac{3844}{20}$$

$$= 276 - 192.2 = 83.8$$

TABLE 4.5
Data from Illumination Experiment: Repeated Measures Design

Subject	High Illumination	Low Illumination	D	D^2
1	26	22	4	16
2	24	21	3	9
3	22	23	−1	1
4	20	14	6	36
5	18	15	3	9
6	18	14	4	16
7	17	14	3	9
8	16	11	5	25
9	15	15	0	0
10	15	12	3	9
11	15	11	4	16
12	15	10	5	25
13	14	12	2	4
14	14	8	6	36
15	12	10	2	4
16	11	13	−2	4
17	11	7	4	16
18	10	6	4	16
19	9	5	4	16
20	7	4	3	9
$n = 20$			$\Sigma D = 62$	$\Sigma D^2 = 276$

Substituting in we have

$$t = \dfrac{\dfrac{62}{20}}{\sqrt{\dfrac{83.8}{20(19)}}} = \dfrac{3.10}{\sqrt{0.221}} = \dfrac{3.10}{0.470} = 6.60$$

With $\alpha = 0.01$, $df = n - 1 = 19$ two-tailed test, our tabled value of $t = 2.861$. We therefore would reject H_0. The greater precision possible with the repeated measures design without losing degrees of freedom resulted in a higher observed t than with the randomized groups or matched groups designs. It should be noted that we actually gained one degree of freedom over the df using the same number of subjects with the randomized group design.

NONPARAMETRIC ANALYSIS

Thus far we have analyzed parametric data (i.e., continuous data), that is on interval or ratio scales and is assumed to be normally distributed with homogeneous variability. The next portion of this chapter presents how to analyze non-parametric data.

Statistical techniques designed for use with interval or higher-level data are called *parametric* statistics. Parametric statistics require the assumptions of normality and homogeneity of variance. *Normality* means that the underlying distribution of the DM is normal in shape. *Homogeneity of variance* is the assumption that the variances of the two samples are equal, that is, they do not differ statistically from one another. Because parametric statistical tests are based on the use of more precise levels of measurement and more stringent assumptions, they are more powerful than statistics requiring less precise levels of measurement. For this reason parametric statistics should be used whenever appropriate. The z-test and t-test are two of the most commonly used parametric statistical tests. Another common parametric test we have mentioned and will discuss in Chapter 6 is the F-test for more complex experimental designs.

Researchers frequently do not have interval level data in their studies. For these situations, a number of statistical techniques require only nominal or ordinal levels of measurement and are known as *nonparametric* statistics. In this chapter, we will discuss the three most common nonparametric statistical tests: (a) the Mann–Whitney U, a nonparametric alternative to the t-test for independent samples; (b) the Wilcoxon matched-pairs signed-ranks test, a nonparametric alternative to the t-test for correlated data; and (c) the Chi-square test that may be used with nominal level data.

MANN–WHITNEY U-TEST

As mentioned above, the Mann–Whitney U-test is a nonparametric alternative to the t-test for independent samples. Accordingly, the Mann–Whitney U is used with the two randomized groups design when the DM does not exceed ordinal level measurement. The U-test is almost as powerful as the t-test (the relative power of U is about 0.95).

Assumptions of Mann–Whitney U-Test

The Mann–Whitney U requires ordinal level data and also requires the assumption that the data are continuously distributed. Accordingly, the power of the test decreases greatly if a large number of tie scores are present. The assumptions of normality of the underlying distribution and homogeneity of variance required by parametric tests are not required by the Mann–Whitney U so it is a good alternative to the t-test when the sample size is very small (even when the researcher has interval level data). With small samples, the underlying assumptions of the t are more likely to be violated and the violations are more likely to grossly affect the test.

Computation of Mann–Whitney U-Test

To illustrate computation of the Mann–Whitney U and explain its underlying rationale, let us return to our illumination study for the two randomized groups design. The first step in computing the Mann–Whitney U is to arrange the DM scores for the two groups into a single ranking as depicted in Column 2 of Table 4.6. To keep track of which ranks refer to high and low illumination scores, we used a and b subscripts, respectively.

The next step is to compute U_a and U_b. U_a represents the number of times a rank from Sample A (high illumination group) is preceded by a rank from Sample B (low illumination group). U_b is the number of times a rank from Sample B is preceded by a rank from Sample A. In our example, $U_a = 18$ and $U_b = 82$. The *smaller* of these two statistics is then designated U. In our example, $U = U_a = 18$. This value is then compared to the value of U in Table D in Appendix A for the desired level of significance. If our observed U of 18 is equal to or smaller than the tabled value, we may reject H_0.

Let us conduct a two-tailed test at the 0.05 level. Turning to Table D, we note that the left-hand column is designated n_1 and the top row is designated n_2. The n_1 symbol refers to the size of the larger sample and the n_2 to the size of the smaller sample. If the sample sizes are the same, the researcher may arbitrarily designate either n_1 or n_2 as the size of Group A and the other as the size of Group B. For our example, consulting the table at the 0.05 level (two-tailed test for $n_1 = 10$, $n_2 = 10$), we find the value of U to be 23. Because our observed U of 18 was smaller than 23, we may reject the null hypothesis. In actual practice, it is frequently easier to compute U_a and U_b as follows:

$$U_a = n_a n_b + \frac{n_b(n_b + 1)}{2} - \Sigma R_b$$

$$U_b = n_a n_b + \frac{n_a(n_a + 1)}{2} - \Sigma R_a$$

TABLE 4.6
Ranking of Data for Computation of Mann–Whitney U

All Scores	All Ranks	A Ranks High Illumination	B Ranks Low Illumination	U_a	U_b
28a	1a	1		0	
27a	2a	2		0	
26a	3a	3		0	
24b	4b		4		3
22a	5a	5		1	
20b	6b		6		4
19a	7a	7		2	
18a	8.5a	8.5		2	
18a	8.5a	8.5		2	
16a	10a	10		2	
15b	11b		11		8
14a	12a	12		3	
13b	13b		13		9
12b	14b		14		9
11b	15b		15		9
10a	16a	16		6	
8b	17b		17		10
7b	18.5b		18.5		10
7b	18.5b		18.5		10
6b	20b		20		10
		$\Sigma R_a = 73$	$\Sigma R_b = 137$	$U_a = 18$	$U_b = 82$

Referring back to Table 4.6 and substituting, we have

$$U_a = 100 + \frac{110}{2} - 137$$

$$= 155 - 137 = 18$$

$$U_b = 100 + \frac{110}{2} - 73$$

$$= 155 - 73 = 82$$

Explanation of U

If our groups displayed no differences, we would have expected the ranking of the scores for the two groups to be heavily intermixed such that $U_a = U_b$. If a real difference existed between our two treatment populations, we would expect the ranks for Sample A to cluster at one end of the distribution and the ranks of Sample B to cluster at the other end. Therefore, the lower the value of the smaller of the two statistics, U_a and U_b, the greater the likelihood that the two treatment populations differ.

Wilcoxon Matched-Pairs Signed-Ranks Test (T)

The Wilcoxon matched-pairs signed-ranks test (T) is a nonparametric alternative to the t-test for correlated data. It therefore is used with the matched groups and repeated measures designs when a researcher has only ordinal level data. As with the Mann–Whitney U, the Wilcoxon test is particularly useful as an alternative to the t-test with small samples because the underlying assumptions of t are more likely to be violated. With small samples, the relative power of the Wilcoxon is about 95%. With larger samples its relative power is lessened.

Assumptions of Wilcoxon Test

In addition to requiring ordinal level of measurement, the primary assumption of the Wilcoxon is that the underlying data be continuously distributed. The Wilcoxon test is particularly sensitive to this assumption and the power of the test will be reduced with tie scores for a given pair. If a number of ties occur in the data, the Wilcoxon test should not be used. Because of the tie scores problem and the fact that the assumptions underlying the t-test for correlated measures are not as likely to be violated as is the case with independent samples, the Wilcoxon test is used much less frequently than the Mann–Whitney U.

Computation of Wilcoxon Test

The data from the illumination experiment using the matched groups design is depicted in Table 4.3. The first step is to rank all the absolute difference scores (disregard sign). These rankings appear in the fifth column of Table 4.7. The ranks then are separated into those corresponding to positive and those corresponding to negative differences as shown in the last two columns of the table. These two columns are then added.

The smaller of the T_a and T_b values is designated T. The value of T is then compared with the value of T shown in Table E of Appendix A for the appropriate value of N (number of pairs of scores) at the selected level of α. If the observed T is equal to or smaller than the tabled value, H_0 may be rejected. In our example $T = T_b = 3$. Assuming a two-tailed test at the 0.01 level, the corresponding tabled value of T is 3. Therefore, we may reject H_0.

TABLE 4.7
Computation of the Wilcoxon Text Using Data From Table 4.3

Pair	High Illumination	Low Illumination	D	Rank	Positive	Negative
1	26	21	5	7.5	7.5	
2	22	19	3	3.5	3.5	
3	18	19	−1	1.0		1.0
4	18	12	6	9.5	9.5	
5	16	10	6	9.5	9.5	
6	14	16	−2	2.0		2.0
7	15	10	5	7.5	7.5	
8	13	10	3	3.5	3.5	
9	11	7	4	5.5	5.5	
10	10	6	4	5.5	5.5	
$N = 10$					$T_a = 52.0$	$T_b = 3.0$

Explanation of Wilcoxon Test

The rationale for the Wilcoxon matched-pairs signed-ranks test is similar to that of the Mann–Whitney U. If two groups are random samples from the same population, the positive ranks will equal the negative ranks ($T_a = T_b$). However, the larger the difference between the two treatment conditions, the greater the difference will be between T_a and T_b. Accordingly, the lower the smaller of these two values is, the greater the likelihood of a real difference between our two treatment populations.

CHI-SQUARE

One of the more powerful and widely used nonparametric statistical tests is the *Chi-square* (χ^2). The test may be used to determine (a) whether two or more measures are related (Chi-square tests of independence) or (b) whether an observed frequency distribution differs significantly from some hypothesized frequency distribution (Chi-square tests of goodness of fit). To use Chi-square, your data need only be nominal in form.

Chi-Square Distribution

Suppose we drew a large number of samples of the same size from a normally distributed population. Next, for each sample we determined the z-score from the raw score and then computed the sum of squares for this relative frequency (z-score) distribution. The sampling distribution of these sums of squares is known as the Chi-square distribution. As was the case with the t distribution, the Chi-square distribution is a function of sample size. Thus, there are as many Chi-square distributions as there are degrees of freedom (df). Unlike the t distribution, the Chi-square distribution is asymmetrical and for small samples is sharply skewed to the right. As N approaches infinity, the distribution of Chi-square approaches the normal distribution (see Figure 4.4). Chi-square distributions with 30 or more degrees of freedom tend to have essentially the same near-normal shape. It may be demonstrated that the 1 df Chi-square = z^2.

All Chi-square values are positive and range from zero to infinity; the mean of any Chi-square distribution is equal to its degrees of freedom. The mode = $df - 2$ except for the distribution for one degree of freedom. With rare exceptions, all usual applications of the Chi-square statistic require a one-tailed test and the format of the Chi-square table (Table F in Appendix A) reflects this. The probabilities recorded across the top of the table are for random Chi-squares *equal to or larger than the tabled values.*

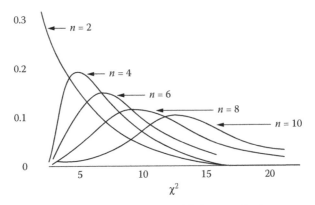

FIGURE 4.4   Representative distributions of Chi-square with five sample sizes.

Chi-Square Tests of Independence

Let us assume that the data from the illumination experiment using 100 subjects had simply been recorded as high performance or low performance. The resulting 2 × 2 *contingency table* might look like Table 4.8a.

To determine what frequencies we could have expected by chance alone, we compute Table 4.8b. As may be noted, the expected frequency for a given cell is computed by multiplying its row by its column marginal values and dividing by the total number of observations. For example, the expected frequency for cell a = column m total × row k total ÷ total number of subjects (T) or (50) (40)/100 = 20. The general formula for Chi-square is

$$\chi^2 = \Sigma \frac{(O-E)^2}{E}$$

where

$\chi^2 = \chi$-square or Chi-square
O = observed frequency in given cell
E = expected frequency for same cell

TABLE 4.8
Data for Computation of the Chi-Square Test for Independence

a. Observed Frequencies

	High Illumination	Low Illumination	
High Performers	30_a	10_b	$_k 40$
Low Performers	20_c	40_d	$_l 60$
	50_m	50_n	T 100

b. Expected Frequencies

	High Illumination	Low Illumination	
High Performers	$\frac{mk}{T} = 20$	$\frac{nk}{T} = 20$	40
Low Performers	$\frac{ml}{T} = 30$	$\frac{nl}{T} = 30$	60
	50	50	100

Substituting, we have

$$\chi^2 = \frac{(30-20)^2}{20} + \frac{(10-20)^2}{20} + \frac{(20-30)^2}{30} + \frac{(40-30)^2}{30}$$

$$= \frac{100}{20} + \frac{100}{20} + \frac{100}{30} + \frac{100}{30}$$

$$= 5 + 5 + 3.33 + 3.33$$

$$= 16.7$$

The test of our hypothesis is outlined below

1. H_0: performance is unrelated to illumination level; H_1: performance is dependent on level of illumination.
2. Table value of χ^2, one-tailed test, $\alpha = 0.01$ level, 1 $df = 6.64$ (see Table F in Appendix A).
3. Because our observed Chi-square is *greater* than the tabled value, reject H_0 and assert H_1.

Computation of Degrees of Freedom for Chi-Square Tests

For Chi-square and other statistical tests whose data may be presented in contingency table form, the degrees of freedom equal the number of rows – 1 multiplied by the number of columns – 1:

$$df = (r - 1)(c - 1)$$

Our previous example had two rows and two columns. Thus, our degrees of freedom = $(2 - 1)(2 - 1) = 1$. Rationally, we may see why this is the case. For a given set of marginal values, as soon as one of the cell values is determined, the frequencies of the other three cells become fixed. Hence, there is only one degree of freedom.

Chi-Square Tests of Goodness of Fit

Chi-square tests of goodness of fit or *one-sample Chi-square tests* are used to determine whether an observed frequency distribution is significantly different from a hypothesized frequency distribution. For example, suppose we presented 100 subjects with 4 cereal boxes, each printed in a different color (red, yellow, green, and blue), and asked them to pick the box they preferred. By chance alone, we would expect each box to be picked one-fourth of the time. Table 4.9 depicts the observed and expected frequency distributions for this experiment.

$$\chi^2 = \Sigma \frac{(O-E)^2}{E}$$

$$= \frac{(15-25)^2}{25} + \frac{(35-35)^2}{25} + \frac{(20-25)^2}{25} + \frac{(30-25)^2}{25}$$

$$= \frac{100}{25} + \frac{100}{25} + \frac{25}{25} + \frac{25}{25} + 4 + 4 + 1 + 1$$

$$= 10$$

The test of our hypothesis is as follows:

1. H_0: color choices are uniformly distributed; H_1: color choices are not uniformly distributed.
2. df = number of cells (k) – 1 = 4 – 1 = 3.
3. Tabled value of Chi-square, one-tailed test, $\alpha = 0.05$ level, 3 $df = 7.81$.
4. Because our observed Chi-square is greater than the tabled value, reject H_0 and assert H_1.

TABLE 4.9
Distribution of Observed and Expected Frequency
Distributions of Different Colored Cereal Boxes

Box Color	Observed	Expected
Red	15	25
Yellow	35	25
Green	20	25
Blue	30	25
Total Observations	100	100

Chi-Square Test for Goodness of Fit to Normal

This is a special case of the Chi-square test of goodness of fit and conducted when you are interested in knowing if your data are normally distributed. Starting with frequency distribution, the first step is to determine expected frequencies. This is accomplished by a process known as *normalizing,* and is represented by the columns numbered (1) through (7) in Table 4.10. The data are organized in frequency intervals from highest to lowest.

1. The first step in normalizing the distribution for the χ^2 goodness of fit test is to collapse the extreme intervals having a frequency of less than 5 into the adjacent intervals as shown in Column (1).
2. Enter the upper limit of each interval in Column (2).
3. Determine the deviation of each interval in Column (2) from $\overline{\chi}$ Column (3).
4. Convert each deviation score to a z-score.
5. Using the table of normal probability (Table B in Appendix A, Column 3, area in larger proportion), convert each z-score to a *proportion below* to represent the portion of a normal distribution that would be found to the left of this point on the curve.

TABLE 4.10
Normalizing a Distribution and Computation of χ^2

		(1)	(2)	(3)	(4)	(5)	(6)	(7)	(8)	(9)	(10)
Interval	f	f_O	Upper Limit	$\overline{X}$	Z	Prop. Below	Prop. Within	f_E	$O - E$	$(O - E)^2$	$\dfrac{(O-E)^2}{E}$
95–99	2										
90–94	5	7	94.5	22.1	1.99	0.98	0.04	5	2	4	0.8
85–89	9	9	89.5	17.1	1.54	0.94	0.08	10	−1	1	0.1
80–84	14	14	84.5	12.1	1.09	0.86	0.12	14	0	0	0.47
75–79	20	20	79.5	7.1	.64	0.74	0.16	19	1	1	0.05
70–74	26	26	74.5	2.1	.19	0.58	0.18	22	4	16	0.73
65–69	18	18	69.5	−2.9	−2.6	0.40	0.16	19	−1	1	0.05
60–64	11	11	64.5	−7.9	−.71	0.24	0.12	14	−3	9	0.64
55–59	8	8	59.5	−12.9	−1.16	0.12	0.07	8	0	0	
50–54	4	7	54.5	−17.9	−1.61	0.05	0.05	7	0	0	
45–49	3										
	$\Sigma f = 120$						1.01	121			$\chi^2 = 2.4$
	$\overline{X} = 72.4$										
	$\sigma = 11.1$										

6. For each interval, subtract the *proportion below* for the interval below from its *proportion below* (for the current interval) to obtain the *proportion within*. For the bottom interval, the proportions within and below will be the same value and will represent the portion of the distribution that falls within the interval scores.
7. For each interval, multiply the *proportion within* by Σf to obtain the expected frequency (f_E).
8. As the first step in computing Chi-square, subtract each (f_E) value from the corresponding (f_O) value.
9. Square all values from Column (8).
10. Divide each Column (9) value by its corresponding Column (7) value.
11. Sum column (10) to yield the observed value for χ^2.

In computing our best fitting normal distribution, we maintained the same M, SD, and n found in the original data. In other words, we placed three restrictions on our data when we normalized them. As a result, we lost one degree of freedom for each restriction because, after collapsing the extreme intervals, we had nine intervals (cells) and $df = 9 - 3 = 6$. Because our objective in making this Chi-square test is to ensure that our data does *not* differ significantly from normal, it is appropriate to guard against Type II error by choosing a numerically large value of α. Entering the (χ^2) table (Table F) for $\alpha = 0.10$ and $df = 6$, we find 10.64. Because our observed χ^2 is less than the tabled value, we would fail to reject H_0 and conclude that our data are normally distributed or that we failed to find differences.

Computation of Chi-Square With Small Expected Frequencies

When any of the *expected* frequencies is less than 10, the observed Chi-square is likely to be an *overestimate*. With $df = 1$, a correction for continuity known as Yates's correction may be applied. The correction is required because the distribution of Chi-square is discrete, whereas the values resulting from the Chi-square formula result in continuous distributions. With large frequencies, this difference has little effect. However, as the expected frequencies become small, a correction for continuity is needed. The revised formula using Yates's correction is

$$\chi^2 = \Sigma \frac{(|O - E| - .5)^2}{E}$$

where $|O - E|$ = the absolute value of $O - E$. Based on the data in Table 4.11 and substituting the formula above, we have

$$\chi^2 = \Sigma \frac{\left(|O - E| - .5\right)^2}{E}$$

$$= \frac{(|11 - 7| - .5)^2}{7} + \frac{(|3 - 7| - .5)^2}{7} + \frac{(|8 - 12| - .5)^2}{12} + \frac{(|16 - 12| - .5)^2}{12}$$

$$= \frac{(3.5)^2}{7} + \frac{(3.5)^2}{7} + \frac{(3.5)^2}{12} + \frac{(3.5)^2}{12}$$

$$= \frac{12.25}{7} + \frac{12.25}{7} + \frac{12.25}{12} + \frac{12.25}{12}$$

$$= 1.75 + 1.75 + 1.02 + 1.02$$

$$= 5.54$$

TABLE 4.11
Data for Computation of Yates's Correction

Observed

11	3	14
8	16	24
19	19	38

Expected

7	7	14
12	12	24
19	19	38

The table value for a Chi-square, one-tailed test, $\alpha = 0.05$ is 3.841. Therefore, reject the H_0.

With larger contingency tables, where the degrees of freedom are greater than 1, Yates's correction is not appropriate. In these cases, the expected frequencies may sometimes be increased by collapsing cells prior to computing Chi-square. For an example, see Table 4.12.

When the expected values are less than 5, even Yates's correction for continuity is inadequate, and Fisher's exact method must be used. Explanations of this method can be found in many textbooks on statistics.

TESTING FOR SIGNIFICANCE OF CORRELATION

TEST FOR SIGNIFICANCE OF PHI (Φ)

By now readers may understand that there is a relationship between correlation and tests of statistical significance. For example, with the Chi-square test for independence, the greater the relatedness or correlation, the greater the size of Chi-square. In fact, the phi coefficient may be tested for significance using Chi-square. The formula is:

$$\chi^2 = N\phi^2$$

TABLE 4.12
Collapsing Cells to Increase Expected Frequencies

	E Performance					
	V. High	High	Medium	Low	V. Low	
Group A	5	10	20	10	5	50
Group B	5	10	20	10	5	50
	10	20	40	20	10	100

Original Table

	E Performance			
	High	Medium	Low	
Group A	15	20	15	50
Group B	15	20	15	50
	30	40	30	100

Collapsed Table

Referring back to the computation of the phi coefficient from the data in Table 4.7 and applying the formula, we have

$$\chi^2 = 100(0.41)^2$$

$$= 100(0.1681)$$

$$= 16.81$$

The tabled value of χ^2, 1 df, one-tailed test, $\alpha_{0.01} = 6.64$. Therefore, we reject the H_0. A phi coefficient of this size could have occurred by chance fewer than 1 of 100 times.

TESTING FOR SIGNIFICANCE OF PEARSON r AND SPEARMAN RHO

In Chapter 3 we discussed correlations among other bivariate statistics. In this chapter we examined hypothesis testing and simple experimental designs. It may be useful at this point to consider what constitutes significance of a *correlation coefficient*. Tests of significance have been computed for various values of the Pearson r and Spearman rho for different sample sizes. This information is available for convenient use in Tables G and H in Appendix A. Both tables are entered at $df = n - 1$ pairs of scores and at the researcher's chosen level of α for one- and two-tailed tests. If the tabled correlation coefficient is smaller than the computed correlation coefficient, one must reject the H_0.

SUMMARY

1. For samples of 100 or more units, the sampling distribution of sample means will be normal in shape regardless of the shape of the distribution for a given trait in the population. This characteristic of the sample means is known as the central limit theorem. With only moderately skewed populations, the central limit theorem holds for samples as few as 30. Because of the operation of the central limit theorem, sample means may be interpreted directly in terms of the normal probability distribution (z-table).

2. The standard deviation of a sampling distribution of sample means is known as the standard error of the mean ($\sigma_{\bar{x}}$). Its size varies as a function of sample size; the larger the samples the smaller the $\sigma_{\bar{x}}$. An unbiased estimate of $\sigma_{\bar{x}}$ ($S_{\bar{x}}$) can be calculated by dividing the standard deviation for a single random sample by $\sqrt{n}$.

3. Of particular importance in statistical hypothesis testing is the fact that the distribution of the differences between sample means also follows the central limit theorem. The standard deviation for a distribution of differences between sample means is called the standard error of the difference between the means ($\sigma_{D_{\bar{x}}}$). An unbiased estimate of $\sigma_{D_{\bar{x}}}$ ($S_{D_{\bar{x}}}$) can be calculated from two random samples.

4. By dividing the difference between two sample means by their $\sigma_{D_{\bar{x}}}$, a z-score is obtained. The z-score is referred to as the critical ratio and forms the basis of statistical hypothesis testing. By referring to the normal probability table, the probability of occurrence of any critical ratio can be obtained. This probability is the likelihood that the particular differences among the means tested could have occurred by chance alone and is also known as the alpha (α) or region of rejection.

5. The null hypothesis (H_0) is a hypothesis of no difference. The alternative (H_1) is the experimental hypothesis. To assert H_1, a researcher must demonstrate statistically that H_0 cannot be supported.

6. A researcher may choose to conduct either a one-tailed or two-tailed test. For a one-tailed test, H_0 states that $(\mu_1 - \mu_2 \leq 0)$; H_1 states that $(\mu_1 - \mu_2 > 0)$. For a two-tailed test, H_0 states that $\mu_1 - \mu_2 = 0$, and H_1 states that $\mu_1 - \mu_2 \neq 0$. Correct use of a one-tailed test requires that the researcher be certain that a significant difference between μ_1 and μ_2 occurs only in one direction. Because this directionality is not always known, most behavioral research utilizes two-tailed tests.

7. Type I error results from rejecting the null hypothesis when in fact it should not have been rejected and the probability of the event is equal to alpha. Type II error arises from not rejecting the null hypothesis when in fact it should have been rejected. As α increases, the probability of a Type II error is decreased but the relationship is nonlinear. The probability of a Type II error is equal to $1 - $ power or beta (β).

8. When the purpose of an experiment is to attempt to assert the alternative hypothesis, α usually is set numerically small (0.05 or smaller) to guard against Type I error. When the purpose is to not reject the null hypothesis, α is set high (0.10 or higher) to guard against Type II error.

9. Power is the ability of a statistical test to reject the null hypothesis. The larger the level of significance chosen and/or the larger the sample, the more powerful is a statistical test. Relative power or power efficiency of a statistical test is the ratio of sample size required to reject a given null hypothesis using either the t- or F-test to the sample size required using the statistic under consideration.

10. For the two randomized groups design, subjects are randomly assigned to one of the two treatment conditions in the experiment. This design assumes that on the average as a result of randomization, the two groups will be equivalent with respect to the various extraneous subject variables that may affect the DM.

11. In a two matched groups design, subjects are initially paired on the basis of their scores on one or more extraneous variables known to be correlated with the DM. Subjects within each matched pair are randomly assigned to one of two treatment conditions. This design has the advantage of assuring equality of the experimental groups on at least one extraneous variable known to be important. At the same time, it relies on randomization to assure equality on the other extraneous subject variables.

12. With the repeated measures design, each subject receives all treatment conditions. In essence, the subject acts as his or her own control and equality of the subject under various treatment conditions is further ensured. Because of its greater precision—particularly with small samples where randomization is not as likely to be an effective control of all the extraneous subject variables—this design is preferred when appropriate. It is not appropriate in studies in which a differential order effect between treatment conditions may be possible (where going from Treatment A to Treatment B is different in performance than going from B to A).

13. The most commonly used parametric statistic for analyzing simple experimental designs is the t-test. For very large samples, $t = z$. For smaller samples, the t distribution is a function of degrees of freedom. The t-test assumes (a) normal distribution of the underlying trait in the population, (b) interval level data, and (c) homogeneity of variance for the two groups under study. The t-test has two computational forms: (a) the t-test for independent samples used with the two randomized groups design and (b) the t-test for correlated data used with the two matched groups and repeated measures designs.

14. A nonparametric alternative to the t-test for independent samples is the Mann–Whitney U-test that assumes ordinal level measurement and an underlying continuous distribution. It is particularly useful with small samples where the assumptions of a t-test are more likely to be violated.

15. A nonparametric alternative to the t-test for correlated data is the Wilcoxon matched-pairs signed-ranks test. Like the Mann–Whitney U, the Wilcoxon test requires data on at least an

ordinal scale and the data are assumed to be continuously distributed. The Wilcoxon test is particularly sensitive to tie scores and should not be used if ties are numerous.

16. The Chi-square is the most commonly used nonparametric statistical test. It can be used to determine (a) whether two (or more) measures are related or (b) whether an observed frequency distribution departs significantly from some hypothesized frequency distribution. Chi-square may be used with nominal or higher level data. Its distribution is asymmetrical, and highly positively skewed with a small df value. As the df increases, the Chi-square distribution approaches normality. Degrees of freedom with Chi-square and other statistics utilizing data in contingency table form equal the number of rows minus one times the number of columns minus one: $(r-1)(c-1)$.

17. Correlation coefficients may be tested for significance. For the Pearson r and the Spearman rho, tables have been prepared for various levels of α. The Φ coefficient may be tested using Chi-square.

KEYWORD DEFINITIONS

Alpha (α): Lower case Greek letter indicating probability of a Type I error; the likelihood that the difference between means tested occurred by chance alone. Also known as the region of rejection.

Alternative Hypothesis (H_1): Opposite of null hypothesis. When testing the hypothesis of difference, the alternative hypothesis states that the differences between means or populations are not due to chance alone.

Area in Smaller Proportion: Probability of obtaining a sample mean this far from the population mean (μ) or further.

Asymmetric Transfer Effect: Experimental design error in which there is a greater influence on behavior of having condition A before B than the reverse; an unbalanced order or sequence effect in which the experience on the first condition that influences subsequent conditions is not the same across conditions.

Beta (β): Lower case Greek letter indicating probability of a Type II error.

Carryover Effects: Temporary or permanent change in a participant's behavior; experience with a prior test may have affected his/her behavior in a later test (e.g., practice, fatigue, drug use).

Central Limit Theorem: Even with a skewed distribution of a given trait in the population, the distribution of sample means for that trait will be normal for samples of 100 or larger.

Chi-Square Test (χ^2): Most common nonparametric statistical test used to determine (a) whether two (or more) measures are related or (b) whether an observed frequency distribution departs significantly from some hypothesized frequency distribution.

Contingency Table: Table showing levels of two or more variables by placing the number of observations in each cell. This number takes into consideration the intersection of the different levels of the variables.

Counterbalancing: Systematic technique for varying the order of conditions to distribute various carryover effects so they are not confounded with different treatment conditions; balances the order in which treatments are experienced.

Critical Ratio: Ratio of the difference between the means to the standard error of the difference.

Degrees of Freedom (df): Number of scores in a distribution that are free to vary.

Experimental Hypothesis: Also called the research hypothesis or working hypothesis; statement that describes the predicted relationship between levels of an independent variable and a dependent measure; same as the statistical alternative hypothesis (H_1).

Homogeneity of Variance: Assumption that the variances of two samples are equal and do not differ statistically.

Independent Data: Requirement for between-groups designs that the selection of one group or subject did not influence the selection of the other group or groups of subjects.

Learning Effect: Treatment or experience that results in a relatively permanent change in behavior.

Mann–Whitney U Test: Nonparametric test of the null hypothesis H_1; uses ranks of various observations for two independent samples; serves as alternative t-test for equality of means.

Matched Groups Design: Experimental design in which pairs of subjects are matched on an extraneous variable known to be correlated with the DM. Subjects in each pair are then randomly assigned to the experimental and control groups. This design is used to increase the precision of control over extraneous variables.

Matching Variable: Dimension on which subjects are paired.

Meaningful Difference: Measured difference that is beyond significant and useful and relevant in a real-world context.

Nonparametric: Statistical technique requiring only nominal or ordinal levels of measurement or for use in cases where interval or ratio data are not normally distributed.

Normality: Degree (measured or assumed) to which the distribution of a dependent measure is normal in shape.

Null Hypothesis (H_0): Hypothesis of no difference; also called statistical hypothesis. Rejection of the null hypothesis provides support for the experimental or working hypothesis.

One-Sample Chi-Square: Test used to determine whether an observed frequency distribution is significantly different from a hypothesized frequency distribution.

One-Tailed Hypothesis: Assumption that any real DM difference between experimental and control groups can only be in one direction, for example, the assumption in our hypothetical illumination experiment that the higher illumination condition had to exert no effect or improve performance over the low illumination condition.

One-Tailed Test: Statistical test for one-tailed hypothesis.

Overestimate: Calculated value exceeds actual (or "real") value.

Parametric: Statistical techniques designed for use with interval or higher level data; requires assumptions of normality and homogeneity of variance.

Pooling: Combining samples together to obtain an unbiased estimate of the common population variance.

Power: Ability of a statistical test to reject the null hypothesis. The larger the level of significance chosen and/or the larger the sample, the more powerful is a statistical test.

Practice Effect: Improvement resulting merely from experience with a task or procedure.

Power Efficiency: See *Relative Power*.

Probability: Proportion or fraction of times expressed in decimal form predicting a certain outcome.

Randomization: Assigning subjects to treatments or groups in such a way that each subject has an equal likelihood of being assigned to any treatment or group, and the assignment of any single subject does not influence the assignment of any other subject.

Region of Rejection: See *Alpha*.

Relative Power: Ratio of sample size required to reject a given null hypothesis using the t- or F-test to the sample size required using the statistic under consideration; also known as power efficiency.

Repeated Measure: Obtaining a measure (DM score) on each subject for more than one treatment condition; also called within-subject measure.

Serendipitous Finding: Good luck; beneficial result that was not the original object of a search.

Sequence Effect: see Practice Effect.

Significant Difference: Difference greater than expected by chance alone.

Spurious Results/Findings: Experimental results that are not genuine; false findings.

Standard Error of Differences Between Means ($\sigma_{D_{\bar{x}}}$): Standard deviation for a distribution of differences between sample means.

Standard Error of Mean ($\sigma_{\bar{x}}$): Standard deviation for a sampling distribution of sample means.

t-Test: Ratio of the difference between two sample means and an estimate of the standard deviation of the distribution of differences; a test of the difference between groups.

Transfer Effect: Change in an organism that is carried from one condition, treatment, or experience to another to influence behavior on subsequent testing.

Type I Error: Rejecting the null hypothesis when it should not have been rejected; probability that an event is equal to alpha (α).

Type II Error: Not rejecting the null hypothesis when it should have been rejected; the probability of that an event is equal to $1 - power$.

Wilcoxon Matched-Pairs Signed-Ranks Test: Nonparametric alternative to t-test for correlated data.

Within-Subject Measure: See *Repeated Measure*.

z-Test: Test to determine whether a sample mean qualifies as a probable outcome under the null hypothesis.

EXERCISES

1. Two randomized groups of students enrolled in a physical education class were taught how to bowl using two different teaching methods. The results were as follows:

Group A	Group B
$\bar{X}_A = 157$	$\bar{X}_B = 149$
$\sigma_A = 8$	$\sigma_B = 7$
$n = 20$	$n = 20$

Test the experimental hypothesis at the 0.05 level that the difference between the means is one that could be expected in the populations represented by these sample groups (i.e., that the teaching method probably does make a real difference in student performance).

2. Two randomized groups of students participated in a laboratory experiment on vigilance. Their performance scores were as follows:

Group				
A		B		
12	8	10	6	$H_0 : \mu_1 - \mu_2 \leq 0$
11	8	9	5	
10	6	9	3	$H_1 : \mu_1 - \mu_2 > 0$
10	5	8	3	
9	3	7	2	$\alpha = .05$

Conduct a t-test to determine whether H_0 should be rejected; based on this test, state your conclusion.

3. Using the data from Exercise 2, conduct at Mann–Whitney U to determine whether H_0 should be rejected. Under what conditions would U be the more appropriate test?

4. Twenty employees were matched on the basis of a job knowledge test and the members of each matched pair were randomly assigned to either the experimental group (A) taught with a self-paced programmed learning method or the control group (B) taught via a

conventional lecture and discussion approach. The following final exam results were obtained. Determine if the difference in final exam scores between groups is significant.

Pairs	Group A	Group B	
1	12	12	Test following at 0.05 level
2	11	9	
3	10	8	$H_0 : \mu_1 - \mu_2 = 0$
4	9	9	
5	8	9	$H_1 : \mu_1 - \mu_2 \neq 0$
6	8	5	
7	8	4	
8	7	5	
9	6	3	
10	5	4	

5. Using the data from Exercise 4, test the same hypothesis using the Wilcoxon T. Under what conditions is the Wilcoxon T the more appropriate test?

6. In a human factors experiment, a group of 14 persons performed a psychomotor task under two different temperature conditions (68°F and 100°F). The results were as follows:

Individual	68°F	100°F	Individual	68°F	100°F
1	10	8	8	6	3
2	9	9	9	6	2
3	8	7	10	5	4
4	7	5	11	5	4
5	7	4	12	4	2
6	6	8	13	4	2
7	6	4	14	3	4

Test the following at the 0.05 point: (a) $H_0 : \mu_1 - \mu_2 \leq 0$ and (b) $H_1 : \mu_1 - \mu_2 > 0$.

7. A group of supervisors were randomly assigned to two different human relations training programs. One group received a lecture–discussion program; the other group participated in a behavior modeling program conducted by the same set of instructors who taught the first group and were not aware of the study. Six months after completing the program, the supervisors were evaluated by their subordinates as to whether they had shown (a) little or no improvement or (b) noticeable improvement in their interpersonal skills since the end of the study. The subordinates' categorical ratings were as follows:

	Traditional Training Group	Behavior Modeling Group
Improved Skills	30	45
No Improvement	30	15

Conduct a Chi-square test for independence to determine whether this outcome represents a real difference ($\alpha = 0.01$).

8. In an attitude survey, a group of consumers were asked to select the U.S. automobile manufacturer that they thought designed the best engineered automobiles. The results were

Manufacturer	N
A	490
B	460
C	440
D	410
Total Responses	1800

(a) Test the hypothesis that the above distribution does not differ from what would be
expected by chance ($\alpha = 0.05$).

(b) Test the hypothesis at the 0.10 point that the following distribution follows the normal curve.

Interval	F
95–99	10
90–94	17
85–89	20
80–84	16
75–79	12
70–74	10
65–69	5
60–64	5
55–59	3
50–54	2

$n = 100$
$\bar{X} = 83$
$\sigma = 10$

9. Determine whether each of the following correlations is probably a real or chance relationship.

a.	$r = .42$	c.	rho = .35	e.	$\psi = .40$
	$n = 20$		$n = 40$		$n = 50$
	$\alpha = .01$		$\alpha = .05$		$\alpha = .01$
b.	$r = .28$	d.	rho = .42	f.	$\psi = .32$
	$n = 50$		$n = 50$		$n = 100$
	$\alpha = .05$		$\alpha = .01$		$\alpha = .01$

10. A mathematics instructor at a small West Coast college wants to know whether the conten-
tion that this year's freshman class is not average has any statistical support. Based on the
following scores from a randomly selected group of students within the class, determine
whether the speculation is accurate.

$\alpha = 0.05$			Previous $\bar{X} = 75$			$n = 33$
51	89	91	89	93	69	81
94	95	46	58	75	69	55
68	49	86	74	79	100	88
62	83	52	63	89	83	
73	81	85	96	89	71	

EXERCISE ANSWERS

1. Test for group differences.

$$H_0 : \bar{X}_A - \bar{X}_B = 0; \qquad H_1 : \bar{X}_A - \bar{X}_B \neq 0$$

$$S_{D_{\bar{x}}} = \sqrt{\frac{8^2}{20} + \frac{7^2}{20}}$$

$$= 2.38$$

$$t = \frac{157 - 149}{2.38}$$

$$= 3.36$$

$t_{.025, 38} = 1.96$. If $df = n_1 - 1 + n_2 - 1 = 20 - 1 + 20 - 1 = 19 + 19 = 38$, enter Table C in Appendix A at 38 df; $t_{.025}$ is between 2.021 and 2.042. $t > t_{.025}$. Therefore reject H_0.

2. Test for group differences with raw data.

$\bar{X}_A = 8.2$ $\qquad\qquad\qquad$ $\sigma_A = 2.82$ $\qquad\qquad\qquad$ $df = 18$

$\bar{X}_B = 6.2$ $\qquad\qquad\qquad$ $\sigma_B = 2.86$

$$S_{D_{\bar{x}}} = \sqrt{\frac{2.82^2}{10} + \frac{2.86^2}{10}}$$

$$= 1.27$$

$$t = \frac{8.2 - 6.2}{1.27}$$

$$= 1.57$$

$$t_{0.05,18} = 1.73$$

$t < t_{0.05,18}$. Therefore do *not* reject H_0.

3. Mann–Whitney U calculations.

All Scores	All Ranks	A Ranks	B Ranks	U_a	U_b
12a	1a	1		0	
11a	2a	2		0	
10a	4a	4		0	
10a	4a	4		0	
10b	4b		4		4
9a	7a	7		3	
9b	7b		7		4
9b	7b		7		4
8a	10a	10		3	
8a	10a	10		3	
8b	10b		10		7

All Scores	All Ranks	A Ranks	B Ranks	U_a	U_b
7b	12b		12		7
6a	13.5a	13.5		6	
6b	13.5b		13.5		7
5a	15.5a	13.5		6	
5b	15.5b		15.5		9
3a	18a	18		9	
3b	18b		18		9
3b	18b		18		9
2b	20b		20		10
		$\Sigma R_a = 85$	$\Sigma R_b = 125$	$\Sigma U_a = 30$	$\Sigma U_b = 70$

$U = U_a = 30$

$U_{0.05,10,10} = 27$

$30 > 27$; do not reject H_0. Note: The larger number of tie scores decreases the power of this test.

By shortcut method: $U_a = (10)(10) + \dfrac{(10)(11)}{2} - 125 = 30$. Because $30 > 27$, do not reject H_0.

4. Test for group differences.

Pair	A	B	D	D^2
1	12	12	0	0
2	11	9	2	4
3	10	8	2	4
4	9	9	0	0
5	8	9	−1	1
6	8	5	3	9
7	8	4	4	16
8	7	5	2	4
9	6	3	3	9
10	5	4	1	1
			$\Sigma D = 16$	$\Sigma D^2 = 48$

$$\Sigma d^2 = 48 - \frac{(16)^2}{10}$$

$$= 22.4$$

$$t = \frac{1.6}{\sqrt{\dfrac{22.4}{(10)(9)}}}$$

$$= 3.21$$

$t_{0.05,9} = 2.262$. Because $3.21 > 2.262$, reject H_0.

Therefore, the experimental group (A) had significantly higher final exam scores than the control group (B).

5. Wilcoxon calculations.

Pair	A	B	D	Rank	Positive	Negative
1	12	12	0	1.5	1.5	
2	11	9	2	6	6	
3	10	8	2	6	6	
4	9	9	0	1.5	1.5	
5	8	9	–1	3.5		3.5
6	8	5	3	8.5	8.5	
7	8	4	4	10	10	
8	7	5	2	6	6	
9	6	3	3	8.5	8.5	
10	5	4	1	3.5	3.5	

$T = T_B = 3.5$ $T_n = 51.5$ $T_b = 3.5$

$T_{.025,10} = 8$

$3.5 < 8$. Therefore reject H_0.

Wilcoxon T is the more appropriate test when sample sizes are small and underlying assumptions of t are likely to be violated. It should not be used when the data include many tie scores.

6. Test for within-subject differences.

Subject	68°F	100°F	D	D²
1	10	8	2	4
2	9	9	0	0
3	8	7	1	1
4	7	5	2	4
5	7	4	3	9
6	6	8	–2	4
7	6	4	2	4
8	6	3	3	9
9	6	2	4	16
10	5	4	1	1
11	5	4	1	1
12	4	2	2	4
13	4	2	2	4
14	4	4	–1	1
			$\Sigma D = 20$	$\Sigma D^2 = 62$

$$\Sigma d^2 = 62 - \frac{(20)^2}{14}$$

$$= 33.43$$

$$t = \frac{\dfrac{20}{14}}{\sqrt{\dfrac{33.43}{(14)(13)}}}$$

$$= 3.33$$

$t_{.05,13} = 1.771$. Because $3.33 > 1.771$, reject H_0.

7. Chi-square calculation (independence). H_0 = performance independent of training. H_1 = performance dependent on training.

	T	B	
I	30	45	75
N	30	15	45
	60	60	120

Observed

	T	B	
I	37.5	37.5	75
N	22.5	22.5	45
	60	60	120

Expected

$$\chi^2 = \frac{(-7.5)^2}{37.5} + \frac{(7.5)^2}{37.5} + \frac{(7.5)^2}{22.5} + \frac{(-7.5)^2}{22.5}$$

$$= 1.5 + 1.5 + 2.5 + 2.5$$

$$= 8$$

$\chi^2_{.01,1} = 6.635$. Because $8 > 6.635$, reject H_0.

8.a. Chi-square calculation (uniformity). H_0 = car choice uniformly distributed. H_1 = car choice not uniformly distributed.

	Observed	Expected
A	490	450
B	460	450
C	440	450
D	410	450
	1800	

$df = 3$

$$\chi^2 = \frac{(490 - 450)^2}{450} + \frac{(460 - 450)^2}{450} + \frac{(440 - 450)^2}{450} + \frac{(410 - 450)^2}{450}$$

$$= 3.56 + .22 + .22 + 3.56$$

$$= 7.56$$

$\chi^2_{.05,3} = 7.815$. Because $7.56 < 7.815$, do not reject H_0.

8.b. Chi-square calculation (normality).

Interval	F	f_o	Upper Limit	X	z	Proportion Below	Proportion Within	f_E	O − E
95–99	10	10	99.5	16.5	1.65	.96	.08	8	2
90–94	17	17	94.5	11.5	1.15	.87	.13	13	4
85–89	20	20	89.5	6.5	.65	.74	.18	18	2
80–84	16	16	84.5	1.5	.15	.56	.20	20	−4
75–79	12	12	79.5	−3.5	−.35	.36	.17	16	−4
70–74	10	10	74.5	−8.5	−.85	.20	.11	11	−1
65–69	5	5	69.5	−13.5	−1.35	.09	.06	6	−1
60–64	5	5	64.5	−18.5	−1.85	.03	.02	2	3
55–59	3	5	59.5	−23.5	−2.35	.01	.01	1	4
50–54	2	—							

$(O - E)^2$	$\dfrac{(O - E)^2}{E}$
4	0.5
16	1.23
4	0.22
16	0.8
16	1.0
1	0.09
1	0.17
9	4.5
16	16.0
	$\chi^2 = 24.51$

$n = 100 \quad \Sigma f = 100$

$\bar{X} = 83$

$\sigma = 10$

$\chi^2_{10.6} = 10.645$

$24.51 > 10.645$
Therefore reject H_0

9. Meaningfulness of correlations.
   a. Chance
   b. Real
   c. Real
   d. Chance
   e. $x^2 = (0.40)^2 (50) = 8.0$ real
   f. $x^2 = (0.32)^2 (100) = 10.24$ real

10. Test of sample uniqueness

$$H_0 : \bar{X} = 75, \text{ the students are average}$$

$$H_1 : \bar{X} \neq 75, \text{ the students are not average}$$

$$x = 2528 \text{ the sum of all test scores}$$

$$x^2 = 201,026 \text{ the sum of squares}$$

$$\bar{X} = \frac{2528}{33} = 76.6$$

$$\sigma^2 = \frac{1}{32} \left(201.026 - \frac{2528}{\sqrt{33}} \right) = 230.2$$

$$\sigma = \sqrt{230.2} = 15.17$$

$$z = \frac{76.6 - 75}{\dfrac{15.17}{\sqrt{33}}} = \frac{1.6}{2.6} = 0.62$$

Table value of $Z = 1.96$. The critical region $Z < -1.96$ and $Z > 1.96$. Because 0.62 does not lie in the critical region do not reject H_0. There is no statistical basis for concluding that the freshman class is not average.

REFERENCES

Keppel, G. (1973). *Design and analysis: A researcher's handbook.* Englewood Cliffs, NJ: Prentice-Hall.
Kerlinger, F. N. (1986). *Foundations of behavioral research* (3rd ed.). New York: Holt, Rinehart & Winston.
Krathwohl, D. R. (1985). *Social and behavioral science research: A new framework for conceptualizing, implementing, and evaluating research studies.* San Francisco: Jossey-Bass.

5 Simple Analysis of Variance

A natural extension of hypothesis testing and simple experimental designs is one-way analysis of variance (ANOVA). This chapter builds on the *t*-test introduced in the last chapter to permit analysis of more than two treatments simultaneously. Most important is learning to distinguish between groups (independent, noncorrelated, or randomized) designs from within subjects (repeated measures, correlated, or dependent) designs. The chapter topics are

More Than Two Treatments Design
Single-Factor (Simple) Analysis of Variance
ANOVA for More Than Two Randomized Groups Design
ANOVA for Repeated Measures Design
Post Hoc Analysis: Multiple Comparisons Among Means
Summary
Keyword Definitions
References
Exercises
Exercise Answers

KEYWORDS

Between Groups	Degrees of Freedom (*df*)	Power
Conservatism	Experimental Error	Sum of Squares
Control Group	Latin Square Design	Treatment Group
Critical Distance	Mean Square	Within Groups
Critical Ratio	Post Hoc Tests	

MORE THAN TWO TREATMENTS DESIGNS

Thus far, we have considered experiments using only two levels or treatments of one independent variable. In this chapter, we shall consider experiments that use more than two treatments. We will also examine two types of experimental design and analysis: the between-groups design and the within-subjects or repeated measures design for these sorts of experiments.

REASONS FOR USING MORE THAN TWO TREATMENTS

Using More Than Two Treatments May Yield a Different Answer

When researchers use a simple two treatments design, they implicitly assume that the relationship of the independent variable (IV) and dependent measure (DM) is linear or at least near linear. Therefore, by picking two levels of the IV having fairly good separation, any difference on the DM attributable to the IV will appear (see Figure 5.1a). Frequently, however, in behavioral research this assumption does not hold. In extreme cases, as depicted in Figure 5.1b, little or no difference will result in the two treatments experiment because of the severe curvilinearity of the relationship.

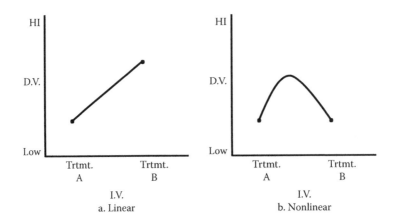

FIGURE 5.1   Relationship between IV and DM in two treatments designs: (a) linearity and (b) nonlinearity.

More commonly, the curvilinearity will be less severe, but still produce too small a difference between treatment conditions to demonstrate a significant relationship between the IV and DM (Type II error). Figure 5.2 depicts a three treatments design. Adding the third treatment (B) demonstrates that the relationship is different from what would be represented if only treatments A and C were plotted.

To Obtain Fairly Precise Knowledge of the IV–DM Relationship

As we already have seen, the two treatments design does not enable us to determine the nature of the IV–DM relationship. The addition of a third treatment in our previous example enabled us to determine that the relationship was nonlinear. By adding additional treatments, we can go beyond determining that the relationship is nonlinear and gain a more precise knowledge of the shape of the relationship. This is illustrated for five treatments in Figure 5.3.

To Study More Than Two Treatment Conditions

Let us assume that an educational researcher is interested in knowing whether (a) small group discussion or (b) lecture and discussion are better techniques for teaching a philosophy course than (c) the conventional lecture method. With a three treatments design our researcher could study and compare the results from each of these methods. Studies of this kind are very common in educational and behavioral science research.

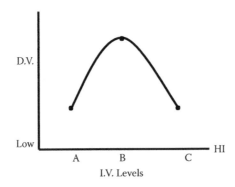

FIGURE 5.2   Relationship between IV and DM in three treatments designs: one factor with three treatment levels.

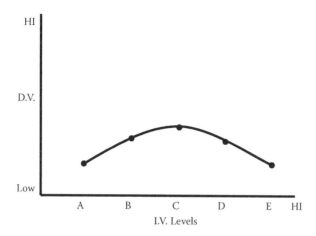

FIGURE 5.3   Relationship between IV and DM in five treatments designs: one factor with five treatment levels.

TYPES OF MORE THAN TWO TREATMENT DESIGNS

As with two treatments designs, more than two treatments designs can be (a) randomized groups, (b) matched groups, or (c) repeated measures (within-subjects) designs. The same advantages and disadvantages covered in the discussion of two treatments also apply here. It should be noted, however, that with the addition of more treatments, the problem of obtaining matched groups becomes much more difficult. As a result, the matched groups design is used infrequently in more than two treatments studies. On the other hand, repeated measures designs are used fairly frequently with more than two treatments.

As mentioned in Chapter 4, the method of incomplete counterbalancing, also called the *Latin square design,* is used to counter the order effects of the experiment. In a simple Latin square design, each treatment occurs equally often in every stage of the experiment; Treatment A occurs first, second, and third as often as Treatments B and C. Figure 5.4a shows a simple Latin square. Note that when a treatment is preceded or followed by another treatment, the same treatment always precedes or follows it.

A balanced Latin square has the additional condition that each treatment be preceded and followed by every other treatment with equal frequency. This design is usually optimal because it minimizes carryover effects that result when a subject's involvement in one treatment affects performance on a subsequent treatment (e.g., fatigue effects on learning). Figure 5.4b illustrates a balanced Latin square. For more information on constructing balanced Latin squares, see Elmes, Kantowitz, and Roediger (1985).

The most frequently used design with more than two treatments is the more than two randomized groups design. The technique is to randomly divide subjects into as many groups as we have treatments, just as we did for the two randomized groups.

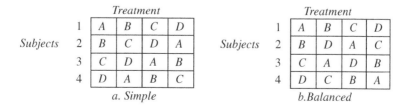

FIGURE 5.4   Latin square design.

SINGLE-FACTOR (SIMPLE) ANALYSIS OF VARIANCE

CONCEPT OF ANALYSIS OF VARIANCE (ANOVA)

In Chapter 4 we noted that the *critical ratio* forms the basis of statistical hypothesis testing. With two treatment groups, this is the ratio of the difference or *variability between* the means to the difference or *variability within* the treatment groups as represented by $S_{D_{\bar{X}}}$. In other words, we can consider the total variation between the two treatment groups to consist of differences caused by (a) the effect of the IV plus (b) differences among the subjects within each group that were not perfectly controlled. These are extraneous variables such as differences among subjects in ability and/or motivation to perform well. Differences due to the effects of extraneous variables are also called *experimental errors,* because they impact the results and are *not* caused by manipulation of the IV.

Analysis of variance may thus be described as the partitioning (analysis) of the total variance into two parts: the variance *between groups* and the variance *within groups*. The between-groups variance measures how much the means of the groups vary from the overall mean, while the within-groups variance indicates how the groups vary about their own means. The within-groups variance is not influenced by the differences between group means; it is treated as an error term. If the null hypothesis is true, the between-groups variance will equal the within-groups variance. That is, the critical ratio will be close to 1.0. As illustrated in Figure 5.5, the greater the difference between the group means and/or the smaller the difference within groups, the greater is the likelihood the null hypothesis is false.

This phenomenon is illustrated in Figure 5.6a and b. The difference between group means is the same, but Figure 5.6b shows a great deal of overlap. In fact, a number of scores in Group 1 were above the mean of Group 2, and some scores in Group 2 were below the mean of Group 1. Thus, it appears much more likely that the differences between the groups in Figure 5.6a are *true* differences resulting from the IV than in Figure 5.6b in which most of the variance is within groups. In Figure 5.6c, the variance within groups is the same as in Figure 5.6b, but the variance between the groups is much greater, resulting in less overlap. In short, the between-groups variation comprises proportionately more of the total variation in both a and c than in b and the critical ratio will

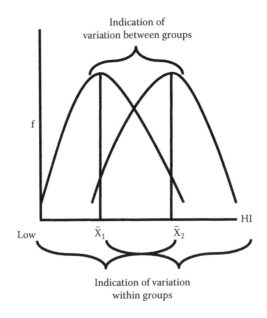

FIGURE 5.5   Between- and within-groups variations.

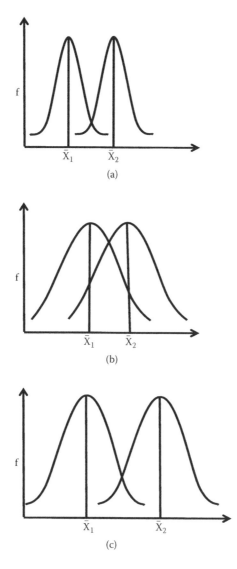

FIGURE 5.6   Different amounts of DM overlap for two groups as a result of differences in both between- and within-groups variation.

therefore be larger. As we know from our discussions of the z- and t-tests, the larger the critical ratio, the greater the likelihood that the difference between groups is a real rather than chance occurrence.

The rationale for partitioning the total variation into between-groups and within-groups variance for the two groups design also applies when more than two treatment groups are used. If the difference among the various group means (i.e., variation of individual treatment means from grand mean or average for the entire group of subjects) is large, the between-groups variance is large. If the difference among all the means is small, the between-groups variation is small. Similarly, if variability of the DM scores within each group is large, the within-groups variation will be large. If the differences within the separate groups are small, the within-groups variance also is small. Next we will examine the principal statistic employed in analysis of variance—the F-test.

F-Test

Rationale for F-Test

The critical ratio used in ANOVA is the F-ratio, developed by one of history's most famous statisticians, Sir Ronald A. Fisher. It can be expressed as

$$F = \frac{between\ groups\ variance}{within\ groups\ variance}$$

$$= \frac{effect\ of\ the\ IV + random\ error}{random\ error}$$

If there are no IV effects, the F-ratio theoretically will be 1.0 (one unbiased estimate of the error variance divided by another unbiased estimate). In practice, we may obtain some apparent effect of the IV by chance (as in the two groups design, the means may differ somewhat by chance). This may result in a numerator that is spuriously larger or smaller than the denominator of the F-ratio. When the F-ratio is smaller than 1, it may be judged not significant simply by inspection. When the F-ratio is greater than 1, the question is how large must it be to be considered significant at a specified level of alpha.

To determine the answer we may use the F distribution. Because the F-ratio is one-tailed, the calculated value is compared to the right tail of the F-distribution as found in Table I in Appendix A. In the two treatments situation, this procedure will yield the identical result as the two-tailed t-test for independent groups.

How large the F-ratio must be to be significant depends on the *degrees of freedom* (*df*) in the experiment. As the *df* increase, the critical F-ratio decreases. The F table requires two *df:* one associated with the numerator (across the top of the table), and one associated with the denominator (down the left side of the table). It should be noted that Table I has two sections. The first gives the F values for the 0.01 level. If the tabled F value for our specified level of significance, numerator *df*, and denominator *df*, is smaller than our computed F-ratio, we may conclude that our F-ratio is significant. The F-ratio will tell us whether to reject or fail to reject the null hypothesis that the measures are equal ($\mu_1 = \mu_2 = \mu_3$), but it does *not* provide information on specific pairs of means. This may be done by using other tests in the *post hoc* analysis.

Assumptions of F-Test

The assumptions underlying the F-ratio are (1) normal distribution of the underlying trait, (2) random samples, and (3) homogeneity of variances (equal variances for the treatment groups). The F-test is highly robust to violations of normality. Even small skewed populations have little effect on ANOVA.

Random sampling is important for all statistical inference and should be controlled by experimental design. Violations of random sampling may invalidate the outcome of any experiment. For the randomized groups design, the random samples are also assumed to be independent. However, in the repeated measures design, variability within treatments may be correlated, because the subjects contribute to more than one treatment measure. The ANOVA for repeated measures takes this into account by using an alternate estimate of the error variance, analogous to the difference between the t-tests for independent data and for correlated data (see McCall, 1970).

Violations of homogeneity of variances have little effect for equal n greater than or equal to 20, but may affect the probability of a Type I error for small or unequal n values. Among the variety of tests for homogeneity of variance are those cited by Bartlett, Scheffe, Hartley, Cochran, and Levene (Glass & Stanley, 1970, p. 374). A more complete discussion of homogeneity of variance can also be found in Games (1984). ANOVA techniques for experiments with unequal n can be found in Myers (1979), Glass and Stanley (1970), Games and Klare (1967), Games (1984), and Hayslett (1968).

WHY MULTIPLE *t*-TESTS SHOULD NOT BE USED

Let us assume we have just completed a five randomized groups experiment and are now ready to test the significance of the differences among our treatment $\bar{X}$s. Our first approach may be to compute *t*-ratios for each of our possible comparisons of group $\bar{X}$s. Because five treatments result in ten possible pairs of $\bar{X}$s, we would have to compute ten *t*-tests. This analytical approach is inefficient simply because of the large amount of numerical work required and will not reveal interactions in multifactor experiments (discussed in the next chapter). However, conducting individual *t*-tests with more than two groups is inappropriate for a more important reason.

If we were to perform a single *t*-test at the 0.05 level ($\alpha = 0.05$), there would be a 5% probability of obtaining a significant *t*-ratio by chance or random sampling alone. With 10 *t*-tests, the probability that at least one was significant at the 0.05 level because of random sampling is much larger. This would not be a problem if it were possible to state how this probability changes as the number of *t*-tests conducted is increased. Unfortunately, this cannot be done because we are comparing all possible pairs in a single experiment. As a result, the same $\bar{X}$ and σ values are used several times and the *t*-ratios are not independent of each other.

Generally speaking, it is inappropriate to use the *t*-ratio to test all possible pairs of $\bar{X}$s in a more than two treatments design. This risk of making one or more Type I errors somewhere within all possible comparisons of means in the experiment is called the *experiment-wise Type I error rate* (EWI) by Games (1984), or the *family-wise Type I error rate* (FWI) after Tukey (cited in Games, 1984) if referring to families of comparisons associated with *F* values. The FWI varies with the design of the experiment, the source of the error term, and the various methods used to control inflation of the FWI.

The most conservative estimate for FWI is the Bonferroni inequality (see Myers, 1979), which states that the FWI is less than or equal to the sums of the α values for each comparison. Thus, for a set of 10 *t*-tests, each at the 0.05 level, the maximum overall probability of a Type I error is 0.50. This means the researcher has a 50/50 chance of rejecting the null hypothesis when it is actually true. This 0.50 represents a maximum probability over all αs using multiple *t*-tests, not the actual α for a particular set of comparisons. Remember: analysis of variance (ANOVA), when appropriate, is a preferred alternative to multiple *t*-tests.

ANOVA FOR MORE THAN TWO RANDOMIZED GROUPS DESIGN

To compute a variance, you may recall from Chapter 3 that it is first necessary to compute the sum of squares ($\sum x^2$ *or* SS) or the sum of squared deviations from the mean. A sample variance is then defined as the sum of squares divided by the size of the sample minus one:

$$\sigma^2 = \frac{SS}{n-1}$$

$n - 1$ may be regarded as the degrees of freedom associated with the sum of squares and the variance may be regarded as a *mean square* (*MS*) or average squared deviation. Therefore:

$$MS = \frac{SS}{df}$$

To compute the necessary variances for a between-groups ANOVA, we must first compute the total sum of squares (SS_{total}), the between groups sum of squares ($SS_{between}$), and the within groups sum of squares (SS_{within}). Mean squares are then calculated using the appropriate degrees of freedom and the *MS*s may then be used to determine the *F*-ratio.

COMPUTATION OF SUMS OF SQUARES

The general formula for computing SS total representing the total deviation from the overall mean is

$$SS_{total} = \left(\sum X_1^2 + \sum X_2^2 + \sum X_3^2 \ldots \sum X_k^2 \right) - \frac{(\sum X_1 + \sum X_2 + \sum X_3 \ldots \sum X_k)}{N}$$

where:

  k = last treatment group

  N = total number of subjects in study

The subscripts 1, 2, 3, ..., k refer tothe treatment groups. Thus $\sum X_1$ is the sum of all scores in Group 1 and $\sum X_1^2$ is the sum of all squared scores in Group 1. Although group scores are kept separate (for further examination), this formula in fact reduces to

$$SS_{total} = \sum X^2 - \frac{(\sum X)^2}{N}$$

Next, we compute $SS_{between}$ using the following general equation:

$$SS_{between} = \frac{(\sum X_1)^2}{n_1} + \frac{(\sum X_2)^2}{n_2} + \frac{(\sum X_3)^2}{n_3} \ldots + \frac{(\sum X_k)^2}{n_k} - \frac{(\sum X_1 + \sum X_2 + \sum X_3 \ldots \sum X_k)}{N}$$

where n is the number of subjects in each treatment group and $SS_{between}$ reflects deviations of the group means from the overall mean. SS_{within} (or SS_w) is the sum of deviations of group members from their respective group means and can be computed by the following formula:

$$SS_{within} = \sum x_1^2 + \sum x_2^2 + \sum x_3^2 \ldots \sum x_k^2 = SS_{erro}$$

However, because $SS_{total} = SS_{between} + SS_{within}$, the SS_{within} can be computed by simple subtraction:

$$SS_{within} = SS_{total} - SS_{between} = SS_{error}$$

SS_{error}, when divided by the df, becomes the denominator of the F-ratio, regardless of the experimental design employed. For between-groups designs, $SS_{error} = SS_{within}$, but this is not true for within-subjects and mixed designs. Until the differences are elaborated, remember that the denominator of the F-ratio is the MS_{error}, which equals the SS_{error} divided by its df.

DEGREES OF FREEDOM, MEAN SQUARES, AND F-RATIO

Having computed our total, between groups, and within groups sums of squares, we may now compute the three corresponding variances. It is customary in ANOVA to refer to these variances as *mean squares*. Recall that they are computed by dividing each SS by its appropriate degree of freedom. Because the total sum of squares involves summing across all scores, total $df = N - 1$. The between groups sum of squares involves summing the squared deviations of the individual group means from the grand mean of all scores. Therefore, the between-groups df equals the number of groups minus 1:

between $df = k - 1$. The within-groups sum of squares involves summing the squared deviations within each group (from their respective group means) and summing across groups. Each group has $n - 1$ degrees of freedom. Summed across the groups, within $df = k(n - 1) = N - k$. In summary:

$$\text{Between } df = k - 1$$

$$\text{Within } df = N - k$$

$$\text{Total } df = \text{between } df + \text{within } df = N - 1$$

We are now in a position to compute our required mean squares, MS between and MS within.

$$MS_{between} = \frac{SS_{between}}{df_{between}}$$

$$MS_{within} = \frac{SS_{within}}{df_{within}}$$

Note that MS_{total} is not needed. SS_{total} was used for computational ease. $MS_{between}$ is our estimate of the between-groups variance, and MS_{within} of the within-groups variance. The F-ratio is determined by these two variances:

$$F = \frac{MS_{between}}{MS_{within}} = \frac{SS_{between} / df_{between}}{SS_{within} / df_{within}}$$

The two df needed to use the F distribution table (Table I in Appendix A) are the between df (numerator) and the within df (denominator).

GENERALIZED ANOVA SUMMARY TABLE

An ANOVA summary table consolidates the aforementioned calculations into a quick reference of the statistical results of an experiment. A generalized summary appears in Table 5.1.

TABLE 5.1
ANOVA Summary Table (Between Groups)

Source	SS[a]	df	MS	F	$F_{critical}$
Between groups (treatments)	$\sum \frac{(\sum X_k)^2}{n_k} - \frac{(\sum X)^2}{n}$	$k - 1$	$\frac{SS_{between}}{df_{between}}$	$\frac{MS_{between}}{MS_{within}}$	[b]
Within groups (error term)	$SS_{total} - SS_{between}$	$N - k$	$\frac{SS_{within}}{df_{within}}$		
Total	$\sum X^2 - \frac{(\sum X)^2}{N}$	$N - 1$			

[a] Specific treatment groups; summation is from 1 to k.
[b] $F_{critical}$ is the tabled F value for chosen level of significance ($\alpha = 0.05$, $\alpha = 0.01$) for dfs $k - 1, N - k$.

TABLE 5.2
**Scores from Three Randomized Groups Design Study
of Teaching Methods**

	Group 1: Lecture Method	Group 2: Discussion Method	Group 3: Lecture–Discussion Method
	0	3	4
	1	5	6
	2	5	7
	4	6	8
	5	6	9
ΣX	12	25	34
ΣX^2	46	131	246
n	5	5	5

COMPUTATIONAL EXAMPLE

Table 5.2 presents the DM scores for a hypothetical three randomized groups design study. Let us assume these are the final exam results from a study of three teaching methods for a philosophy course mentioned earlier in this chapter.

$$SS_{total} = \left(\Sigma X_1^2 + \Sigma X_2^2 + \Sigma X_3^2\right) - \frac{(\Sigma X_1 + \Sigma X_2 + \Sigma X_3)}{N}$$

$$= (46 + 131 + 246) - \frac{(12 + 25 + 34)^2}{15}$$

$$= 86.93$$

$$SS_{between} = \frac{(\Sigma X_1)^2}{n_1} + \frac{(\Sigma X_2)^2}{n_2} + \frac{(\Sigma X_3)^2}{n_3} - \frac{(\Sigma X_1 + \Sigma X_2 + \Sigma X_3)}{N}$$

$$= \frac{144}{5} + \frac{625}{5} + \frac{1156}{5} - 336.07$$

$$= 28.8 + 125.0 + 231.2 - 336.07$$

$$= 385 - 336.07 = 48.93$$

$$SS_{within} = SS_{total} - SS_{between}$$

$$= 86.93 - 48.93 = 38.0$$

Because the total number of observations (N) is 15 and the number of groups (k) is 3, the degrees of freedom for our example are:

Total $df = N - 1 = 15 - 1 = 14$
Between $df = k - 1 = 3 - 1 = 2$
Within $df = N - k = 15 - 3 = 12$

TABLE 5.3
ANOVA Summary Table (Between Groups)

Source	SS	df	MS	F	$F_{.05}$	$F_{.01}$
Between groups	48.93	2	24.47	7.72	3.89	6.93
Within groups	38	12	3.17			
Total	86.93	14				

The mean squares may now be computed:

$$MS_{between} = \frac{SS_{between}}{df_{between}} = \frac{48.93}{2} = 24.465$$

$$MS_{within} = \frac{SS_{within}}{df_{within}} = \frac{38}{12} = 3.17$$

The F-ratio for this example is

$$F = \frac{MS_{between}}{MS_{within}} = \frac{24.46}{3.17} = 7.72$$

Entering the data into an ANOVA summary table yields Table 5.3. Because our calculated F of 7.72 exceeds the table value at the 0.01 level of significance (6.93), we would reject the H_0 and conclude that there *is* a significant difference between the three teaching methods.

ANOVA FOR REPEATED MEASURES DESIGN

Calculations for the within-subjects design differ somewhat from those for the randomized groups design. To clarify the structure of the repeated measures framework, we may diagram it as a matrix of subjects (rows), and treatments (columns) as shown in Table 5.4. The sums of squares required are SS_{total}; $SS_{between}$ or $SS_{treatment}$; $SS_{subjects}$ or SS_{within}; and SS_{error} or $SS_{within \times between}$.

COMPUTATION OF SUMS OF SQUARES

SS_{total} is the deviation of all scores from the overall mean and is calculated the same way as in the between-groups example:

$$SS_{total} = \Sigma X^2 - \frac{(\Sigma X)^2}{N}$$

TABLE 5.4
Repeated Measures Framework

		Treatments				
		k_1	k_2	k_3	Σ_{row}	$(\Sigma_{row})^2$
	s_1					
	s_2					
Subjects	s_3					
	ΣX					
	ΣX^2					

where N is the total number of scores in the study (# subjects × # scores per subject); $SS_{treatment}$ reflects deviations of the treatment means from the overall mean and is also calculated like the $SS_{between}$ in the between-groups design:

$$SS_{treatment} = \frac{(\sum X_1)^2}{n_1} + \frac{(\sum X_2)^2}{n_2} + \frac{(\sum X_3)^2}{n_3} \cdots \frac{(\sum X_k)^2}{n_k} - \frac{(\sum X)}{N}$$

where 1, 2, 3, ..., k denotes the specific treatment; n is the number of subjects under that treatment (number of rows). $SS_{subjects}$ reflects the variability among individuals in the experiment and involves adding the subjects' scores across the rows, then summing the squares of the sums and dividing by the number of scores per subject (number of columns).

$$SS_{subjects} = \frac{\sum (\sum \text{rows})^2}{k} - \frac{(\sum X)^2}{N}$$

where k is the number of treatments. $SS_{subjects}$ is of little interest other than for its use in calculating the SS_{error}. SS_{error} ($SS_{subjects \times treatment}$) is the random variability expected of the same subject under the same conditions and is a different error term from that used in the between-groups design. The error term is calculated by subtraction:

$$SS_{SXT} = SS_{total} - SS_{treatment} - SS_{subjects}$$

Degrees of Freedom, Mean Squares, and F-Ratio

The repeated measures design total df and treatment df are the same ones used in the between-groups design:

$$\text{Total } df (df_{total}) = N - 1$$
$$\text{Treatment } df (df_{treatment}) = k - 1$$
$$\text{Subjects } df (df_{subjects}) = \text{number of subjects} - 1$$
$$\text{Subjects X treatment } df (df_{SXT}) = \text{subjects } df \times \text{treatment } df:$$
$$\text{SXT } df = (n - 1)(k - 1)$$

Again, the mean squares are calculated by dividing the SS by their respective df. The F-ratio needed for repeated measures ANOVA is, as before, the treatment variance divided by the error variance, but the error variance is different from the between groups error term. Note that the variance due to differences among subjects has been removed from the denominator of the F-ratio, thereby making it smaller compared to the randomized groups denominator and increasing the sensitivity of the test. The degrees of freedom used for the table F value are the treatment df (numerator) and the subject X treatment df (denominator).

Generalized ANOVA Summary Table

Our repeated measures calculations are summarized in Table 5.5.

TABLE 5.5
ANOVA Summary Table (Repeated Measures)

Source	SS[a]	df	MS	F	F$_{critical}$
Treatments	$\Sigma \dfrac{(\Sigma X_k)^2}{nk} - \dfrac{(\Sigma X)^2}{N}$	$(k-1)$	$\dfrac{SS_{treatment}}{df_{treatment}}$	$\dfrac{MS_{treatment}}{MS_{SXT}}$	[b]
Subjects	$\Sigma \dfrac{(\Sigma_{rows})^2}{k} - \dfrac{(\Sigma X)^2}{N}$	$(n-1)$			
SXT (error term)	$SS_{total} - SS_{treat} - SS_{subjects}$	$(n-1)(k-1)$	$\dfrac{SS_{SXT}}{df_{SXT}}$		
Total	$\Sigma X^2 - \dfrac{(\Sigma X)^2}{N}$	$(N-1)$			

[a] Specific treatment groups; summation is from 1 to k.
[b] $F_{critical}$ is tabled F value of chosen level of significance for df's $k-1$, $(n-1)(k-1)$.

COMPUTATION EXAMPLE

We could have performed the fictitious experiment presented in the randomized groups section as a repeated measures experiment with three treatments and five subjects. The data from Table 5.2 are shown in Table 5.6, in the repeated measures format.

$$SS_{total} = \Sigma X^2 - \frac{(\Sigma X)^2}{N}$$

$$= (46 + 131 + 246) - \frac{(12 + 25 + 34)^2}{15}$$

$$= 423 - 336.07 = 86.93$$

$$SS_{treatment} = \frac{(\Sigma X_1)^2}{n_1} + \frac{(\Sigma X_2)^2}{n_2} + \frac{(\Sigma X_3)^2}{n_3} - \frac{(\Sigma X)^2}{N}$$

$$= \frac{144}{5} + \frac{625}{5} + \frac{1156}{5} - 336.07 = 48.93$$

$$SS_{subjects} = \frac{\Sigma (\Sigma rows)^2}{k} - \frac{(\Sigma X)^2}{N}$$

$$= \frac{(49 + 144 + 196 + 324 + 400)}{3} - 336.07$$

$$= \frac{1113}{3} - 336.07$$

$$= 371 - 336.07 = 34.93$$

$$SS_{SXT} = SS_{total} - SS_{treatment} - SS_{subjects}$$

$$= 86.93 - 48.93 - 34.93 = 3.07$$

TABLE 5.6
Scores from Repeated Measures Design Study of Teaching Methods

		Treatment				
		1 **Lecture** **Method**	**2** **Discussion** **Method**	**3 Lecture–** **Discussion** **Method**	Σ_{row}	$(\Sigma_{row})^2$
Subjects	**1**	0	3	4	7	49
	2	1	5	6	12	144
	3	2	5	7	14	196
	4	4	6	8	18	324
	5	5	6	9	20	400
	ΣX	12	25	34		
	ΣX^2	46	131	246		
	N	5	5	5		

The degrees of freedom for this example are

$$\text{Total } df = N - 1 = 15 - 1 = 14$$

$$\text{Treatment } df = k - 1 = 3 - 1 = 2$$

$$\text{Subjects } df = n - 1 = 5 - 1 = 4$$

$$\text{SXT } df = (n - 1)(k - 1) = (4)(2) = 8$$

The mean squares are

$$MS_{treatment} = \frac{SS_{treatment}}{df_{treatment}} = \frac{48.93}{2} = 24.465$$

$$MS_{SXT} = \frac{SS_{SXT}}{df_{SXT}} = \frac{3.07}{8} = 0.38 = \text{Error term}$$

The F-ratio is

$$F = \frac{MS_{treatment}}{MS_{SXT}} = \frac{24.465}{0.38} = 64.39$$

Table 5.7 is the ANOVA summary for this example. Note that the F-ratio df decreased compared to the randomized groups design, with corresponding increases in the tabled F values. But also note

TABLE 5.7
ANOVA Summary Table (Repeated Measures)

Source	SS	df	MS	F	$F_{.05}$	$F_{.01}$
Treatment	48.93	2	24.47	64.39	4.46	8.65
Subjects	34.93	4				
SXT	3.07	8	0.38			
Total	86.93	14				

how much larger this F-ratio is than in the between-groups analysis. Of course, we have assumed that our experiment was properly counterbalanced, there were no differential order effects, and no other major assumptions of the F-test were violated.

POST HOC ANALYSES: MULTIPLE COMPARISONS AMONG MEANS

Just knowing that there is a significant difference among the treatment means is not very informative. For example, in our study of three different methods for teaching a philosophy course, we would probably want to know whether both or only one of the experimental methods were really better than the traditional lecture method. To isolate the particular means that contributed to our rejection of the null hypothesis, we must do a post hoc analysis for which several procedures allow multiple comparisons among means.

Some examples are multiple t-tests or the test of least significant difference (LSD); the Neuman–Keuls test; Tukey's WSD (wholly significant difference) test; the Bonferroni t-test; and the Scheffe test. Dunnett's test compares a control group with all experimental groups and will not be discussed here. For more information on these tests, see Myers (1979), Glass and Stanley (1970), and Games (1984).

The tests for multiple comparisons among means were listed in order of increasing *conservatism* or decreasing *power*. As the LSD method is the least conservative, it requires the smallest difference between means to show significance. The Scheffe test is the most conservative—a larger difference between means is required to indicate significance. Conversely, the LSD method is the most powerful and the Scheffe is the least powerful in terms of rejecting null hypotheses. All tests can be viewed as special cases of the t-test, differing in their abilities to control the family-wise error rate (FWI) mentioned earlier. The use of multiple t-tests has already been discussed as undesirable because it increases the FWI with each successive comparison. The four other tests are briefly discussed in the following section.

Tukey's WSD (Wholly Significant Difference)

$$WSD = q_{\alpha \, , \, k \, , \, df_{denom}} \sqrt{\frac{MS_{error}}{n}}$$

Tukey's WSD method is based on the sampling distribution of q, the *Studentized range* statistic; q represents the range of a set of observations divided by an estimate of the standard deviation of the population from which the observations were drawn. For comparisons among means, the calculated q value is

$$q = \frac{\left| \bar{X}_1 - \bar{X}_2 \right|}{\sqrt{\dfrac{MS_{error}}{n}}}$$

where:

n = number of observations in each treatment

$\sqrt{\dfrac{MS_{error}}{n}}$ = the estimate of the sampling error

The WSD method sets the family-wise probability of a Type I error (FWI) at α and decreases the comparison error rates in contrast to multiple t-tests that set each contrast error rate at α, thereby increasing FWI. The WSD requires equal ns, independent and randomized groups, and homogeneous variances.

The q value can be obtained for all comparisons and evaluated against the critical q from Table J in Appendix A that shows the number of means across the top and the df for the error term down the left side. Alternatively, a *critical distance* can be calculated and the means judged significant by inspection. If

$$\left|\bar{X}_1 - \bar{X}_2\right| > (q_{critical})\sqrt{\frac{MS_{error}}{n}}$$

then there is a significant difference between the two means. The data from our teaching example (randomized groups) are as follows:

$\bar{X}_1$ Lecture method = 2.4
$\bar{X}_2$ Discussion method = 5.0
$\bar{X}_3$ Lecture–discussion method = 6.8
MS_{error} = 3.17
n (number of observations per mean) = 5
df_{error} = 12
q (0.5, 3, 12) = 3.77 for α = 0.05, 3 means compared, and a df for MS_{error} = 12

The critical distance is

$$WSD = q_{(\alpha,\ \#\ \text{of means},\ df_{MS_{error}})}\sqrt{\frac{MS_{error}}{\#\ \text{obs per mean}}}$$

$$= q_{(.05,\ 3,\ 12)}\sqrt{\frac{3.17}{5}}$$

$$= (3.77)\sqrt{\frac{3.17}{5}}$$

$$= 3.00$$

Arranging the means from lowest to highest, we see differences between $\bar{X}_1$ and $\bar{X}_2$ and between $\bar{X}_2$ and $\bar{X}_3$. The difference between $\bar{X}_1$ and $\bar{X}_2$ (2.6) is not greater than our computed WDS (3.0), so it is not significant. The same is true for the difference between $\bar{X}_2$ and $\bar{X}_3$. The only significant difference is between $\bar{X}_1$ and $\bar{X}_3$. To illustrate this using the Duncan underline technique (1955), simply underscore the differences which are not significant.

$\bar{X}_1$	$\bar{X}_2$	$\bar{X}_3$
2.4	5.0	6.8

Although the WSD method assumes independent groups, it may be used for a repeated measures design via use of the appropriate error term and error df. However, the repeated measures experiment presents more of a risk of heterogeneous variances and the q statistic may only approximate the true relationship. Tukey's WSD is used frequently because it falls in the middle of the range of post hoc tests because of its conservatism and power. If the WSD is larger than the difference between the lowest mean and the highest mean and the omnibus F-test shows the difference is significant, the WSD is too conservative and should be replaced with a more liberal test, possibly the Neuman–Keuls or LSD test.

NEUMAN–KEULS TEST

The Neuman–Keuls test uses the same q statistic as the WSD, subject to the same assumptions. However, it is performed sequentially, testing the largest range first and then continuing with the smaller ranges if the first test proves significant, changing the critical q for each test. As the range decreases, the number of means decreases, and the table q value that depends on the number of means also decreases. With each comparison, the critical distance for significance becomes smaller. The first test performed exactly like the WSD; the last test between two means is the same as the LSD test. Therefore, the Neuman–Keuls test is more powerful (and less conservative) than the WSD. From our teaching example (randomized groups):

First comparison: $q_{.05,3,\,12} = 3.77$; d = 3.00
Second (last) comparison: $q_{.05,\,2,\,12} = 3.08$; d = 2.45

where d is the critical difference. This test yields different results from the WSD in that the distance between X_1 and X_2 exceeds the critical distance (e.g., $|\bar{X}_2 - \bar{X}_1| > $ d) :

$$2.4 \quad\quad 5.0 \quad\quad 6.8$$

Now the difference between $\bar{X}_1$ and $\bar{X}_2$ is also judged to be significant.

BONFERRONI t-TEST

The Bonferroni t-test is fairly conservative; it places an upper limit on FWI and uses a smaller α for the comparisons. It employs the t table with error df and may be used for independent or dependent means and with unequal n. For a FWI of $\alpha = 0.05$ for five comparisons, use the critical t value for $\frac{\alpha}{k} = .01,$ and the df_{error} (two-tailed). The calculated t value is:

$$t = \frac{\bar{X}_1 - \bar{X}_2}{\sqrt{\frac{2MS_{error}}{n}}}$$

Or the critical distance is:

$$|\bar{X}_1 - \bar{X}_2| > \left(\frac{a}{k}, df_{error}\right)\sqrt{\frac{2MS_{error}}{n}}$$

Doubling the MS_{error} term increases the critical distance compared to the WSD, thereby making this test less powerful than the WSD.

SCHEFFE TEST FOR ALL POSSIBLE COMPARISONS

The Scheffe test sets the upper limit of the FWI at the overall α and is the most conservative (least powerful) post hoc test. Like the ANOVA, the Scheffe is fairly insensitive to violations of the assumptions of normality and homogeneity of variance. The formula for any comparison using the Scheffe is as follows:

$$F = \frac{(\bar{X}_1 - \bar{X}_2)^2}{MS_{error}\left(\frac{1}{n_1} + \frac{1}{n_2}\right)(k-1)}$$

where $df = k - 1, N - k$, or $(n - 1)(k - 1)$. Because the numerator and denominator df for the Scheffe are the same as for the overall F in the ANOVA, the tabled value of F also is identical for the specified level of significance (i.e., $F_{.05 \, for \, 2, \, 12 \, df} = 3.89$). For the data in Table 5.1, the treatment means are as follows:

$\bar{X}_1$ Lecture method = 2.4
$\bar{X}_2$ Discussion method = 5.0
$\bar{X}_3$ Lecture–discussion method = 6.8

The Scheffe tests of all possible comparisons are as follows:

$$F_{3,2} = \frac{(6.8 - 5.0)^2}{3.17\left(\frac{1}{5} + \frac{1}{5}\right)(3 - 1)} = \frac{(1.8)^2}{2.54} = \frac{3.24}{2.54} = 1.28$$

$$F_{3,1} = \frac{(6.8 - 2.4)^2}{2.54} = \frac{(1.8)^2}{2.54} = \frac{3.24}{2.54} = 7.62$$

$$F_{2,1} = \frac{(5.0 - 2.4)^2}{2.54} = \frac{(2.6)^2}{2.54} = \frac{6.76}{2.54} = 2.66$$

As may be noted, the only F-ratio reaching significance at the 0.05 level is for the comparison of the lecture–discussion method with the lecture method (7.62). Thus, we may conclude that the lecture–discussion method was found superior to the lecture method based on final exam scores. However, the superiority of the discussion method over the lecture method and of the lecture–discussion method over the discussion method cannot be asserted based on this study. When a number of treatment groups of the same size are used in a simple ANOVA design study, it is more efficient to use the Scheffe procedure to derive the critical distance between the means:

$$d = \sqrt{\frac{2(k - 1)(F_{critical})(MS_{error})}{n}}$$

where k represents the number of means compared. Substituting in the data from our teaching techniques study we have the following:

$$d = \sqrt{\frac{2(2)(3.89)(3.17)}{5}} = 3.14$$

The difference between means 1 and 2 is 2.6, and between means 2 and 3 is 1.8. Accordingly, because each difference is less than 3.14, the null hypothesis cannot be rejected. The difference between means 1 and 3 is 4.4. Thus, because 4.4 is greater than our calculated critical distance, the null hypothesis may be rejected. These findings are identical to those obtained previously by computing three separate F-ratios.

The Scheffe is not as powerful a statistical procedure as the ANOVA. Because of this, it is possible to obtain a significant F from the ANOVA and yet not have any two means differ significantly based on the Scheffe procedure. When this occurs, it is acceptable to assume that the largest mean is significantly larger than the smallest mean.

In addition to making simple comparisons of pairs of means, the Scheffe procedure also may be used to test more complex relationships. For example, in our study, suppose we wanted to compare the mean for the traditional lecture method with the mean for the two experimental methods combined. First, we would pool our two experimental group means.

$$\bar{X}_{2+3} = \frac{(n_2)(\bar{X}_2) + (n_3)(\bar{X}_3)}{n_2 + n_3} = \frac{(5)(5.0) + (5)(6.8)}{10}$$

$$= \frac{25 + 34}{10} = \frac{59}{10}$$

$$= 5.9$$

The computation of the Scheffe test would be as follows:

$$F = \frac{\left(\bar{X}_{2+3} - \bar{X}_1\right)^2}{MS_{error}\left(\frac{1}{n_2+n_3} + \frac{1}{n_3}\right)(k-1)} = \frac{(5.9 - 2.4)^2}{3.17\left(\frac{1}{10} + \frac{1}{5}\right)(2}$$

$$= \frac{(3.5)^2}{1.902} = \frac{12.25}{1.902}$$

$$= 6.44$$

Because the tabled value of $F_{.05}$ for 2 means, 12 df is 3.89, we may conclude that our F-ratio is significant. Accordingly, we may reject our null hypothesis at the 0.05 level. Our conclusion would be that based on the final exam scores, the experimental teaching methods together resulted in better overall student performance than the lecture method.

SUMMARY

1. More than two treatments designs are used in educational and behavioral research when researchers (a) suspect the relationship between the IV and DM is not linear and using more treatments may yield a different result, (b) wish to obtain a precise knowledge of the nonlinear relationship, or (c) have more than two treatment conditions they wish to study. As with two treatment designs, more than two treatment designs may be (a) randomized groups, (b) matched groups, or (c) repeated measures designs. The most frequently used is the randomized groups design. Because the problem of obtaining matched groups is more difficult, matched groups designs are used infrequently with more than two treatments.
2. Counterbalancing with more than two groups repeated measures designs may be accomplished with the Latin square. Each treatment condition appears once and only once in each row and column of the design matrix.
3. In analysis of variance (ANOVA), the total variation of the DM scores is partitioned into two orthogonal parts: (a) the variance between groups and (b) the variance within groups. In general, the larger the variance between groups and/or the smaller the variance within groups, the greater is the likelihood that the between-groups variation is a real difference resulting from manipulation of the IV (rather than simply a fluctuation resulting from experimental error).
4. The F-ratio is an estimate of the variance due to random error plus an estimate of the effect of the IV divided by an independent estimate of the random error variance. Theoretically, because of this, an F-ratio may never be less than 1; thus, the F-ratio is one-tailed. Accordingly, it is compared to the right tail of the F-distribution in testing for significance. If the calculated F-ratio is *larger* than the table F value for the specified level of significance, the F-ratio is significant. In the two treatments situation, this procedure will yield the same result as the two-tailed t-test.

5. The assumptions underlying the F-ratio calculated in the simple ANOVA for randomized groups are essentially the same as for the t-test for independent samples: (a) interval level data, (b) independent random samples, (c) normal distribution of the underlying trait, and (d) homogeneity of variance. Like the t-test, the F-test is insensitive to minor violations of the assumptions of normality and homogeneity of variance. For repeated measures, a different error term compensates for nonindependence of samples.

6. It is inappropriate to use the t-ratio to test all possible pairs of means in a more than two treatments study because (1) multiple t-tests cannot reveal interactions between facts and (2) as the number of t-ratios calculated increases, the true level of α increases to a value less than or equal to the sum of α values for all comparisons. By far the most common procedure for testing significance in more than two treatments studies is ANOVA.

7. In ANOVA, the degrees of freedom for the total variation is total $N - 1$ because all scores vary about a grand mean. The various treatment means also vary about the grand mean; hence, between-groups df equal $k - 1$. For randomized groups, the scores within each treatment group vary about their respective group means; accordingly, within-groups df are $n_1 - 1 + n_2 - 1 \ldots n_k - 1$ or $N - k$. For repeated measures, the subjects df is $n - 1$ and the error df is $(n - 1)(k - 1)$.

8. Dividing sums of squares *(SSs)* by their respective degrees of freedom yields the corresponding variances or mean squared deviations, i.e., MS between, MS within, and MS error. Dividing the treatment (or between) MS by the error (or within) MS yields the F-ratio.

9. A repeated measures design usually yields a smaller error term and a larger F-ratio than a randomized groups design. However, because the error df is also smaller, the critical F (table value) is correspondingly larger; accordingly, a larger calculated F-ratio is required to show significance for a repeated measures design.

10. If researchers need to know which specific group means differ significantly from each other, they may make multiple comparisons among the group means using post hoc tests. Most of the available post hoc tests ensure that the probability of Type I error for any comparison does not exceed the level of significance specified in the ANOVA for the overall hypothesis. However, the actual α level for any given comparison may be much smaller than that specified for the overall test. Post hoc tests vary in conservatism and power.

11. Arranged from the most to the least powerful, the common post hoc tests are multiple t-tests or the test of least significant difference (LSD); the Neuman–Keuls test; Tukey's WSD (wholly significant difference) test; the Bonferroni t-test; and the Scheffe test. The Scheffe is one of the most commonly used post hoc tests for all possible comparisons. Like the ANOVA, it is (a) insensitive to violation of the assumptions of normality and homogeneity of variance and (b) utilizes the F distribution for significance testing. The Scheffe is less powerful than ANOVA. As a result, it is possible to obtain a significant ANOVA and not have any two means differ significantly using the Scheffe procedure. In this event, the difference between the largest and smallest means is considered significant at the specified α level. The Scheffe also may be used for more complex post hoc comparisons. Tukey's WSD method is also frequently used because it falls midway in the range of power and conservatism.

KEYWORD DEFINITIONS

Between Groups: Variance that measures how much the means of groups vary from the overall mean.

Conservatism: Cautious analysis and interpretation; biased toward maintaining status quo.

Control Group: Sample of subjects designated not to receive the experimental treatment condition.

Critical Distance: Minimum difference between two numbers required for significance.

Critical Ratio: Basis of statistical hypothesis testing. With two treatment groups, critical ratio indicates the difference or variability between the means to the difference or variability within the treatment groups, as represented by $S_{D_{\bar{x}}}$.

Degrees of Freedom (*df*): Term that distinguishes among members in families of *t*, Chi-square, and *F* distributions. One *df* is gained for every independent observation or data point gathered, and one is lost for every population parameter estimated using the data collected.

Experimental Error: Differences due to effects of extraneous variables.

Latin Square Design: Method of incomplete counterbalancing used to counterorder effects of an experiment.

Mean Square (*MS*): Variance or average squared deviation.

Post Hoc Tests: Pair-wise comparisons of more than two means conducted after an omnibus *F*-test has shown significance to determine whether differences are significant.

Power: Measure of sensitivity of a test to detect differences that actually exist.

Sum of Squares (Σx^2 or *SS*): Sum of squared deviations from the mean.

Treatment Group: Sample of subjects receiving condition involving experimental manipulation.

Within Groups: Variance indicating how groups vary about their own means.

EXERCISES

1. Given the following data from a four randomized groups study, determine at the 0.05 level (a) whether a real difference in criterion performance exists among the treatment groups and (b) if so, between which specific treatments does the real difference lie? Use both Tukey's WSD method and the Scheffe method to determine critical distances; illustrate results with Duncan underlining. (c) State your conclusion based upon your analysis of the results.

Treatment Group 1	Treatment Group 2	Treatment Group 3	Treatment Group 4
2	4	4	6
3	5	6	7
3	5	7	9
4	6	8	9
5	6	8	10
6	7	10	10
6	7	10	12
7	8	12	14
9	9	15	15

2. Analyze the data in Exercise 1 as a repeated measures design. Determine at the 0.05 level (a) whether a real difference in performance of treatments exists and (b) if so, between which specific treatments does the real difference lie? Use Tukey's WSD method and the Scheffe method to determine critical distances; illustrate results with Duncan underlining. (c) State your conclusion.

3. A construction company has three suppliers from which to choose in deciding which one will supply building products for a major high-rise project. To determine the supplier of building products, the construction company purchased five samples from each supplier. Each sample contained the same type and number of items. The table below indicates the number of defects found in each sample. Determine at the 0.01 level of significance, the following: (a) Based on the number of defects per sample, would you conclude that real differences exist among the three suppliers? (b) Conduct a post hoc test utilizing Tukey's WSD method and illustrate the results using the Duncan underline procedure. (c) State your conclusion.

Sample	Supplier X	Supplier Y	Supplier Z
1	30	25	40
2	35	15	35
3	35	15	25
4	20	30	30
5	25	20	40

4. A researcher at a local university wishes to determine whether a person's ability to remember is affected by age. Four age brackets are determined and randomized groups are selected for each. Subjects in each group have been matched on other important variables such as IQ and sex. Each subject is shown a series of random three-letter combinations. The series is then repeated and the subjects are asked to write as many combinations as they can remember. The number of combinations remembered for each group is shown in the following table. (a) Conduct an analysis of variance with a 0.05 level of significance. Prepare an ANOVA summary table. (b) Conduct a post hoc test using Tukey's WSD. (c) State your conclusions.

25–34	35–44	45–54	55–64
17	11	9	8
10	18	15	9
14	12	17	13
12	12	13	6
13	15	14	10
15	16	14	7

EXERCISE ANSWERS

1.a. H_0 = no real difference in performance among groups ($\mu_1 = \mu_2 = \mu_3 = \mu_4$); H_1 = significant difference in performance among groups ($\mu_1 \neq \mu_2 \neq \mu_3 \neq \mu_4$). The following table refers to Exercise 1.a:

Group	1	2	3	4
ΣX	45	57	80	92
ΣX^2	265	381	798	1012
N	9	9	9	9
$\bar{X}$	5	6.3	8.9	10.2

$$SS_{Total} = (265 + 381 + 798 + 1012) - \frac{(45 + 57 + 80 + 92)^2}{36}$$

$$= 2456 - 2085.4 = 370.6$$

$$SS_{Between} = \frac{(45)^2}{9} + \frac{(57)^2}{9} + \frac{(80)^2}{9} + \frac{(92)^2}{9} - 2085.4$$

$$= 2237.6 - 2085.4 = 152.2$$

$$SS_{Within} = 370.6 - 152.2$$

$$= 218.4$$

Source	SS	df	MS	F	$F_{.05, 3, 32}$
Between	152.2	3	50.73	7.43	2.92
Within	218.4	32	6.82		
Total	370.6	35			

$$F_{.05, 3, 32} = 2.92; 7.43 > 2.92. \text{ Reject } H_0.$$

1.b. Scheffe method.

$$d = \sqrt{\frac{2(3)(2.92)(6.82)}{9}}$$

$$= 3.64$$

$\bar{X}_2 - \bar{X}_1 = 1.3$

$\bar{X}_3 - \bar{X}_1 = 3.9$

$\bar{X}_4 - \bar{X}_1 = 5.2$

$\bar{X}_3 - \bar{X}_2 = 2.6$

$\bar{X}_4 - \bar{X}_2 = 3.9$

$\bar{X}_4 - \bar{X}_3 = 1.3$

	$\bar{X}_1$	$\bar{X}_2$	$\bar{X}_3$	$\bar{X}_4$
	5	6.3	8.9	10.2

There are real differences between Groups 1 and 3, 1 and 4, and 2 and 4.
Tukey's WSD.

$$q_{.05, 4, 32} = 3.845$$

$$d = 3.845\sqrt{\frac{6.82}{9}} \quad 5 \quad 6.3 \quad 8.9 \quad 10.2$$

$$= 3.35$$

1.c. There are real differences between Groups 1 and 3, 1 and 4, and 2 and 4. Both post hoc tests confirm the results of ANOVA, so reject H_0.

2.a. Treatment differences.

	Treatment 1	Treatment 2	Treatment 3	Treatment 4
ΣX	45	57	80	92
ΣX^2	265	381	798	1012
$\bar{X}$	5	6.3	8.9	10.2

$$\Sigma(\Sigma row)^2 = 9142$$

$$SS_{Total} = 370.6$$

$$\frac{(\Sigma X)^2}{N} = 2085.4$$

$$SS_{Treatment} = 152.2$$

$$SS_{Subject} = \frac{9142}{4} = 2085.4 = 200.1$$

$$SS_{error} = 370.6 - 152.2 - 200.1$$

$$= 18.3$$

Source	SS	df	MS	F	$F_{.05,\,3,\,324}$
Treatments	200.1	8			3.01
Subjects	152.2	3	50.73	66.75	
Error	18.3	24	.76		
Total	370.6	35			

2.b. Tukey WSD.

$$q_{.05,4,24} = 3.90$$

$$d = 3.90\sqrt{\frac{.76}{9}}$$

$$= 1.13$$

5	6.3	8.9	10.2

All differences are significant.

Scheffe method.

$$d = \sqrt{\frac{(2)(3)(3.01)(.76)}{9}}$$

$$= 1.23$$

5	6.3	8.9	10.2

2.c. All differences are significant; therefore no underscore. Reject H_0.

3.a. H_0 = no real differences among suppliers in average number of defects per sample; H_1 = real differences among suppliers in average number of defects per sample.

$$F = \frac{215}{42.5} = 5.06$$

$$F_{.05,2,12} = 6.93$$

Source	SS	df	MS	F_{calc}	$F_{0.01}$
Between	430	2	215	5.06	6.93
Within	510	12	42.5		
Total	940	14			

Because $5.06 < 6.93$, do not reject H_0.

3.b.

$$WSD = q(\alpha, k, df_{error})\sqrt{\frac{MS_{error}}{\# \ obs \ per \ \overline{X}}}$$

$$WSD = (5.05)\sqrt{\frac{42.5}{5}} = 14.72$$

$$\overline{X}_1 \ (\text{supplier } Y) = 21$$

$$\overline{X}_2 \ (\text{supplier } X) = 29$$

$$\overline{X}_3 \ (\text{supplier } Z) = 34$$

Arrange the means from lowest to highest and compare their differences using the WSD value as the standard against which they are measured. Normally this is only done if the F is significant. The means in order were 21, 29, 34. Because 14.72 is required for a difference between any two means to be significant, a Duncan underline would be drawn under all three to indicate no significant differences among the means.

3.c. Conclude that there are no real differences among suppliers in average number of defects per sample because $|X_1 - X_2|$, $|X_2 - X_3|$, and $|X_3 - X_1|$ all are less than the WSD values.

4.a. H_0 = no real effect of age on memory; H_1 = real effect of age on memory.

Source	SS	df	MS	F_{calc}	$F_{.05}$
Between	108.33	3	36.11	5.45	3.10
Within	132.67	20	6.63		
Total	241.00	23			

$$WSD = q(\alpha, \ k, \ df_{error})\sqrt{\frac{MS_{error}}{\# \ obs \ per \ \overline{X}}}$$

4.b. Tukey's WSD.

$$WSD = (3.958)\sqrt{\frac{6.63}{6}}$$

$$= (3.958)(1.105) = 4.180$$

$$\overline{X}_1 = 55\text{--}64 \text{ age bracket}$$

$$\overline{X}_2 = 25\text{--}34 \text{ age bracket}$$

$$\overline{X}_3 = 45\text{--}54 \text{ age bracket}$$

$$\overline{X}_4 = 35\text{--}44 \text{ age bracket}$$

Arrange the means from lowest to highest and compare their differences using the WSD value. A WSD of 4.18 indicates a significant difference between the upper age bracket and the other three.

There are no significant differences among the younger three age groups. (Note: If a difference occurred between $\bar{X}_2$, and $\bar{X}_4$, we would have to show the Duncan underline as in the table below.)

$\bar{X}_1$	$\bar{X}_2$	$\bar{X}_3$	$\bar{X}_4$
8.83	13.5	13.7	14.0

4.c. Conclusion. Because 5.45 > 3.10, reject H_0 and conclude that there is a real effect of age on memory.

REFERENCES

Duncan, D. B. (1955). Multiple range and multiple *F*- tests. *Biometrics, 11*(1), 1–42.

Elmes, D. G., Kantowitz, B. H., & Roediger, H. L. (1985). *Research methods in psychology.* St. Paul: West Publishing.

Games, P. A. (1984). *Textbook for educational psychology 506.* University Park, PA: Department of Educational Psychology, Pennsylvania State University.

Games, P. A., & Klare, G. R. (1967). *Elementary statistics: Data analysis for the behavioral sciences.* New York: McGraw-Hill.

Glass, G. V., & Stanley, J. C. (1970). *Statistical methods in education and psychology.* Englewood Cliffs, NJ: Prentice-Hall.

Hayslett, H. T. (1968). *Statistics made simple.* Garden City, NY: Doubleday.

Jaeger, R. M. (1990). *Statistics: A spectator sport* (2nd ed.). Newbury Park, CA: Sage Publications.

Keppel, G. (1973). *Design and analysis: A researcher's handbook.* Englewood Cliffs, NJ: Prentice-Hall.

Kerlinger, F. N. (1986). *Foundations of behavioral research* (3rd ed.). New York: Holt, Rinehart & Winston.

McCall, R. B. (1970). *Fundamental statistics for psychology.* New York: Harcourt, Brace & World.

Myers, J. L. (1979). *Fundamentals of experimental design* (3rd ed.). Boston: Allyn & Bacon.

6 Multifactor Analysis of Variance

Chapter 5 introduced analysis of variance (ANOVA) on a single factor (i.e., independent variable). This procedure permits comparisons across more than two treatment conditions, but does not allow examination of two (or more) factors simultaneously or interactions of factors. Thus, this chapter builds on the last one to introduce multifactor univariate analysis using between, within, and mixed designs.

KEYWORDS

Between-Groups Design	Level	Single IV Experiment
df	Main Effect	*SS*
Factorial Design	Mixed Design	Within-Subjects Design
Fully Crossed Design	*MS*	
Interaction Effect	Multiple IV Experiment	

RATIONALE FOR FACTORIAL DESIGNS

Thus far, we have considered experiments having only one independent variable (IV). However, in the real world it is rare to find an IV that influences criterion measures in isolation from other IVs. Usually two or more IVs contribute to changes in dependent measures. *Factorial designs* permit examination of the effects of two or more IVs separately (*main effects*) or in combinations (*interactions*). An interaction occurs when the effects of one IV vary according to the level of another IV.

For example, in one of our fictitious experiments in Chapter 5, we investigated the effects of different instructional methods on student performance in a philosophy course. The question of the most effective method may be answered by saying that effectiveness *depends* on the small or large size of the class. The *depends* verb implies an interaction between the *class size* and *teaching method* IVs. To determine whether an interaction is significant, the IVs must be manipulated simultaneously in the same experiment. While the appropriate *F*-tests from an ANOVA provide an objective basis of whether interactions and main effects are significant, it is often difficult to

TABLE 6.1
Effects of Teaching Method and Class Size on
Student Performance

Class Size	Teaching Method B1 (Lecture)	Teaching Method B2 (Lecture–Discussion)
A1 (Large)	$\bar{X} = 10$	$\bar{X} = 8$
	$n = 10$	$n = 10$
A2 (Small)	$\bar{X} = 8$	$\bar{X} = 10$
	$n = 10$	$n = 10$

Note: $\bar{X}$ = mean. n = Number of classes taught under each method.

understand and to interpret the outcomes of factorial designs, especially those involving interactions. Hopefully, some examples will help improve reader understanding.

In our teaching methods study, we could have tried each instructional method with both large and small class sections as shown in Table 6.1.

Judging from the means of each group, we appear to have an interaction between two independent variables: with large classes the lecture method appears to work better; with small classes, the lecture–discussion method was the more effective approach. Thus, the effect of a given teaching method on performance depends on class size. If we plot our results, they would appear as in Figure 6.1. Note that the plots for A1 and A2 are *not* parallel. IVs may interact if the plotted lines converge, diverge, or cross. However, ANOVA is required to determine whether the nonparallel lines represent a significant interaction of the IVs. *Standard error bars* project from the means to illustrate the variability about the means. Standard error (of the mean) bars are used instead of standard deviation bars because they more closely represent the within-cell variance that is greatly influenced by sample size. (Remember, $\sigma_{\bar{X}} = \frac{\sigma}{\sqrt{n}}$.) Some other possible outcomes of this study are shown in Figure 6.2.

In Figure 6.2a through d, the plots of A1 and A2 are parallel, indicating no interaction. Because the plots in 6.2a and c are horizontal rather than sloping, no differences of student performance between teaching methods are apparent. That means the students performed equally

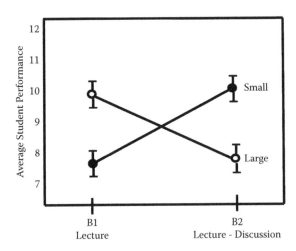

FIGURE 6.1 Effects of teaching method and class size on student performance (means and standard errors of means).

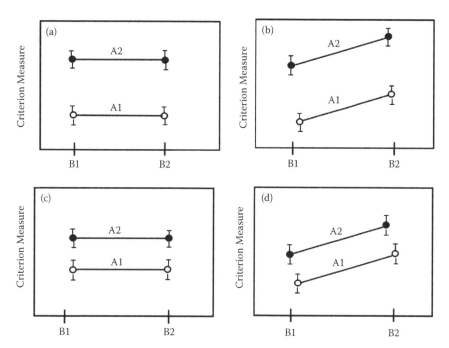

FIGURE 6.2   Other possible outcomes from a study of the effects of teaching methods and class size on student performance.

well under both methods. However, plot A2 is considerably higher than A1, indicating that small classes performed better than large classes. In Figure 6.2b, the A plots also slope, indicating that, in addition to better performance of smaller classes, all students performed better under the lecture–discussion method (B2) than under the lecture method (B1). Because the lines were parallel, there was no interaction.

The plot lines of Figure 6.2c are horizontal and at about the same height. This indicates that manipulating both teaching method and class size had no effect on student performance. In Figure 6.2d, the results indicate no performance effects from class size, but do indicate that the lecture–discussion method (B2) resulted in better performance than the lecture method (B1). Again, because the plot lines are parallel, the interaction of teaching method and class size probably was not significant.

From inspection of the possible outcomes illustrated above, it should be apparent now that an experiment involving two IVs can be analyzed three ways. First, we can ignore the effects of the A variable and simply compare all the results under the B1 treatment with all the results under treatment B2 (i.e., treat the results as though they arose from only one teaching method IV, the B variable). In multiple IV experiments, this result is known as the *main effect* of variable B. Second, we could ignore variable B and treat the study as though the only IV was variable A, class size; this is called the main effect of variable A. Third, we can analyze the differences among all four treatment conditions (A1–B1, A2–B1, A1–B2, and A2–B2) to determine whether interaction effect is present. This ability to determine whether an interaction is present gives a multiple IV experiment an advantage over several single IV experiments.

FACTORIAL DESIGNS

Studies in which two or more independent variables are manipulated simultaneously are known as *factorial designs*. The different treatment values for a given IV or factor are *levels*.

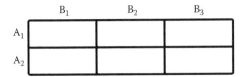

FIGURE 6.3   Diagram of 2 × 3 factorial design.

Two-Factor Designs

The previous experiment included two factors: class size (A) and teaching method (B). Each factor had two levels: A1 and A2, B1 and B2. This type of study is referred to as a two by two (2 × 2) factorial design. If variable B had three levels, the study would have a 2 × 3 factorial design and could be diagrammed as in Figure 6.3. Note that there are 3 × 2 or 6 possible treatment conditions. If factor A had three levels, the study would have had a 3 × 3 factorial design. This is illustrated by the addition of a third row to the Figure 6.3 diagram (see Figure 6.4).

Three-Factor Designs

A third factor (for example, different courses such as philosophy and English) added to the experiment (3 × 3 × 2) would yield a diagram as shown in Figure 6.5. Note that adding a third variable with two levels doubled the size of the experiment. To diagram this addition, we simply repeated the Figure 6.4 diagram. Note that adding a third variable produces an additional main effect (main effect of variable C) and also adds a number of possible interactions. It is now possible to have a three-factor interaction, A × B × C. It also is possible to have a total of three two-factor interactions, A × B, A × C, and B × C.

Four-Factor Designs

If we were to add a fourth variable with two levels, D1 and D2, the size of our experiment would again be doubled as shown in Figure 6.6. The experiment would involve the following 15 possible main effects and interactions (four main effects + six 2-way interactions + four 3-way interactions + one 4-way interaction):

A	B × C
B	B × D
C	C × D
D	A × B × C
A × B	A × B × D
A × C	A × C × D
A × D	B × C × D
	A × B × C × D

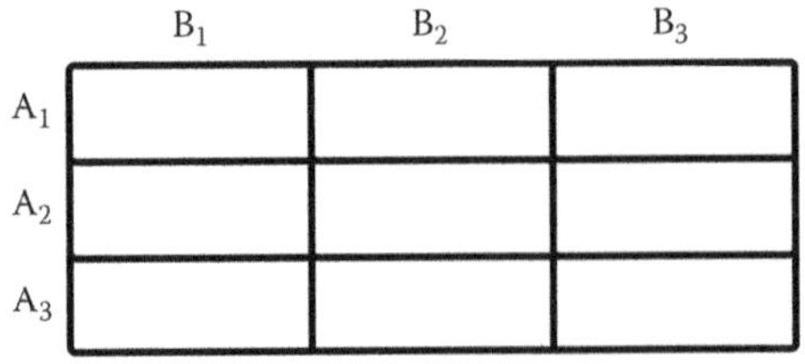

FIGURE 6.4   Diagram of 3 × 3 factorial design.

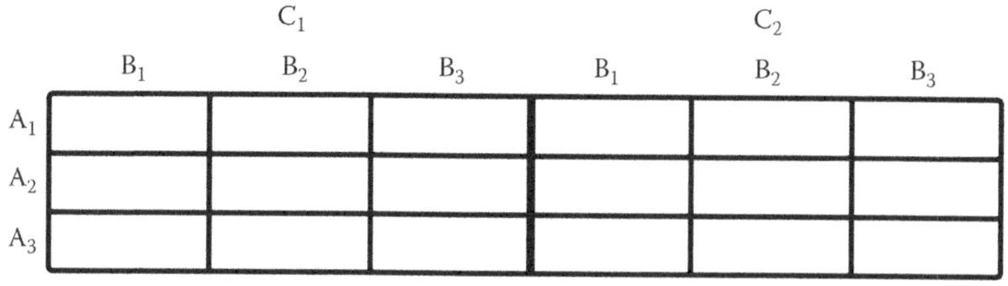

FIGURE 6.5   Diagram of $3 \times 3 \times 2$ factorial design.

As can be seen, each time an additional independent variable is added to an experiment, the number of possible interactions increases algebraically. Experiments with many factors also result in many more treatment conditions and thus require many more subjects. As a result, experiments with more than four factors are fairly uncommon.

NESTED DESIGNS

FULLY CROSSED DESIGNS

In multiple IV designs like those mentioned in the previous section, each level of a factor was crossed (combined) with each level of every other factor to form a treatment condition. Thus, level A1 (class size) of our two-factor design was crossed with both levels of variable B to form treatment conditions A1–B1 (large class, lecture method) and A1–B2 (small class, lecture method). Similarly level B2 (lecture–discussion method) also was crossed with levels A1 and A2. Designs in which each level of every IV is crossed with all levels of all other IVs are referred to as *fully crossed* factorial designs.

NESTED DESIGNS

As noted above, one disadvantage of fully crossed designs with many factors is that they require many treatment conditions. One way of reducing the number of treatment conditions is *not* to cross every level of one factor with every level of another factor in an experiment. In this case, we refer to the factor that is not fully crossed as "nested" under the other factor. For example, assume we have a $2 \times 4 \times 2$ fully crossed factorial design (Figure 6.7) that results in 16 treatment conditions.

Let us assume this is a teaching methods study. Variable B is class size, A is teaching method, and C consists of four different courses (philosophy, English, history, and biology). If we want 10 class sections under each treatment condition, we must have 160 class sections for the experiment. Figure 6.8 shows a diagram of the study in which variable C (different courses) is nested

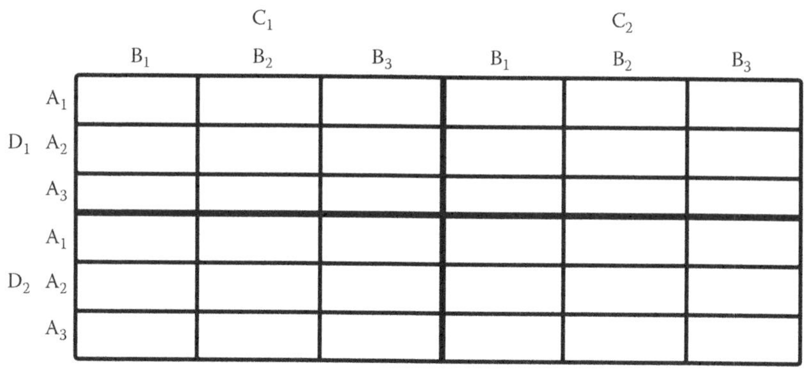

FIGURE 6.6   Diagram of $3 \times 3 \times 2 \times 2$ factorial design.

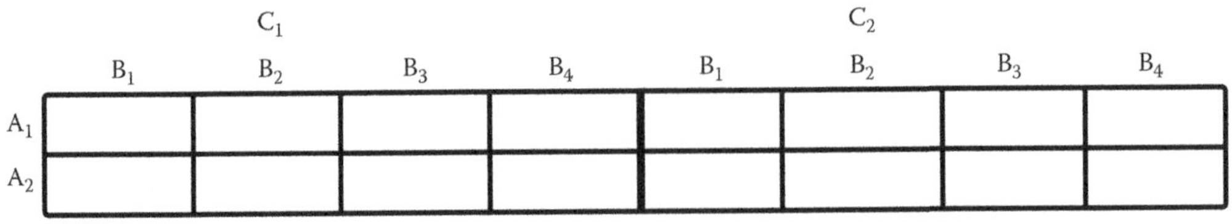

FIGURE 6.7   Diagram of fully crossed $2 \times 4 \times 2$ factorial design.

under variable B (class size). Note that by not crossing all courses with both teaching methods, we cut the size of our experiment in half. Clearly this method is more advantageous than a fully crossed design for the same study.

Limitation of Nested Designs

One major limitation of nested designs is the inability to analyze all possible interactions because the IVs are not fully crossed. For instance, in our example we cannot look at the A × B × C or the B × C interactions because all levels of A are not crossed with all levels of C. Thus, in addition to the A, B, and C main effects, we can look at only the A × B interaction. Because of this limitation, a nested design is only appropriate when a researcher has a sound basis for assuming that the interactions that cannot be analyzed are not pertinent to the research question or do not really exist.

Thus, in the above nested design study, a researcher would have to assume that using the different courses would not interact with the teaching methods in determining performance. Similarly, he also would have to assume that using different courses would not interact with teaching methods and class size in combination (i.e., the A × B × C interaction). When assumptions of this kind cannot be made, a nested design is inappropriate.

TYPES OF ANALYSIS OF VARIANCE DESIGNS

In conducting statistical analyses, one of the most crucial problems is choosing the correct error term for the denominator of the *F*-ratio. The choice of the correct error term will depend on the type of ANOVA design: between-groups, within-subjects (repeated measures), or mixed (i.e., combination of both). Characteristics of factorial ANOVA designs are presented in Table 6.2.

BETWEEN-GROUPS DESIGNS

Two-factor between-groups designs contain two independent variables, each with mutually exclusive levels. The key issue to remember with this design is that each subject receives only one treatment condition. An example for one factor might be gender; obviously, one subject can be of only one sex, and hence, can only be correctly placed on one level of that factor. The appropriate error term for this design is the within-subjects mean square (MS_{within}). We would look for two main effects (one for each variable) and an interaction of the two variables. The convention established for the labeling of between factors in this chapter is to use the letters A, B, C, and D.

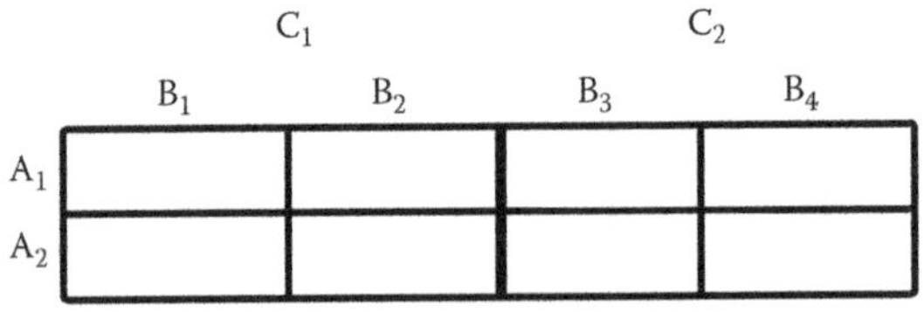

FIGURE 6.8   Diagram of $2 \times 2 \times 2$ factorial design with factor C nested under factor B.

TABLE 6.2
Characteristics of Factorial ANOVA Designs

Between-
Groups Design	Matrix	Definition	Features	Problems
2 × 3 factorial		Two independent variables, one with 2 levels and one with 3 levels yielding 6 independent groups; each subject receives only one treatment	Two main effects and interaction of 2 variables	Must control confounding variables of subjects and groups

B

		1	2	3
A	1	S1	S3	S5
		S2	S4	S6
	2	S7	S9	S11
		S8	S10	S12

Observations = 12
Subjects = 12
Observations/subject = 1

Within-Subjects Design

2 × 3 within factorial (treatments × subjects)	Matrix	Each subject receives all levels of each independent variable (6 treatments in 2 × 3)	More efficient than between designs; each subject is own control; 2 main effects and interaction	Use Latin square design, counterbalance, or randomize to minimize confounding from carryover effects

K

		1	2	3
J	1	S1	S1	S1
		S2	S2	S2
	2	S1	S1	S1
		S2	S2	S2

Observations = 12
Subjects = 2
Observations/subject = 6

Mixed Design

Combination of independent variables manipulated between and within subjects	Matrix	Within each independent group, each subject receives each level of within-subjects variable	Has virtues of repeated measures design; especially good for analyzing interaction of practice (trials) or learning with another independent variable	All controls above are necessary

J

		1	2	3
A	1	S1	S1	S1
		S2	S2	S2
	2	S3	S3	S3
		S4	S4	S4

Observations = 12
Subjects = 4
Observations/subject = 3

COMPLETELY WITHIN-SUBJECTS (REPEATED MEASURES) DESIGNS

In a two-factor completely within-subjects design, each subject receives each level of each independent variable. The levels of each variable are therefore mutually exclusive. The appropriate error term is the within-subjects mean square for each variable. Hence, for a two-factor application of this design we would have three error terms: one of each of the two variables and a third for the interaction of the two variables. As before, we would look for two main effects and an interaction of the two variables. Hereafter the within-subjects factors will be labeled J, K, L, and M.

MIXED DESIGNS

In a *mixed design*, each subject receives each level of the variable which is applied within subjects plus one level of the between-groups variable. Hence, as its name implies, this design is a combination of the two previous designs. The appropriate error term for this design is the between-groups residual sum of squares for the within-subjects factor and the interaction of the two factors. In describing mixed designs, the between-groups factors will be labeled A, B, C, and D; the within-subjects factors will be J, K, L, and M.

BETWEEN-GROUPS (RANDOM BLOCKS) TWO-FACTOR ANOVA DESIGNS

RATIONALE FOR BETWEEN-GROUPS ANOVA DESIGNS

The two-dimensional analysis of variance (ANOVA) is an extension of the simple analysis of variance (Chapter 5) to studies involving two IVs. Table 6.3 depicts fictitious data for our teaching methods study. It shows final exam scores for each class section used in the study. In partitioning our total sum of squares (SS_{total}) we can do each of the following:

- Ignore the row variable (*A*) and compute the sum of squares between the *B1* and *B2* columns (SS_B).
- Ignore the column variable (*B*) and compute the sum of squares between the *A1* and *A2* rows (SS_A).
- Treat each cell or treatment condition as a separate group and compute the sum of squares between cells or interaction (SS_{AB} or SS_{AXB}).
- Compute the sum of squares within cells or treatment conditions (SS_{within}, SS_w, or SS_{error}).

Thus, we have partitioned the total sum of squares in our two-factor design into its four component parts. Three parts (SS_A, SS_B, and SS_{AB}) actually represent a partitioning of the overall $SS_{between}$ into its component parts. Similarly, the degrees of freedom for each component part also add up to the total *df*:

$$\text{Total } df = \text{row } df + \text{col } df + \text{interaction } df + \text{error } df$$

or

$$(N-1) = (r-1) + (c-1) + (r-1)(c-1) + (N-rc)$$

or

$$df_{total} = (a-1) + (b-1) + (a-1)(b-1) + (N-ab)$$

Because the A variable consists of two subjects (rows) that vary about the grand mean, we subtract one *df* from the total number of rows. A similar rationale is used for the B variable. The A × B interaction refers to the variations between cells of the matrix. Thus, the interaction

$$df_{AB} = (r-1)(c-1)$$

or

$$df_{AB} = (a-1)(b-1)$$

TABLE 6.3
A 2 × 2 Between-Groups Factorial Design Study

Class Size	B1 Lecture Method	B2 Lecture–Discussion Method
A1 Large	6, 7, 7, 8, 8	6, 7, 7, 8, 8
	8, 8, 9, 9, 10	8, 8, 9, 9, 10
A2 Small	5, 6, 6, 7, 7	8, 8, 9, 10, 10
	7, 7, 8, 8, 9	10, 10, 11, 12, 12

Because the within-cells (treatment) variation refers to the variation of scores about their respective treatment means, the within $df = N - 1$ df for each treatment condition mean or $N - rc$. Dividing each of the four-component sum of squares by its respective df yields the corresponding mean squares: MS_A, MS_B, MS_{AB}, and MS_W. To determine whether either main effect or the interaction is significant, we divide each of these mean squares by MS_W (i.e., MS_{error}) to calculate the corresponding F-ratio. These results can then be tested for significance at a specified level of α by referring to the F table (Table I in Appendix A) as we did for the simple ANOVA in Chapter 6.

COMPUTATIONAL EXAMPLE

Using the data presented in Table 6.3, we must first determine the ΣX and ΣX^2 for each of our treatment conditions or cells. These data are presented in Table 6.4. Next, we compute each of our sums of squares. In each case a correction term (C) will be used: $\frac{(\Sigma X_i)^2}{N}$

$$SS_{total} = \left(\Sigma X_1^2 + \Sigma X_2^2 + \Sigma X_3^2 + \Sigma X_4^2 \right) - \frac{(\Sigma X_1 + \Sigma X_2 + \Sigma X_3 + \Sigma X_4)^2}{N}$$

$$= (652 + 652 + 502 + 1018) - \frac{(80 + 80 + 70 + 100)^2}{40}$$

$$= 101.5$$

$$SS_{between} = \frac{(\Sigma X_1)^2}{n_1} + \frac{(\Sigma X_2)^2}{n_2} + \frac{(\Sigma X_3)^2}{n_3} + \frac{(\Sigma X_4)^2}{n_4} - \frac{(\Sigma X_1 + \Sigma X_2 + \Sigma X_3 + \Sigma X_4)^2}{N}$$

$$= 640 + 640 + 490 + 1000 - 2722.5$$

$$= 47.5$$

Partitioning the overall $SS_{between}$ into its three component parts, we have:

$$SS_A = \frac{(\Sigma X_1 + \Sigma X_2)^2}{n_1 + n_2} + \frac{(\Sigma X_3 + \Sigma X_4)^2}{n_3 + n_4} + \frac{(\Sigma X_1 + \Sigma X_2 + \Sigma X_3 + \Sigma X_4)^2}{N}$$

$$= 1280 + 1445 - 2722.5$$

$$= 2.5$$

$$SS_B = \frac{(\Sigma X_1 + \Sigma X_3)^2}{n_1 + n_3} + \frac{(\Sigma X_2 + \Sigma X_4)^2}{n_2 + n_4} - \frac{(\Sigma X_1 + \Sigma X_2 + \Sigma X_3 + \Sigma X_4)^2}{N}$$

$$= 1125 + 1620 - 2722.5$$

$$= 22.5$$

$$SS_{AB} = SS_{between} - (SS_A + SS_B)$$

$$= 47.5 - (2.5 + 22.5)$$

$$= 22.5$$

TABLE 6.4
Summary of Between-Groups ANOVA Data from Table 6.3

		B_1 Lecture		B_2 Lecture–Discussion		Row Totals	
		6	$\bar{X} = 8.0$	6	$\bar{X} = 8.0$	12	$\bar{X} = 8.0$
		7	$\sigma = 1.1$	7	$\sigma = 1.1$	14	$\sigma = 1.1$
		7	$\sigma_{\bar{X}} = 0.3$	7	$\sigma_{\bar{X}} = 0.3$	14	$\sigma_{\bar{X}} = 0.2$
		8	$n_{1 = 10}$	8	$n_{2 = 10}$	16	
		8		8		16	
Class Size		8		8		16	
	A_1 Large	8		8		16	
		9		9		18	
		9		9		18	
		10		10		20	
		$\Sigma X_1 = 80$		$\Sigma X_2 = 80$		$\Sigma X_1 + \Sigma X_2 = 160$	
		$\Sigma X_1^2 = 652$		$\Sigma X_2^2 = 652$		$\Sigma X_1^2 + \Sigma X_1^2 = 1304$	
		5	$\bar{X} = 7.0$	8	$\bar{X} = 10.0$	13	$\bar{X} = 8.5$
		6	$\sigma = 1.1$	8	$\sigma = 1.3$	14	$\sigma = 1.9$
		6	$\sigma_{\bar{X}} = 0.3$	9	$\sigma_{\bar{X}} = 0.4$	15	$\sigma_{\bar{X}} = 0.4$
		7	$n_{3 = 10}$	10	$n_{4 = 10}$	17	
		7		10		17	
	A_2 Small	7		10		17	
		7		10		17	
		8		11		19	
		8		12		20	
		9		12		21	
		$\Sigma X_3 = 70$		$\Sigma X_4 = 100$		$\Sigma X_3 + \Sigma X_4 = 170$	
		$\Sigma X_3^2 = 502$		$\Sigma X_4^2 = 1018$		$\Sigma X_3^2 + \Sigma X_4^2 = 1520$	
	Column totals	$\bar{X} = 7.53 \ \sigma = 1.2$		$\bar{X} = 9.0 \ \sigma = 1.2$		Grand total $= \Sigma(\Sigma X) = 330$	
		$\sigma_{\bar{X}} = 0.3$		$\sigma_{\bar{X}} = 0.4$		Grand total $= \Sigma(\Sigma X^2) = 2824$	
		$\Sigma X_1 + \Sigma X_3 = 150$		$\Sigma X_2 + \Sigma X_4 = 180$		$N = 40$	
		$\Sigma X_1^2 + \Sigma X_3^2 = 1154$		$\Sigma X_2^2 + \Sigma X_4^2 = 1670$			

As in the simple ANOVA in Chapter 5, SS_W (SS_{within}) is obtained by subtracting the overall $SS_{between}$ from SS_{total}:

$$SS_{within} = SS_{total} - SS_{between}$$

$$= 101.5 - 47.5$$

$$= 54.0$$

Entering our various sums of squares into our ANOVA summary table, we can determine df, calculate mean squares, and compute three F-ratios. These data appear in Table 6.5. Note that for both the sum of squares and df, adding values for A, B, and A × B yields the total sum of squares and total df.

TABLE 6.5
A 2 × 2 Between-Groups ANOVA Summary Table

Source	SS	df	MS	F_{calc}	$F_{1,36,.05}$	$F_{1,36,.01}$
A	2.5	1	2.5	1.67	4.17	7.56
B	22.5	1	22.5	15.00[a]	4.17	7.56
AB	22.5	1	22.5	15.00[a]	4.17	7.56
Within	54.0	36	1.5			
Total	101.5	39				

[a] $p < 0.01$.

To determine whether the three F-ratios are significant, let us test the null hypotheses at the 0.01 level. Three hypotheses are to be tested: (1) there are no differences between rows, (2) there are no differences between columns, and (3) there are no differences between cells. For all three tests in the present study, we enter the F table (Table I in Appendix A) with 1 and 36 df. The tabled value of F with 1, 36 df at the 0.01 level is 7.56 (do not interpolate; when the actual df falls between two points on the table, go with the more conservative, e.g., use 30 df instead of 40). Note that the F-ratios for the B main effect and the A × B interaction exceed the tabled value of 7.56. Thus, the null hypothesis for these two outcomes can be rejected. The A main effect F-ratio did not reach significance and therefore the corresponding null hypothesis would not be rejected. These outcomes can clearly be seen by plotting the results as shown in Figure 6.9.

We can clearly see a strong interaction between teaching method and class size. We also can see that the single most effective approach was use of the lecture–discussion method with small classes. With the lecture method, class size made little or no real difference.

Normally, when an interaction is significant, it is meaningless to interpret the significance of any main effect that is represented in that interaction. This can be seen in our example. Even though the B main effect was significant, it has little meaning because, in fact, teaching method depends on class size. Thus, the A × B interaction was important in interpreting our results. Therefore, it is wise to read ANOVA summary tables from the bottom up, i.e., beginning with the interactions.

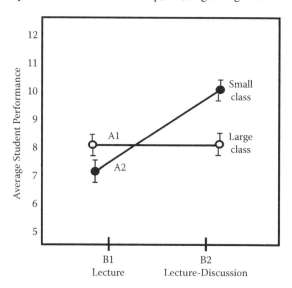

FIGURE 6.9 Graph of results from Table 6.2 as a polygon using means and standard errors of the mean.

WITHIN-SUBJECTS (REPEATED MEASURES) TWO-FACTOR ANOVA DESIGNS

RATIONALE FOR WITHIN-SUBJECTS ANOVA DESIGNS

To look at the within-subjects design we can still use the basic matrix shown in Table 6.4, but the two variables must be modified slightly. Instead of looking at four groups of subjects (two different classes—large and small—and two teaching methods), we will consider the *same* class of students under two different types of teaching methods and two different levels of task difficulty. Each class will receive all levels of each factor, resulting in a two-factor repeated measures design. The study data matrix is shown in Table 6.6.

The same procedure outlined in the between-groups design can be used for computing SS_J, SS_K, SS_{JK}, and their respective degrees of freedom. However, because we must compute three error terms, we must also:

1. Compute the sum of squares of the subject's cell values (SS_S), ignoring the row and column variables.
2. Compute the sum of squares of the subject's every row values ($SS_{J/S}$), ignoring the column variables only.

TABLE 6.6
Matrix for Within-Subjects ANOVA Data from Table 6.3

		K_1 Lecture		K_2 Lecture–Discussion		Row Totals	
		6	$\bar{X} = 8.0$	6	$\bar{X} = 8.0$	12	$\bar{X} = 8.0$
		7	$\sigma = 1.1$	7	$\sigma = 1.1$	14	$\sigma = 1.1$
		7	$\sigma_{\bar{x}} = 0.3$	7	$\sigma_{\bar{x}} = 0.3$	14	$\sigma_{\bar{x}} = 0.2$
		8		8		16	
		8		8		16	
Task		8		8		16	
Difficulty	J_1 Hard	8		8		16	
		9		9		18	
		9		9		18	
		10		10		20	
		$\Sigma X_1 = 80$		$\Sigma X_2 = 80$		$\Sigma X_1 + \Sigma X_2 = 160$	
		$\Sigma X_1^2 = 652$		$\Sigma X_2^2 = 652$		$\Sigma X_1^2 + \Sigma X_1^2 = 1304$	
		5	$\bar{X} = 7.0$	8	$\bar{X} = 10.0$	13	$\bar{X} = 8.5$
		6	$\sigma = 1.1$	8	$\sigma = 1.3$	14	$\sigma = 1.9$
		6	$\sigma_{\bar{x}} = 0.3$	9	$\sigma_{\bar{x}} = 0.4$	15	$\sigma_{\bar{x}} = 0.4$
		7		10		17	
		7		10		17	
	J_2 Easy	7		10		17	
		7		10		17	
		8		11		19	
		8		12		20	
		9		12		21	
		$\Sigma X_3 = 70$		$\Sigma X_4 = 100$		$\Sigma X_3 + \Sigma X_4 = 170$	
		$\Sigma X_3^2 = 502$		$\Sigma X_4^2 = 1018$		$\Sigma X_3^2 + \Sigma X_4^2 = 1520$	
		$\bar{X} = 7.53 \ \sigma = 1.2$		$\bar{X} = 9.0 \ \sigma = 1.6$		Grand Total $= \Sigma(\Sigma X) = 330$	
	Column	$\sigma_{\bar{x}} = 0.3$		$\sigma_{\bar{x}} = 0.4$		Grand Total $= \Sigma(\Sigma X^2) = 2824$	
	totals	$\Sigma X_1 + \Sigma X_3 = 150$		$\Sigma X_2 + \Sigma X_4 = 180$		$N = 10$	
		$\Sigma X_1^2 + \Sigma X_3^2 = 1154$		$\Sigma X_2^2 + \Sigma X_4^2 = 1670$			

---

3. Compute the sum of squares of the subject's every column values ($SS_{K/S}$), ignoring the row variables only.
4. Compute the sum of squares of the subject's every cell values ($SS_{JK/S}$), taking into account both the row and column variables.

The degrees of freedom for these four additional component parts of the total sum of squares are computed as follows:

$$df_S = n - 1$$

$$df_{K/S} = (k-1)(n-1)$$

$$df_{J/S} = (j-1)(n-1)$$

$$df_{JK/S} = (j-1)(k-1)(n-1)$$

Dividing each of the component sum of squares by its respective df yields the corresponding mean squares: MS_J, MS_K, MS_{JK}, MS_S, $MS_{J/S}$, $MS_{K/S}$, $MS_{JK/S}$. To determine whether the main effects and interactions are significant, we divide the mean squares as follows:

$$\text{For J main effect: } \frac{MS_J}{MS_{J/S}}$$

$$\text{For K main effect: } \frac{MS_K}{MS_{K/S}}$$

$$\text{For J} \times \text{K interaction: } \frac{MS_{JK}}{MS_{JK/S}}$$

Each of these calculations yield an F-ratio that can be compared against a table F value (see Table I in Appendix A) to test for significance.

COMPUTATIONAL EXAMPLE

First we compute each of our sum of squares (where $C = \frac{(\sum X_i)^2}{N}$, j = number of rows, and k = number of columns):

$$SS_{\text{total}} = \left(\sum X_1^2 + \sum X_2^2 + \sum X_3^2 + \sum X_4^2\right) - C = 2824 - 2722.5$$

$$= 101.5$$

$$SS_J = \frac{(\sum X_1 + \sum X_2)^2 + (\sum X_3 + \sum X_4)^2}{nk} - C$$

$$= \frac{160^2 + 170^2}{20} - 2722.5$$

$$= 2.5$$

$$SS_K = \frac{(\sum X_1 + \sum X_3)^2 + (\sum X_2 + \sum X_4)^2}{nj} - C = \frac{150^2 + 180^2}{20} - 2722.5$$

$$= 22.5$$

$$SS_S = \frac{(\sum S_1)^2 + (\sum S_2)^2 \dots (\sum S_{10})^2}{jk} - C$$

$$= \frac{25^2 + 28^2 + 29^2 + 33^2 + 33^2 + 33^2 + 33^2 + 37^2 + 38^2 + 41^2}{4} - 2722.5$$

$$= 52.5$$

$$SS_{JK} = \frac{(\sum X_1)^2 + (\sum X_2)^2 + (\sum X_3)^2 + (\sum X_4)^2}{n} - C - SS_J - SS_k$$

$$= \frac{80^2 + 80^2 + 70^2 + 100^2}{10} - 2722.5 - 2.5 - 22.5$$

$$= 22.5$$

$$SS_{JS} = \frac{\sum (\text{Each Subject's Row Total})^2}{k} - C - SS_J - SS_S$$

$$= \frac{(12^2 + 14^2 + 14^2 + 16^2 + 16^2 + 16^2 + 16^2 + 18^2 + 18^2 + 20^2 + 13^2 + 14^2 + 15^2 + 17^2 + 17^2 + 17^2 + 17^2 + 19^2 + 20^2 + 21^2}{2}$$

$$-C - SS_J - SS_S$$

$$= 2778 - 2722.5 - 2.5 - 52.2$$

$$= 0.5$$

$$SS_{KS} = \frac{\sum (\text{Each Subject's Column Total})^2}{j} - C - SS_K - SS_S$$

$$= \frac{(11^2 + 13^2 + 13^2 + 15^2 + 15^2 + 15^2 + 15^2 + 17^2 + 17^2 + 19^2 + 14^2 + 15^2 + 16^2 + 18^2 + 18^2 + 18^2 + 18^2 + 20^2 + 21^2 + 22^2}{2}$$

$$-C - SS_K - SS_S$$

$$= 2798 - 2722.5 - 2.5 - 52.5$$

$$= 0.5$$

$$SS_{JKS} = SS_{total} - SS_J - SS_K - SS_S - SS_{JK} - SS_{JS} - SS_K$$

$$= 101.5 - 22.5 - 2.5 - 52.5 - 22.5 - 0.5 - 0.5$$

$$= 0.5$$

Each sum of square value is entered into the ANOVA summary table along with *df*, *MS*, and calculated and table *F*-ratios. These data are shown in Table 6.7. Note that the sum of squares and *df* for J, K, S, JK, JS, K/S, and JK/S sum to the total sum of squares and df_{total}.

To determine whether the obtained *F*-ratios are significant, each *F* value is compared with the critical *F* values from the *F* distribution table (I in Appendix A). For all three *F*-ratios, we enter the *F*-table with one *df* in the numerator and 9 *df* in the denominator. Hence, the tabled value for *F* at the 0.01 level with 1, 9 *df* is 10.56. Because our calculated *F*-ratios all exceed the tabled or critical *F*-ratio, the null hypothesis for each of the two main factors and the interaction of the two factors may be rejected. As before, these outcomes can be graphically depicted as in Figure 6.9.

TABLE 6.7
A 2 × 2 Repeated Measure ANOVA Summary Table

Source	SS	df	MS	F_{Calc}	$F_{1,9,.05}$	$F_{1,9,.01}$
J	2.5	1	2.50	41.7	5.12	10.56
K	22.5	1	22.50	375.0	5.12	10.56
S	52.5	9	5.83			
JK	22.5	1	22.50	375.0	5.12	10.56
J/S	0.5	9	0.06			
K/S	0.5	9	0.06			
JK/S	0.5	9	0.06			
Total	101.5	39				

Again, we can clearly see a strong interaction between the two factors, indicating that the effects of teaching method on student performance depend on the level of task difficulty. As before, it is possible to determine the effects of teaching method and task difficulty, but the results may be misleading because the significant effect observed for each factor *depends* on the influence of the other factor.

MIXED TWO-FACTOR ANOVA DESIGNS

RATIONALE FOR MIXED ANOVA DESIGNS

To present an example of the mixed design for two-factor ANOVA, we can use the same matrix used for the between-groups example shown in Table 6.4. Close inspection of that matrix shows that if we can assume that one factor (class size) is mutually exclusive (i.e., students can be in only one class, never both); that assumption is the between-groups factor (A). The other factor, however, is a within-subjects (repeated measure) because each subject receives all levels of this (J) factor. All students from both classes will experience both teaching methods. The mixed ANOVA design will compute results different from those of the between-groups and within-subjects designs shown earlier because each design employs different error terms (Table 6.8).

The same procedure as shown in the between-groups design may be used to compute SS_A, SS_J, and SS_{AJ}, and their respective degrees of freedom. However, we must also compute two new error terms as follows:

1. Compute the overall between-groups sum of squares ($SS_{between}$).
2. Compute the between-groups residual sum of squares ($SS_{A/S}$).
3. Compute the within-subjects residual sum of squares ($SS_{AJ/S}$).
4. Compute the overall within-subjects sum of squares (SS_{within}).

The degrees of freedom for these additional component parts of the total sum of squares can be computed as follows, where lower case letters refer to the number of levels of that factor:

$$df_{A/S} = a(n-1)$$

$$df_{between} = an - 1$$

$$df_{AJ/S} = a(n-1)(j-1)$$

$$df_{within} = an(j-1)$$

TABLE 6.8
Matrix for Mixed ANOVA Design Data from Table 6.3

		J_1 Lecture		J_2 Lecture–Discussion		Row Totals	
		7	$\sigma = 1.1$	7	$\sigma = 1.1$	14	$\sigma = 1.1$
		7	$\sigma_{\bar{X}} = 0.3$	7	$\sigma_{\bar{X}} = 0.3$	14	$\sigma_{\bar{X}} = 0.2$
		8	$n_1 = 10$	8	$n_2 = 10$	16	
		8		8		16	
Class Size		8		8		16	
	A_1 Large	8		8		16	
		9		9		18	
		9		9		18	
		<u>10</u>		<u>10</u>		<u>20</u>	
		$\Sigma X_1 = 180$		$\Sigma X_2 = 80$		$\Sigma X_1 + \Sigma X_2 = 160$	
		$\Sigma X_1^2 = 652$		$\Sigma X_2^2 = 652$		$\Sigma X_1^2 + \Sigma X_1^2 = 1304$	
		5	$\bar{X} = 7.0$	8	$\bar{X} = 10.0$	13	$\bar{X} = 8.5$
		6	$\sigma = 1.1$	8	$\sigma = 1.3$	14	$\sigma = 1.9$
		6	$\sigma_{\bar{X}} = 0.3$	9	$\sigma_{\bar{X}} = 0.4$	15	$\sigma_{\bar{X}} = 0.4$
		7	$n_3 = 10$	10	$n_4 = 10$	17	
		7		10		17	
	A_2 Small	7		10		17	
		7		10		17	
		8		11		19	
		8		12		20	
		<u>9</u>		<u>12</u>		<u>21</u>	
		$\Sigma X_3 = 70$		$\Sigma X_4 = 100$		$\Sigma X_3 + \Sigma X_4 = 170$	
		$\Sigma X_3^2 = 502$		$\Sigma X_4^2 = 1018$		$\Sigma X_3^2 + \Sigma X_4^2 = 1520$	
	Column totals	$\bar{X} = 7.53 \; \sigma = 1.2$		$\bar{X} = 9.0 \; \sigma = 1.6$		Grand Total $= \Sigma(\Sigma X) = 330$	
		$\sigma_{\bar{X}} = 0.3$		$\sigma_{\bar{X}} = 0.4$		Grand Total $= \Sigma(\Sigma X^2) = 2824$	
		$\Sigma X_1 + \Sigma X_3 = 150$		$\Sigma X_2 + \Sigma X_4 = 180$		$N = 40$	
		$\Sigma X_1^2 + \Sigma X_3^2 = 1154$		$\Sigma X_2^2 + \Sigma X_4^2 = 1670$			

Dividing each of these component sum of squares by their respective df will yield the corresponding mean squares: MS_A, $MS_{A/S}$, MS_J, MS_{AJ}, $MS_{AJ/S}$ (note that $SS_A + SS_{A/S} = SS_{between}$ and $SS_J + SS_{AJ} + SS_{AJ/S} = SS_{within}$). Finally, to determine significance, the F-ratios may be calculated as follows:

Between-groups:

$$\text{For A main effect: } \frac{MS_A}{MS_{A/S}}$$

Within-subjects:

$$\text{For J main effect: } \frac{MS_J}{MS_{AJ/S}}$$

$$\text{For A} \times \text{J interaction: } \frac{MS_{AJ}}{MS_{AJ/S}}$$

Each of these F-ratios can then be compared against the table F-ratios (see Table I in Appendix A) to test for significance.

COMPUTATIONAL EXAMPLE

Using the data from Table 6.4, we first compute the sums of squares:

$$SS_{total} = (\sum X_1^2 + \sum X_2^2 + \sum X_3^2 + \sum X_4^2) - C = 2824 - 2722.$$

$$= 101.5$$

$$C = \frac{(\sum X_i)^2}{N} = \frac{(330)^2}{40}$$

$$= 2722.5$$

$$SS_{between} = \frac{\sum(\sum Rows)^2}{J} - C$$

$$= \frac{\left(\begin{array}{c} 12^2 + 14^2 + 14^2 + 16^2 + 16^2 + 16^2 + 16^2 + 18^2 + 18^2 + 20^2 + \\ 13^2 + 14^2 + 15^2 + 17^2 + 17^2 + 17^2 + 17^2 + 19^2 + 20^2 + 21^2 \end{array} \right)}{2} - 2722.5$$

$$= 2778 - 2722.5$$

$$= 55.5$$

$$SS_A = \frac{(\sum X_1 + \sum X_2)^2 + (\sum X_3 + \sum X_4)^2}{nj} - C = \frac{160^2 + 170^2}{20} - 2722.5$$

$$= 2.5$$

$$SS_{A/S} = SS_{between} - SS_A = 55.5 - 2.5$$

$$= 53.0$$

$$SS_{within} = SS_{total} - SS_{between} = 101.5 - 55.5$$

$$= 46.0$$

$$SS_J = \frac{(\sum X_1 + \sum X_3)^2 + (\sum X_2 + \sum X_4)^2}{na} - C = \frac{150^2 + 180^2}{20} - 2722.5$$

$$= 22.5$$

$$SS_{AJ} = \frac{(\sum X_1)^2 + (\sum X_2)^2 + (\sum X_3)^2 + (\sum X_4)^2}{n} - C - SS_A - SS_J$$

$$= \frac{(80^2 + 80^2 + 70^2 + 100^2)}{10} - 2722.5 - 2.5 - 22.5$$

$$= 22.5$$

$$SS_{AJ/S} = SS_{within} - SS_J - SS_{AJ} = 46.0 - 22.5 - 22.5$$

$$= 1.0$$

TABLE 6.9
A 2 × 2 Mixed Design ANOVA Summary Table

Source	SS	df	MS	F	$F_{1,18,.05}$	$F_{1,18,.01}$
Between groups	55.5	19				
A	2.5	1	2.50	0.85	4.41	8.20
A/S	53.0	18	2.94			
Within subjects	46.0	20				
J	22.5	1	22.50	40.18	4.41	8.20
AJ	22.5	1	22.50	40.18	4.41	8.20
AJ/S	1.0	18	0.56			
Total	101.5	39				

Each sum of squares value is then entered into our ANOVA summary table along with *df, MS,* and calculated and tabled *F*-ratios. These data are shown in Table 6.9. Note that the sums of squares and *df* for A and A/S add up to the sum of squares for between groups, and the sums of squares and *df* for J, AJ, and AJ/S add up to the sum of squares for within subjects. Also, the sums of squares and *df* for between and within subjects add up to the total sum of squares and *df.*

To determine whether our *F*-ratios are significant at the 0.01 level, we enter the table with one *df* in the numerator and 18 *df* in the denominator. We can see that at the 0.01 level, the critical value above which *F*-ratios are significant is 8.20. Starting at the bottom of the summary table, we can see that the calculated *F*-ratio for the A × J interaction (AJ) and the J main effect exceed the tabled *F*-ratio, but the same does not hold true for the A main effect. Hence, we may reject the null hypothesis for the interaction of A and J, and also for J itself. However, we cannot reject the null hypothesis for the A factor. Thus, we can still conclude that the effect of teaching method on student performance is significant depending on class size. The Figure 6.9 graph may be used again to depict the interaction.

It is important to note that although we used the same data for each of the three ANOVA designs, we found different *F*-ratios for each design and, in some cases, different decisions as to rejecting the null hypothesis. This is, of course, due to the use of different error terms in each design. Hence, the reader is advised to be sure to choose the appropriate ANOVA design prior to making the calculations and drawing conclusions.

SUMMARY

1. In practice, the effect of an IV on a DM often depends on the degree of presence or absence of one or more additional IVs. When this is the case, there is an interaction between the IVs.
2. A multiple IV study has an advantage over separate single IV studies in that we can determine interactions between the IVs. Studies in which two or more IVs are manipulated simultaneously are referred to as factorial designs.
3. In a two-factor study, the overall $MS_{between}$ may be partitioned into three parts: MS_A, MS_B, and MS_{AB}. These, together with MS_{within} comprise the MS_{total}. As the number of factors in a study increases arithmetically, the number of possible interactions increases algebraically.
4. Studies with many factors also result in many treatment conditions and thus require large numbers of subjects. As a result, studies with more than four factors are less common.

5. Studies in which each level of every factor is crossed with each level of every other factor are called fully crossed factorial designs. Two factors are "crossed" when each possible combination of treatment levels represents a treatment condition in the experiment.

6. One way of reducing the number of treatment conditions is to "nest" the levels of one factor under another factor. Nesting also has a disadvantage when compared with fully crossed designs in that any interaction involving the nested factors cannot be tested.

7. In a factorial study, each partitioned portion of the overall $MS_{between}$ forms the numerator of an F-ratio. As was the case with the simple ANOVA, an F-ratio is tested by comparing it with the tabled value of F for the chosen significance level and numerator and denominator df. Normally, if the interaction F-ratio is significant, the associated main effects must be cautiously interpreted.

8. ANOVA designs may be classified as (a) between groups, (b) within subjects (repeated measures), or (c) mixed designs. In a between-groups design, the MS_A, MS_B, and MS_{AB} are divided by the within-groups error term MS_{within} for all F-ratios. In the within-subjects (repeated measures) design, the MS values are divided by the within-groups error term for each variable. Thus, MS_J is divided by $MS_{J/S}$, MS_K is divided by $MS_{K/S}$, and MS_{JK} is divided by $MS_{JK/S}$ to yield the corresponding F-ratios. In the mixed design, the MS terms are divided by the residual MS values. Thus, MS_A is divided by $MS_{A/S}$, MS_J is divided by $MS_{AJ/S}$, and MS_{AJ} is divided by $MS_{AJ/S}$ to yield the corresponding F-ratios.

KEYWORD DEFINITIONS

Between-Groups Design: Statistical procedure to determine whether differences between independent groups of subjects are significant.

df: Degrees of freedom; number of independent observations on which a statistic is based minus the number of restrictions placed on the freedom of those observations.

Factorial Design: Study in which two or more independent variables are manipulated simultaneously.

Fully Crossed Design: Characteristic of designs in which each level of every IV is crossed with every level of all other IVs.

Interaction Effect: Circumstance in which the combination of two or more IVs produces an effect that is not simply an additive representation of each factor's effects singly.

Level: Different treatment values for a given IV or factor.

Main Effect: Treating results in multiple IV experiments as though there was only one IV, variable B, or other factor.

Mixed Design: Statistical procedure with at least one between-groups factor and one within-subjects factor.

MS: Mean square; sum of squares of the differences between the observations and the mean divided by its degrees of freedom.

Multiple IV Experiment: Study employing more than one factor to examine the interaction of experimental variable on the dependent measure(s).

Single IV Experiment: Study employing only one experimental factor with multiple levels.

SS: Sum of squares; sum of squared deviations for some set of scores about their mean.

Within-Subjects Design: Statistical procedure to determine whether differences between repeated tests on the same subjects are significant.

EXERCISES

1. In a 2×2 between-groups factorial study, the following data were obtained. The specific levels of the two treatment dimensions were deliberately selected by the researcher.

	B$_1$	**B$_2$**
A$_1$	1 2 3 3 4	2 4 4 4 5
	5 5 6 7 8	5 5 6 7 8
A$_2$	3 4 6 7 7	5 7 8 9 9
	8 8 8 9 10	10 10 11 12 12

a. Conduct an analysis of variance for these data at the $\alpha = 0.01$ and 0.05 levels of significance.
b. Graph the treatment means.
c. Based on the ANOVA and on inspection of the graph, describe the outcome of this study.

2. A researcher wishes to investigate the effects of the following variables on subject accuracy in a tracking task.

Times of day: 8–10 a.m.; 2–4 p.m.; 8–10 p.m.; 2–4 a.m.
Training: basic education level; advanced education level
Noise: loud; quiet
Directional changes per unit of time: 10; 7; 4; 1

a. Diagram this study as a fully crossed factorial design.
b. List all main effects and possible interactions.
c. Diagram this study with the time-of-day factor nested under the noise factor.
d. List all possible interactions for the nested design.

3. Analyze the following 2×4 data matrix as a between-groups design, completely within-subjects design, and a 2×4 mixed design with repeated measures on the second factor. (*Source:* Adapted from Freedman, 1958, pp. 585–586; McCall, 1970, pp. 272–280.)

		Second Factor			
		1	**2**	**3**	**4**
		1	7	6	9
		4	10	9	7
	1	3	10	7	10
		1	9	8	10
		2	6	5	8
First Factor		2	8	10	9
		5	9	1	2
		1	9	0	6
	2	4	8	3	3
		1	10	1	4
		2	5	2	5
		3	8	4	3

a. Show the data matrix ΣX, ΣX^2, $\bar{X}$, σ, $\sigma_{\bar{X}}$.
b. Show the 2×4 between-groups ANOVA summary table.
c. Show the 2×4 completely within-subjects ANOVA summary table.

    d. Show the 2 × 4 mixed ANOVA summary table in which the first factor is the between-groups factor (i.e., the second factor is a repeated measure for each subject within the first factor groups).

    e. Illustrate the main effect and interaction means and standard errors.

    f. Perform Tukey's WSD post hoc test on the data analyzed in the between-groups design.

    g. Illustrate the WSD results using the Duncan underline procedure.

    h. Interpret the results of the between-groups ANOVA design.

    i. Describe the differences in the finding of each design.

4. A researcher wants to determine whether differences exist between the amount of physical fitness improvement that can be achieved between males and females after members of each sex go through one of two types of training regimens. One group of each sex participated in a running program and the other group in a swimming program. Fitness was determined by measuring the number of minutes it took to return the individuals to their resting heart rates.

	B_1 Swimming	B_2 Running
A_1 Males	8	8
	7	9
	8	7
	5	6
	6	7
A_2 Females	9	9
	6	7
	8	8
	8	8
	7	7

    a. Is this ANOVA a between, within, or mixed design? Why?

    b. Conduct an ANOVA for these data at the 0.05 level of significance.

    c. Construct an ANOVA summary table.

    d. Interpret the findings.

5. Given the following 2 × 3 data matrix:

    a. Conduct a fully within-subjects ANOVA.

    b. Construct an ANOVA summary table.

    c. Interpret the results.

	K_1	K_2	K_3
J_1	3,4,7,5,3, 6,8,8,7,4	9,8,8,7,8, 7,5,10,9,6	1,3,4,3,6, 2,2,4,3,4
J_2	6,4,7,7,8, 7,3,5,5,9	10,8,7,6,8, 8,9,7,8,9	3,5,2,2,1, 1,4,3,3,2

EXERCISE ANSWERS

1.a. Between-groups ANOVA.

	A_1-B_1	A_1-B_2	A_2-B_1	A_2-B_2
n	10	10	10	10
ΣX	44	70	50	93
ΣX^2	238	532	276	909
$\bar{X} d$	4.4	7.0	5.0	9.3
σ	2.1	2.0	1.6	2.1
$\sigma_{\bar{X}}$	0.7	0.6	0.5	0.7

$$SS_{total} = (238 + 532 + 276 + 909) - \frac{(44 + 70 + 50 + 93)^2}{40}$$

$$= 1955 - 1651.23$$

$$= 303.77$$

$$SS_{between} = \frac{(44)^2}{10} + \frac{(70)^2}{10} + \frac{(50)^2}{10} + \frac{(93)^2}{10} - 1651.2$$

$$= 1798.5 - 1651.2$$

$$= 147.3$$

$$SS_A = \frac{(44 + 70)^2}{20} + \frac{(50 + 93)^2}{20} - 1651.2$$

$$= 649.8 + 1022.45 - 1651.2$$

$$= 21.05$$

$$SS_B = \frac{(44 + 50)^2}{20} + \frac{(70 + 93)^2}{20} - 1651.2$$

$$= 441.8 + 1328.45 - 1651.2$$

$$= 119.05$$

$$SS_{AB} = 147.3 - (21.05 + 119.05)$$

$$= 7.2$$

$$SS_{within} = 303.8 - 147.3$$

$$= 156.5$$

Source	SS	df	MS	F	$F_{.05}$	$F_{.01}$
A	119.05	1	119.05	27.37	4.17	7.56
B	21.05	1	21.05	4.84	4.17	7.56
AB	7.2	1	7.2	1.66	4.17	7.56
Within	156.5	36	4.35			
Total	303.8	39				

At the 0.05 level, only the A main effect is significant.
1.b. Plot of means and standard errors:.

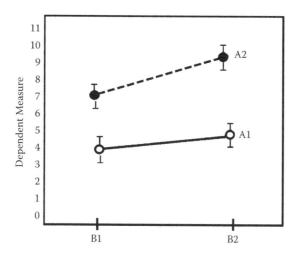

1.c. Interpretation: The interaction of A and B was not significant ($p > 0.05$) nor was the B main effect ($0.05 > p > 0.01$). However, A2 was significantly greater than A1 ($F_{1,36} = 27.37$; $p < 0.01$). The significant A effect may also be illustrated as follows:

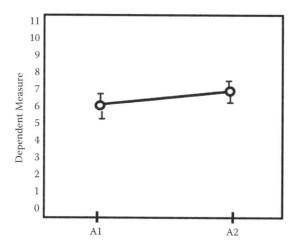

2. Factors and levels:

Time of day: T_1, T_2, T_3, T_4
Training: E_1, E_2,
Noise: N_1, N_2
Directional changes: D_1, D_2, D_3, D_4

2.a. Fully crossed factorial design:

	E_1				E_2			
	T_1	T_2	T_3	T_4	T_1	T_2	T_3	T_4
N_1 D_1								
D_2								
D_3								
D_4								
N_2 D_1								
D_2								
D_3								
D_4								

2.b. Main effects and interactions. Main effects = T, D, E, and N. Interactions are shown below:

$$T \times D \times E \times N$$
$$T \times D \times E$$
$$T \times D \times N$$
$$T \times E \times N$$
$$D \times E \times N$$
$$T \times D$$
$$T \times E$$
$$T \times N$$
$$D \times E$$
$$D \times N$$
$$E \times N$$

2.c. Partial design.

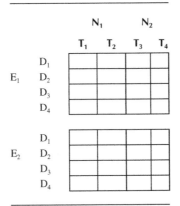

2.d. Nested interactions.

$$T \times D \times E$$
$$N \times D \times E$$
$$T \times D$$
$$T \times E$$
$$D \times E$$
$$D \times N$$
$$E \times N$$

3.a. Data matrix statistics.

	1	2	3	4	Row Effects
1	$n = 6$ $\Sigma X = 13$ $\Sigma X^2 = 35$ $\bar{X} = 2.17$ $\sigma_n = 1.07$ $\sigma_{\bar{x}} = 0.44$	$n = 6$ $\Sigma X = 50$ $\Sigma X^2 = 430$ $\bar{X} = 8.33$ $\sigma_n = 1.49$ $\sigma_{\bar{x}} = 0.61$	$n = 6$ $\Sigma X = 45$ $\Sigma X^2 = 355$ $\bar{X} = 7.50$ $\sigma_n = 1.71$ $\sigma_{\bar{x}} = 0.70$	$n = 6$ $\Sigma X = 53$ $\Sigma X^2 = 475$ $\bar{X} = 8.33$ $\sigma_n = 1.07$ $\sigma_{\bar{x}} = 0.44$	$n = 24$ $\Sigma X = 161$ $\Sigma X^2 = 1295$ $\bar{X} = 6.71$ $\sigma_n = 2.99$ $\sigma_{\bar{x}} = 0.61$
2	$n = 6$ $\Sigma X = 16$ $\Sigma X^2 = 56$ $\bar{X} = 2.67$ $\sigma_n = 1.49$ $\sigma_{\bar{x}} = 0.61$	$n = 6$ $\Sigma X = 49$ $\Sigma X^2 = 415$ $\bar{X} = 8.17$ $\sigma_n = 1.57$ $\sigma_{\bar{x}} = 0.64$	$n = 6$ $\Sigma X = 11$ $\Sigma X^2 = 31$ $\bar{X} = 1.83$ $\sigma_n = 1.34$ $\sigma_{\bar{x}} = 0.55$	$n = 6$ $\Sigma X = 23$ $\Sigma X^2 = 99$ $\bar{X} = 3.83$ $\sigma_n = 1.34$ $\sigma_{\bar{x}} = 0.55$	$n = 24$ $\Sigma X = 99$ $\Sigma X^2 = 601$ $\bar{X} = 4.12$ $\sigma_n = 2.83$ $\sigma_{\bar{x}} = 0.58$
Column Effects	$n = 12$ $\Sigma X = 29$ $\Sigma X^2 = 91$ $\bar{X} = 2.42$ $\sigma_n = 1.32$ $\sigma_{\bar{x}} = 0.38$	$n = 12$ $\Sigma X = 99$ $\Sigma X^2 = 845$ $\bar{X} = 8.25$ $\sigma_n = 1.53$ $\sigma_{\bar{x}} = 0.44$	$n = 12$ $\Sigma X = 56$ $\Sigma X^2 = 386$ $\bar{X} = 4.6$ $\sigma_n = 3.22$ $\sigma_{\bar{x}} = 0.93$	$n = 12$ $\Sigma X = 76$ $\Sigma X^2 = 574$ $\bar{X} = 6.33$ $\sigma_n = 2.78$ $\sigma_{\bar{x}} = 0.80$	$n = 48$ $\Sigma X = 260$ $\Sigma X^2 = 1896$ $\bar{X} = 5.42$ $\sigma_n = 3.19$ $\sigma_{\bar{x}} = 0.46$

3.b. 2 × 4 between-groups ANOVA summary table.

Source	SS	df	MS	F	F	F	Assumptions Met P
A	80.09	1	80.08	33.94	4.08	7.31	< .0001
B	221.17	3	73.72	31.24	2.84	4.31	< .0001
AB	92.08	3	30.69	13.00	2.84	4.31	< .0001
Error	94.33	40	2.36				
Total	487.67	47					

3.c. Calculations:

$$C = \frac{(\Sigma X_i)^2}{N}$$

$$= \frac{(\Sigma X_1 + \Sigma X_2 + \Sigma X_3 + \Sigma X_4 + \Sigma X_5 + \Sigma X_6 + \Sigma X_7 + \Sigma X_8)^2}{N}$$

$$= \frac{(13 + 50 + 45 + 53 + 16 + 49 + 11 + 23)^2}{48}$$

$$= \frac{260^2}{48} = \frac{67600}{48} 1408.33$$

$$SS_{total} = \left(\sum X_1^2 + \sum X_2^2 + \sum X_3^2 + \sum X_4^2 + \sum X_5^2 + \sum X_6^2 + \sum X_7^2 + \sum X_8^2 \right) - C$$

$$= 1896 - 1408.33 = 487.67$$

$$SS_J = \frac{(\sum X_1 + \sum X_2 + \sum X_3 + \sum X_4)^2 + (\sum X_5 + \sum X_6 + \sum X_7 + \sum X_8)^2}{nk} - C$$

$$= \frac{161^2 + 99^2}{(6)(4)} - 1408.33 = 1488.42 - 1408.33 = 80.09$$

$$SS_K = \frac{(\sum X_1 + \sum X_5)^2 + (\sum X_2 + \sum X_6)^2 + (\sum X_3 + \sum X_7)^2 + (\sum X_4 + \sum X_8)^2}{nj} - C$$

$$= \frac{29^2 + 99^2 + 56^2 + 76^2}{(6)(2)} - 1408.33 = 1629.5 - 1408.33 = 221.17$$

$$SS_S = \frac{(\sum S_1)^2 + (\sum S_2)^2 + (\sum S_3)^2 + (\sum S_4)^2 + (\sum S_5)^2 + (\sum S_6)^2}{jk} - C$$

$$= \frac{40^2 + 46^2 + 48^2 + 44^2 + 35^2 + 47^2}{(2)(4)} - 1408.33$$

$$= 1423.75 - 1408.33 = 15.42$$

$$SS_{JK} = \frac{(\sum X_1)^2 + (\sum X_2)^2 + (\sum X_3)^2 + (\sum X_4)^2 + (\sum X_5)^2 + (\sum X_6)^2 + (\sum X_7)^2 + (\sum X_8)^2}{n} - C - SS_J - SS_K$$

$$= \frac{13^2 + 50^2 + 45^2 + 53^2 + 16^2 + 49^2 + 11^2 + 23^2}{6} - 1408.33 - 80.09 - 221.17$$

$$= 1801.67 - 1408.33 - 80.09 - 221.17$$

$$= 92.08$$

$$SS_{J/S} = \frac{\sum (Subjects' Row Total)^2}{k} - C - SS_J - SS_S$$

$$= \frac{23^2 + 30^2 + 30^2 + 28^2 + 21^2 + 29^2 + 17^2 + 16^2 + 18^2 + 16^2 + 14^2 + 18^2}{4} - C - SS_J - SS_S$$

$$= 1510.00 - 1408.33 - 80.09 - 15.42$$

$$= 34.08$$

$$SS_{K/S} = \frac{\sum (Subjects' Column Total)^2}{J} - C - SS_K - SS_S$$

$$= \frac{\left(\begin{array}{c} 6^2 + 5^2 + 7^2 + 2^2 + 4^2 + 5^2 + 16^2 + 19^2 + \\ 18^2 + 19^2 + 11^2 + 16^2 + 7^2 + 9^2 + 10^2 + 9^2 + \\ 7^2 + 14^2 + 11^2 + 13^2 + 13^2 + 14^2 + 13^2 + 12^2 \end{array} \right)}{2}$$

$$-C - SS_K - SS_S$$

$$= 1976.00 - 1408.33 - 221.17 - 15.42$$

$$= 175.16$$

$$SS_{JK/S} = SS_{total} - SS_J - SS_K - SS_S - SS_{JK} - SS_{J/S} - SS_{K/S}$$

$$= 487.67 - 80.09 - 221.17 - 15.42 - 92.08 - 6.16 - 34.08$$

$$= 38.67$$

2×4 within-subjects ANOVA summary table.

Source	SS	df	MS	F	$F_{.05}$	$F_{.01}$	Assumptions Method P	Conservative Adjustment P
S	15.42	5	3.08					
J	80.09	1	80.09	65.11	6.61	16.26	< .0005	< .0005
J/S	6.16	5	1.23					
K	221.17	3	73.72	32.48	3.29	5.42	< .0005	.002
K/S	34.08	15	2.27					
JK	92.08	3	30.69	11.90	3.29	5.42	< .0005	.018
JK/S	38.67	15	2.58					
Total	487.67	47						

3.d. Calculations:

$$C = \frac{(\sum X_i)^2}{N}$$

$$= \frac{(\sum X_1 + \sum X_2 + \sum X_3 + \sum X_4 + \sum X_5 + \sum X_6 + \sum X_7 + \sum X_8)^2}{N}$$

$$= 1408.33$$

$$SS_{total} = \left(\sum X_1^2 + \sum X_2^2 + \sum X_3^2 + \sum X_4^2 + \sum X_5^2 + \sum X_6^2 + \sum X_7^2 + \sum X_8^2 \right) - C$$

$$= 487.67$$

$$SS_{between} = \frac{\sum (\sum Rows)^2}{J} - C$$

$$= \frac{23^2 + 30^2 + 30^2 + 28^2 + 21^2 + 29^2 + 17^2 + 16^2 + 18^2 + 16^2 + 14^2 + 18^2}{4} - C$$

$$= 101.67$$

$$SS_A = \frac{(\sum X_1 + \sum X_2 + \sum X_3 + \sum X_4)^2 + (\sum X_5 + \sum X_6 + \sum X_7 + \sum X_8)^2}{nj} - C$$

$$= 80.09$$

$$SS_{A/S} = SS_{between} - SS_A$$

$$= 21.58$$

$$SS_{within} = SS_{total} - SS_{between}$$
$$= 386.00$$

$$SS_J = \frac{(\sum X_1 + \sum X_5)^2 + (\sum X_2 + \sum X_6)^2 + (\sum X_3 + \sum X_7)^2 + (\sum X_4 + \sum X_8)^2}{na} - C$$
$$= 221.17$$

$$SS_{AJ} = \frac{(\sum X_1)^2 + (\sum X_2)^2 + (\sum X_3)^2 + (\sum X_4)^2 + (\sum X_5)^2 + (\sum X_6)^2 + (\sum X_7)^2 + (\sum X_8)^2}{n} - C - SS_A - SS_J$$
$$= 92.08$$

$$SS_{AJ/S} = SS_{within} - SS_J - SS_{AJ}$$
$$= 386.00 - 221.17 - 92.08$$
$$= 72.75$$

2×4 mixed ANOVA summary table:

Source	SS	df	MS	F	$F_{.05}$	$F_{.01}$	Assumptions Method P	Conservative Method P	Adjustment P
Between	101.67								
A	80.08	1	80.08	37.10	4.96	10.04	< .0005		
A/S	21.58	10	2.16						
Within	386.00								
J	221.17	3	73.72	30.40	2.92	4.51	< .0005	< .0005	< .0005
AJ	92.08	3	30.69	12.66	2.92	4.51	< .0005	< .0005	< .0005
AJ/S	72.75	30	2.42						
Total	487.67	47							

3.e. See following illustrations:

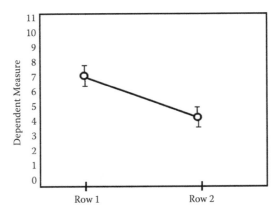

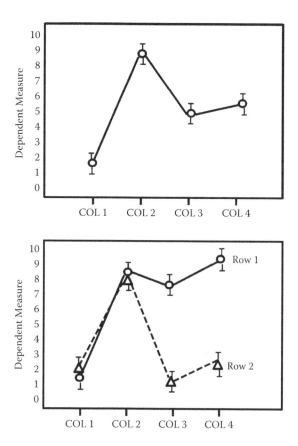

3.f. Tukey's WSD calculations:

$$WSD_{interaction} = q\left(\alpha, k, df_{MSE}\right)\sqrt{\frac{MS_E}{n}}$$

$$= q(.05, 8, 40)\sqrt{\frac{2.36}{6}}$$

$$= (4.521)(0.627)$$

$$= 2.85$$

3.g. Duncan underline (underscore indicates no significant difference):

<u>1.83 2.17 2.67 3.83</u>   <u>7.50 8.17 8.33 8.83</u>

3.h. Interpretation: The effects of the column factor depend on the level of the row factor such that there are no differences between Row 1 and Row 2 at the levels of Column 1 and Column 2, but at Column 3 and Column 4, these differences are significant (p ≤ .002). Within Row 1, the only significant difference was in the increase from Column 1 to Column 2. For Row 2, Column 2 was significantly higher than Columns 1, 3, and 4.

3.i. Design differences: Although all the *F*-ratios were different in this case, the decisions were identical across designs. Regardless of analysis method, the main and interaction effects were significant at least at the 0.0025 level. It is important to note the required *F* values from the *F* table required for significance. Despite the fact that the data were identical across designs, as degrees of freedom change, dramatic changes occur in the critical values used in decision making. (Note: This problem is based on the example used in Vercruyssen and Edwards, 1988, originally based on a between-groups design described in Freedman, 1958, pp. 585–586, and McCall, 1970, pp. 272–280.)

4.a. This is a between-groups ANOVA because each group is fully independent from the other.

4.b. Between-groups ANOVA:

	A1–B1	A1–B2	A2–B1	A2–B2
N	5	5	5	5
ΣX	34	37	38	39
ΣX^2	238	279	294	307

$$SS_{total} = (238 + 279 + 294 + 307) - \frac{(34 + 37 + 38 + 39)^2}{20}$$

$$= 1118 - 1095.2$$

$$= 22.8$$

$$SS_{between} = \frac{(34)^2}{5} + \frac{(37)^2}{5} + \frac{(38)^2}{5} + \frac{(39)^2}{5} - 1095.2$$

$$= 1098 - 1095.2$$

$$= 2.8$$

$$SS_A = \frac{(34 + 38)^2}{10} + \frac{(38 + 39)^2}{10} - 1095.2$$

$$= 1096 - 1095.2$$

$$= 0.8$$

$$SS_B = \frac{(34 + 37)^2}{10} + \frac{(38 + 39)^2}{10} - 1095.2$$

$$= 1097 - 1095.2$$

$$= 1.8$$

$$SS_{AB} = 2.8 - (0.8 + 1.8)$$

$$= 0.2$$

$$SS_{within} = 22.8 - 2.8$$

$$= 20$$

4.c. ANOVA summary table:

Source	SS	df	MS	F_{calc}	$F_{0.05}$
A	0.8	1	0.8	0.64	4.49
B	1.8	1	1.8	1.44	4.49
AB	0.2	1	0.2	0.16	4.49
Within	20.0	16	1.25		
Total	22.8	19			

4.d. In each case $F_{calc} < 4.49$, and therefore not significant; do not reject H_0. Conclude that neither a swimming nor running program is better for either sex.

5.a. Within-subjects ANOVA:

	J_1–K_1	J_2–K_1	J_3–K_1	J_1–K_2	J_2–K_2	J_3–K_2
n	10	10	10	10	10	10
X	55	77	32	61	80	26
X^2	337	613	120	403	652	82

$$SS_{total} = (337 + 613 + 120 + 403 + 652 + 82) - \frac{(55 + 77 + 32 + 61 + 80 + 26)^2}{60}$$

$$= 2207 - 1862.02$$

$$= 380.98$$

$$SS_J = \frac{(164)^2 + (167)^2}{30} - 1826.02$$

$$= 1826.17 - 1826.02$$

$$= 0.15$$

$$SS_K = \frac{(116)^2 + (157)^2 + (58)^2}{20} - 1826.02$$

$$= 2073.45 - 1826.02$$

$$= 247.43$$

$$SS_S = \frac{(32)^2 + (32)^2 + (35)^2 + (30)^2 + (34)^2 + (31)^2 + (31)^2 + (37)^2 + (35)^2 + (34)^2}{6} - 1826.02$$

$$= 1833.5 - 1826.02$$

$$= 7.48$$

$$SS_{JK} = \frac{(55)^2 + (77)^2 + (32)^2 + (61)^2 + (80)^2 + (26)^2}{10} - 1826.02 - 0.15 - 247.43$$

$$= 2077.5 - 1826.02 - 0.15 - 247.43$$

$$= 3.9$$

$$SS_{J/S} = \left[\frac{\begin{bmatrix} (13)^2 + (15)^2 + (19)^2 + (15)^2 + (17)^2 + (15)^2 + (15)^2 + \\ (22)^2 + (19)^2 + (14)^2 + (19)^2 + (17)^2 + (16)^2 + (15)^2 + \\ (17)^2 + (16)^2 + (16)^2 + (15)^2 + (16)^2 + (20)^2 \end{bmatrix}}{2} \right]$$

$$-1826.02 - 0.15 - 7.48$$

$$= \frac{5573}{3} - 1826.02 - 0.15 - 7.48$$

$$= 24.02$$

$$SS_{K/S} = \left[\frac{\begin{bmatrix} (9)^2 + (8)^2 + (14)^2 + (12)^2 + (11)^2 + (13)^2 + (11)^2 + \\ (13)^2 + (12)^2 + (13)^2 + (19)^2 + (16)^2 + (15)^2 + (13)^2 + \\ (16)^2 + (15)^2 + (14)^2 + (17)^2 + (17)^2 + (15)^2 + (4)^2 + \\ (8)^2 + (6)^2 + (5)^2 + (7)^2 + (3)^2 + (6)^2 + (7)^2 + \\ (6)^2 + (6)^2 \end{bmatrix}}{2} \right]$$

$$-1826.02 - 247.43 - 7.48$$

$$= \frac{4225}{2} - 1826.02 - 247.43 - 7.48$$

$$= 31.57$$

$$SS_{JK/S} = 380.98 - 0.15 - 247.43 - 7.48 - 3.9 - 24.02 - 31.57$$

$$= 66.43$$

5.b. ANOVA summary table:

Source	SS	df	MS	F_{calc}	$F_{0.05}$	$F_{0.01}$
J	0.15	1	0.15	0.06	5.12	10.56
K	247.43	2	123.72	70.55	3.55	6.01
S	7.48	9	0.83			
JK	3.9	2	1.95	0.53	3.55	6.01
J/S	24.02	9	2.67			
K/S	31.57	18	1.75			
JK/S	66.43	18	3.69			
Total	380.98	59				

5.c. Because $70.70 > 3.55$ (or 6.01 at a 0.01 level of significance), reject H_0 for variable K. Because 0.06 (J value) and 0.53 (JK value) are less than all significant levels of F, do not reject H_0 for factors J and JK.

REFERENCES

American Psychological Association. (1982). *Ethical principles in conduct of research with human participants.* Washington, D.C.: Author.

Campbell, D. T., & Stanley J. C. (1966). *Experimental and quasi-experimental designs for research.* Chicago: Rand McNally.

Cozby, P. C. (1984). *Using computers on the behavioral sciences.* Palo Alto, CA: Mayfield.

Edwards, A. L. (1985). *Multiple regression and the analysis of variance and covariance.* New York: Freeman.

Herrnstein, R. J., & Boring, E. G. (1966). *A source book in the history of psychology.* Cambridge, MA: Harvard University Press.

Keppel, G. (1973). *Design and analysis: A researcher's handbook.* Englewood Cliffs, NJ: Prentice-Hall.

Kerlinger, F. N. (1986). *Foundations of behavioral research* (3rd ed.). New York: Holt, Rinehart & Winston.

Malmo, R. B. (1959). Activation: A neurophysiological dimension. *Psychological Reviews, 66,* 367–368.

Neale, J. M., & Liebert, R. M. (1973). *Science and behavior: An introduction to methods of research.* Englewood Cliffs, NJ: Prentice-Hall.

Neale, J. M., & Liebert, R. M. (1986). *Science and behavior: An introduction to methods of research* (3rd ed.). Englewood Cliffs, NJ: Prentice-Hall.

Rosenthal, R. (1963). Experimenter attitudes as determinants of subjects' responses. *Journal of Protective Techniques and Personality Assessment, 67,* 324–331.

Rosenthal, R. (1966). *Experimental bias in behavioral research.* New York: Appleton-Century-Crofts.

Rosenthal, R. (1967). Covert communication in the psychological experiment. *Psychological Bulletin, 67,* 356–367.

Rosenthal, R. (1978a). Combining the results of independent studies. *Psychological Bulletin, 85,* 185–193.

Rosenthal, R. (1978b). How often are our numbers wrong? *American Psychologist,* November, 1005–1008.

Rosenthal, R. (1979). The "file drawer problem" and tolerance for null results, *Psychological Bulletin, 86,* 638–641.

Rosenthal, R., & Fode, K. L. (1963a). The effects of experimenter bias on the performance of the albino rat. *Behavioral Science, 8,* 183–189.

Rosenthal, R., & Fode, K. L. (1963b). Three experiments in experimenter bias. *Psychological Reports, 12,* 491–511.

Vercruyssen, M., & Edwards, J. C. (1988). ANOVA/TT: Analysis of variance teaching template for Lotus 1-2-3. *Behavioral Research Methods, Instruments, and Computers, 20*(3), 349–354.

7 Planning, Conducting, and Reporting Research

This final chapter was written to help readers plan research studies, avoid mistakes in the control of intervening variables, conduct experiments, and prepare research reports. For some this may involve thesis or dissertation research while for others it may involve development of guidelines for completing a term project to demonstrate the capability of completing a larger document. This chapter is about research communication. The topics presented are

Planning and Conducting Study
Techniques for Controlling Extraneous Variables
Conducting Experiment
Writing Research Report
Summary
Keyword Definitions
References
Exercises
Exercise Answers

KEYWORDS

Abscissa	Informed Consent	Randomization
Abstract Section	Intervening Variable	Reference Section
Apparatus	Introduction Section	Results Section
Balancing	Latin Square Design	Seriation
Confounding Variable	Main Heading	Side Heading
Control Group	Method Section	Subjects
Counterbalancing	Nuisance Variable	Summary Section
Design	Ordinate	Survey of Literature
Discussion Section	Paragraph Heading	Table
Extraneous Variable	Problem	Treatment of Data
Figure	Procedure	

PLANNING AND CONDUCTING STUDY

SURVEYING LITERATURE

Beginning with a general idea of the problem or area you wish to investigate, the first step you, as a researcher, should undertake is to survey all the previous work that is relevant to your planned study. Undertaking a literature survey early in your planning will allow you to

Clearly identify and formulate the specific problem to be studied — By reviewing the research of others (literature) you may gain knowledge of past discoveries, present positions, and what next needs to be studied. In some cases, you may find that the necessary studies already have been conducted and the information you need is readily available in the literature.

Identify and gain information about extraneous variables that may contaminate your study if not controlled — You may also gain information about how to control for them.

Gain insight into various design methods — You may also identify designs to avoid because they do not adequately control for contaminating extraneous variables.

Tentatively determine how an independent variable (IV) is likely to affect your dependent measure (DM) — This information may be particularly useful for formulating your experimental hypothesis.

An excellent place to begin your survey of the behavioral and social sciences literature is *PsychINFO* (http://www.apa.org/psychinfo). PsychINFO contains bibliographic citations, abstracts, cited references, and descriptive information to help you find what you need across a wide variety of scholarly publications in the behavioral and social sciences. It covers more than 2,450 titles, of which 99% are peer-reviewed journals (for a listing of the journals, see http://www.apa.org/psychinfo/covlist.html). Also included are selected chapters from books and selected dissertations of psychological relevance included in the Dissertation Abstracts International (A and B) database that contains more than 2.7 million records and is updated weekly (retrieved from http://www.apa.org/psychinfo).

STATING PROBLEM AND HYPOTHESIS

By the time you complete your literature survey you should be in a position to state the problem you will attempt to answer in a clear, concise manner. Your statement should be a single sentence, preferably in the form of a question that can be answered in a positive or negative fashion. For example, "The purpose of this experiment is to determine whether increasing the illumination level in grade school classrooms improves student performance." Occasionally, however, experiments and research projects are undertaken to answer many questions and therefore require a single sentence that is more open-ended. For instance, "The purpose of this experiment is to determine the effects of age, sex, and health status on the speed of central nervous system processing as measured by reaction time."

Having both reviewed the literature and identified the problem, you should now be in a position to formulate the hypothesis. It may be helpful to keep in mind three types of hypotheses: (a) the *theoretical hypothesis* that uses experimental outcomes or findings to support or refute a theoretical position, (b) the *experimental hypothesis* that explicitly states the expected outcome of the experiment, and (c) the *statistical (null) hypothesis* (discussed in Chapter 4) that simply states that differences between treatment conditions will not be statistically significant.

The statistical hypothesis is merely a necessary assumption for conducting behavioral statistical tests and is seldom mentioned in describing most research. An example of the theoretical hypothesis might be, "If reaction time slows exponentially with advancing age, there is support for the Cerella neural network hypothesis and evidence against Birren's generalized slowing hypothesis." An example of the experimental hypothesis is, "Reaction time will slow with advancing age." Usually, such hypotheses are so explicitly stated only in theses and dissertations. Formulating a hypothesis forces a researcher to understand and focus on questions the research is supposed to answer. Refereed journals typically do not emphasize wording to this level of detail. The form of the hypothesis section (theoretical or experimental) of a thesis is determined by the student's mentor and academic program.

Defining Variables

Having developed the hypothesis, the next step is to *operationally* define the independent variable (IV) and dependent variable or measure (DM). As we noted in discussing the principles of science in Chapter 1, the principle of operational definition requires that the variables be defined in terms of how they will actually be *measured* in the study. The decision of how you operationally define the

IV and DM necessarily involves consideration of any *apparatus* that you plan to use in the study. This is the case because an apparatus will aid the experimenter to (a) manipulate the IV and (b) precisely measure subjects' responses (DM).

For example, assume we decided to carry out an experiment in response to our illumination hypothesis in grade school classrooms cited earlier. We could use a special apparatus to control the amount of electrical energy input to the lighting system or alternately use a light meter to measure the amount of illumination falling on students' desks. The two devices lead to two different operational definitions of illumination. Similarly, for our DM we could have used scores on a paper and pencil comprehension test or a special apparatus to record how much prescribed printed material the students read.

Selecting Design

The information gained and decisions made in the foregoing steps will play a strong role in determining how specifically to design your study. In your literature review, you may have identified a good matching variable that may lead you to decide to use a matched groups design. Identification of a suspected differential interaction effect may rule out the use of a repeated measures design (see Chapter 4 for explanations of both of types of designs).

If you planned your study as a learning experiment, you probably would rule out the use of a repeated measures design just to avoid potential differential learning effects. If you suspected a curvilinear relationship between the IV and DM, you would want to use a more-than-two-treatments design (see Chapter 5). Had you found from your literature review that any effect of the IV on the DM could depend on the presence of another IV (i.e., an interaction effect), you probably would resort to a multivariate design, as described in Chapter 6.

Based on a literature review or a pilot study, you may have some idea of the size difference you can expect between your treatment groups and how variable the subject responses will be within each treatment group. This information may be helpful in deciding how many subjects you will need (statistical power and determining appropriate sample sizes are discussed in Chapter 4).

Perhaps your first decision is which scientific approach to use. Will you conduct a laboratory experiment, a field study, or simply determine the ability of one variable to predict another through use of the correlation approach? In deciding which approach to take, you will have to consider the advantages and disadvantages of each method as outlined in Chapter 1. When these are considered in light of your statement of the problem, you should be able to reach a good decision regarding the best scientific approach to take.

Developing Experimental Procedure

The importance of specifying the procedure you will use during the data collection phase of your study cannot be overemphasized. You should plan, step by step, everything that will take place from the time of initial contact with the subjects until they complete their involvement in the study. Included in the experimental procedure are preparation of the instructions to subjects, method of administration of the independent variable, and recording of results. As noted in Chapter 1, it always is a good idea to conduct a pilot study with several subjects from your study population to examine your procedures. Rarely does a pilot study not result in significant improvement in the procedures and refinement of the instructions to subjects. Failure to adequately plan and try out your procedures usually leads to contamination of the data because the procedure lacks uniform procedures or administration of the procedures to subjects is inconsistent. Additionally, your subjects may not understand the directions as to what is required of them in the study (even though the directions may seem very clear to you). Consequently, the data collected may not result from a change in the intended IV.

Analyzing Results

From the material in Chapter 4, you know that your experimental design and the nature of the data collected play large roles in determining how the data are to be analyzed. Also involved are the kinds of assumptions you can make regarding the distribution of the DM in your population because some statistical tests require more stringent assumptions than others. Your literature review may shed some light on analysis and data analysis.

All too often inexperienced researchers fail to consider how they will statistically analyze their data until after the data already are collected. As a result, they may find that they are unable to conduct the kind of statistical analysis required to adequately test their hypothesis.

TECHNIQUES FOR CONTROLLING EXTRANEOUS VARIABLES

Perhaps the most important aspects of planning and conducting scientific research are identifying and controlling for the effects of contaminating variables. Rarely is research initiated in some new frontier of behavioral science without even the most skilled and experienced of researchers failing to identify and control some extraneous variable. We suggested that a survey of the literature may be an excellent source for identifying important extraneous variables. It also is a good idea to talk with other researchers—particularly those experienced in the area you are investigating. For purpose of clarification, it may be useful to review definitions of nuisance or unwanted variables that may seriously affect the interpretability of your experiment.

- An extraneous variable is of no particular interest to the investigator but represents a potential source of confounding or error variance. It must be controlled or otherwise discounted to allow the investigator to ascribe any changes in the DM to the IV.
- A confounding variable that varies systematically with the IV. When a second variable varies simultaneously with the IV, the researcher cannot determine whether a change noted in the DM was produced by the IV, the potential confounding variable, or both variables acting together. In short, confounding variables make it difficult or impossible to determine the effect of the IV on the DV.
- An intervening variable (also known as a mediating variable) is an unnoticed or unobservable process or entity hypothesized to explain relationships between antecedent conditions and consequent events (like a hypothetical construct). Such variables may have no effect on the experiment or may act as unwanted or extraneous variables in confounding the research. An intervening variable may be postulated as a predictor of one or more dependent variables and simultaneously be predicted by one or more independent variables.
- A nuisance variable is unwanted, unplanned, or uncontrolled and makes interpretation of the results more difficult.

All extraneous variables are nuisances and may compromise study effort, integrity, or findings. Unexpected extraneous variables also can occur during an experiment. For example, if an investigator wanted to determine the effect of a unique teaching method on learning in children, an assumption is made that differences in test–retest scores will be due to the new teaching method. However, the cause–effect relationship of the teaching method with performance scores may be influenced by a number of extraneous variables including (a) attrition of subjects due to illness or move to another school; (b) changes of investigators, assistants, or testing instruments; (c) improved performance due simply to practice or impaired performance due to fatigue in repeated testing situations; (d) maturation of subjects that systematically changes performance between tests; and (e) historical events inside or outside the testing environment (epidemic, adverse weather, disaster, teacher strike, student protest, etc.).

Variables are considered extraneous only if they exert contaminating effects on dependent measures. Because the effects are unwanted, they must be discounted. The effects of most extraneous variables may be discounted by one of five general techniques: (a) *eliminating conditions*, (b) *holding conditions constant*, (c) *balancing conditions*, (d) *counterbalancing conditions*, and (e) *randomizing conditions*.

ELIMINATING CONDITIONS

If at all practical, the best way by far for controlling for the effect of an extraneous variable is simply to eliminate it. This technique is most often possible with variables external to the subject. For example, in our illumination experiment, even low levels of background noise could potentially interfere with performance. We therefore would want to eliminate noticeable ambient noise from an experimental situation. This could be done by conducting our experiment in a quiet classroom. In some experiments, it even would be desirable to use a soundproof room.

HOLDING CONDITIONS CONSTANT

When a variable cannot easily be eliminated, the next best approach is to hold it constant during the experiment. Many extraneous variables readily lend themselves to this approach. For example, in our illumination study we could hold constant such variables as the time of day the experiment begins, experimental setting, apparatus used to present the IV, or recording subject responses, ages, IQs, visual acuity levels, and socioeconomic status, just to mention a few. When appropriate, using a repeated measures design (see Chapter 4) can enable a researcher to hold constant a large number of extraneous subject variables.

BALANCING CONDITIONS

When it is not practical or possible to hold the level of an extraneous variable constant, you may be able to balance out its effect in a study. This technique may be used in situations in which (a) you are unable to or uninterested in identifying the extraneous variables or (b) you have identified an extraneous variable that does not readily lend itself to control by elimination or holding its level constant.

This technique requires the use of a *control group* in the experiment. For example, suppose we were to conduct an experiment to determine whether a new teaching method would improve student performance over the present method. If subject performance improved under the new method, it could result from additional exposure to the material in this (or some related) course, further maturation of the students, or additional motivation provided by being selected to participate in the study. To rule out these and other possibilities, we could set up a control group of similar students and have them exposed to the same experimental conditions except for the new teaching method. We then could compare any change in the experimental group's performance to that in the control group. If the experimental group's performance improved (differed significantly) while the control group's performance stayed the same (or decreased), we could feel more confident in ascribing the improvement to the new teaching method.

Balancing also is an excellent technique of control when there is more than one experimenter. To control for any effects due to differences among experimenters, one-half of the experimental group and one-half of the control group could be randomly assigned to Experimenter A and the remaining halves of both groups to Experimenter B. Thus any effects due to differences between the experimenters can be balanced out. Balancing may be used whenever multiple factors are involved in an experimental situation (e.g., devices, experimental locations, times of day when treatments are administered, and homogeneous populations represented by the subjects [such as males and females]).

Another important use of control groups is to evaluate the influence of one or more extraneous variables that cannot be completely removed from the study. When this occurs, we merely need

TABLE 7.1
Experimental Design With Second Control Group to Evaluate Effect of Pre-Testing

	Received Pre-Test	Received Instruction	Received Post-Test
Experimental Group	Yes	Yes	Yes
Control Group 1	Yes	No	Yes
Control Group 2	No	No	Yes

to add additional control groups in which the extraneous variable of interest is absent. One situation in which this problem frequently occurs is in educational studies designed to evaluate the effectiveness of additional instruction in improving performance. To measure the amount of additional knowledge or skill gained during an experimental treatment, a pre-test and post-test must be administered. The differences between the pre- and post-test scores thus become the DM values of our study. However, it is well known that the experience of taking a pre-test often leads to increased performance on the post-test even without intervening instruction. To evaluate the effect of the pre-testing experience on final performance, a second control group not subjected to the pre-test may be added to the study. Assuming random assignment of subjects to the three groups, differences in post-test scores between the two control groups may be interpreted to represent the effects of pre-testing. The resulting experimental design may resemble Table 7.1.

COUNTERBALANCING CONDITIONS

Counterbalancing is used in conjunction with repeated measures designs to distribute the effects of going from one treatment condition to another equally among all conditions. Counterbalancing is accomplished in a two treatments design by assigning one half of the subjects to Treatment A first and the other half to Treatment B first. This principle can be expanded to any number of treatment conditions. For example, in a four-treatments repeated measures design, the sequential assignment of subjects to treatments may be as shown in Table 7.2. The general principles of counterbalancing are

- Each condition is presented to each subject an equal number of times.
- Each condition must occur an equal number of times at each experimental session.
- Each condition must precede and follow all other conditions an equal number of times.

Our counterbalanced design example in Table 7.2 is known as an incomplete or simple Latin square design. It is incomplete in that the third condition above is missing. When the third condition is met, the design is a complete or balanced Latin square (see Chapter 5).

TABLE 7.2
Counterbalancing in Four-Treatments Repeated Measures Design

	Experimental Session				
	1	2	3	4	
1/4 Subjects	A	B	C	D	
1/4 Subjects	B	C	D	A	Treatment Conditions
1/4 Subjects	C	D	A	B	
1/4 Subjects	D	A	B	C	

RANDOMIZING CONDITIONS

For any variables that cannot readily be identified and/or controlled using one of the four foregoing techniques, you must resort to randomization. Of particular note here are a wide variety of subject variables that frequently cannot be specifically identified and measured, much less controlled. These include a broad spectrum of personality traits and factors relating to the subjects' physiological conditions at the time of the experiment. By randomly assigning subjects to treatment conditions, we expect differences on these numerous variables to average out. What differences appear between groups on such variables are considered by our statistical analysis. This technique is much more effective in precisely controlling for the effects of these numerous extraneous variables when the subject groups are large.

CONDUCTING AN EXPERIMENT

ETHICS

Conducting an experiment requires knowledge of the ethical principles of research because all investigators are obligated to minimize harm to their subjects. Three of the basic principles typically found in the codes of ethics of behavioral science professional societies are to (a) protect from harm, (b) undo any harm, and (c) ensure confidentiality.

- You should design the experiment to protect the subject from danger and discomfort including the ill effects of stress and psychological conditioning. Explore alternative experimental designs and inform the subjects of potential harm. Never ask a subject to do anything illegal like jaywalking or stealing. You also are responsible for the ethical conduct of your collaborators.
- At the end of the experiment, explain the purpose of the research and the nature of the manipulation to the subjects. Remove any misconceptions and any ill effects you detect. Ensure that self-esteem is restored, especially if the study involved negative psychological conditioning. Give the subjects a list of people to contact if they experience any problems or have additional questions. You may want to follow up with participants by phone or a written survey.
- The experimental results for subjects should be confidential. Do not release data of a personal nature without a subject's consent. One way to ensure confidentiality is to keep personal identities and data separate, using code numbers to identify subjects. Another step is to destroy the personal information after the data are analyzed. If you plan to keep all personal data, inform the subjects beforehand of the safeguards you will use to protect their privacy.

Most professional society codes of ethics cover a number of additional topics. Accordingly, you should contact the national professional society of your discipline to obtain its ethics code. For example, the American Psychological Association's code can readily be downloaded from http://www.apa.org/ethics/code.html.

INFORMED CONSENT

The idea of informed consent is to give a subject enough information to make a judgment about participating in an experiment. You must establish a fair and clear agreement with participants that explains all obligations and responsibilities on your part and theirs. Ensure that participation is voluntary and involves no coercion or pressure. Inform subjects of all possible risks. Make clear to all subjects that they are free to withdraw from the experiment at any time. Describe your procedure for ensuring confidentiality. Answer all questions of participants. After the study, provide a complete debriefing as soon as possible and take all steps necessary to uncover harmful effects.

The use of deception in an experiment requires special safeguards to ensure the welfare of the subjects. When full disclosure of information would produce invalid results, deception may be used, but you must first determine whether it is really justified and consider alternative experimental designs. Deception in experiments is a gray ethical area. If possible, consult other researchers about the ethics of using deception in your experiment.

ETIQUETTE

In the clinical framework of conducting an experiment, remember that you are dealing with people. Treat your subjects with respect and courtesy. Ask them whether they have questions; ask them whether the instructions are clear. Thank them for participating in your study. Remember that they have the freedom to withdraw at any time; treating them well will make study participation easier for all involved.

WRITING RESEARCH REPORT

After you complete a scientific study, if it is to have value in furthering knowledge in even a small way, you must record the information and communicate it to others. While the specific details may differ, in efforts to facilitate communication, the general format of research reports has become universal across scientific disciplines including the behavioral and social sciences, languages, and locations. The following instructions for preparing research reports reflect those universal guidelines as generally used in the behavioral and social sciences. However, you should check with your instructor or thesis advisor for specific guidelines. If you are submitting a paper for publication, check on the journal's requirements since they may deviate from the guidelines in this chapter. A good general reference book with far more detail is the American Psychological Association's *Concise Rules of APA Style* (6th ed., 2009). This style guide is followed by many of the behavioral sciences. An even more extensive reference book is the *Publication Manual of the American Psychological Association* (6th ed.) from which the style manual was culled.

GENERAL COMMENTS

All report should be typed or computer produced in double-spaced format on white bond paper. In general, the quality and style of writing should be precise, unambiguous, and economical. Keep in mind that the goal of scientific writing is effective communication.

Title — A title should not be excessively long but must convey the exact topic of the study. It should be sufficiently unique to distinguish this study from similar investigations. It may identify the IV and DM used in the experiment. Meaningless introductory phrases such as "A Study of…" and "An Investigation of…" should be avoided.

Name and institutional affiliation — The author's name should be centered below the title on the title page. His or her institutional connection should appear centered below the name. For multiple authors, this sequence is repeated for each author. The name of the principal author appears first. If all authors are from the same institution, the name of the institution should be noted only once, after the name of the last author. The name of the department of the institution and professional titles of the author are not listed. For student experiments where in a sense the entire class conducts the experiment, the usual practice is for students to include only their own names on written reports of their studies.

Headings — The heading structure of a research paper enables readers to understand the order of ideas in the report. Professional journals and research reports use as many as five orders of subordination in headings. In most cases, however, only the first three levels are needed unless a report covers multiple experiments. The heading styles recommended by the American Psychological Association (APA) changed recently to make planning simpler and better accommodate electronic formats (APA, 2009a, 2009b). The five heading levels or types now recommended by the APA are as follows:

- *Level 1*: Main headings are centered, in boldface, and the first letter of each major word is capitalized. No concluding punctuation is used. The text follows on the next double-spaced line. (Note: A number of behavioral science journals print main headings in all caps and do not use boldface.)
- *Level 2*: Side headings are typed flush to the left margin. Like Level 1 headings, they are boldface, the first letters (only) of major words are capitalized, and no concluding punctuation is used. The text following a side heading starts on the next double-spaced line and is indented. (Note: A number of journals do not use boldface.)
- *Level 3*: Paragraph headings are indented, boldface; only the first letter of the first word is capitalized and the last word is followed by a period. The text begins on the same line without extra spacing. (Note: A number of journals do not use boldface. Some capitalize the first letters of all main words, do not use periods, begin the text on the next double-spaced line with a paragraph indent, and reserve paragraph headings for Level 4.)
- *Level 4*: Paragraph headings are indented, boldface, italicized, and only the first letter of the first word is capitalized. The last word is followed by a period. The text begins on the same line without extra spacing. Normally, use of a fourth level will be required only when covering several studies in a single report. (Note: Some journals do not use boldface.)
- *Level 5*: Paragraph headings are indented, italicized, and only the first letter of the first word is capitalized. No boldface is used. The text begins on the same line without extra spacing. (Note: Not all journals use this level.)

Seriation — This is a technique to help readers understand the organization of key points within a section or paragraph. Separate paragraphs, sentences, or items within a series are identified by Arabic numerals followed by periods. Alternately, to avoid the implication that the numbers indicate a priority listing, you can use bullets instead. Word processing programs, such as MS Word, are set up to do this automatically.

Example of Numbered Seriation	Example of Bulleted Seriation
1. Control Group 1 subjects … [sentence or paragraph continues].	• Highlands Ranch students … [sentence or paragraph continues].
2. Control Group 2 subjects … [sentence or paragraph continues].	• Arvada students … [sentence or paragraph continues].
3. Experimental group subjects … [sentence or paragraph continues].	• Denver students … [sentence or paragraph continues].

Within a paragraph or sentence, identify items in a series by lower case letters in parentheses (e.g., "The three leadership theories described in the lecture were (a) contingency, (b) path–goal, and (c) situational."). Also, use commas to separate three or more items if the individual items do not contain commas; use semicolons to separate the elements if they contain commas (e.g., "Subjects were assigned to one of three treatment groups, based on their scores on a mechanical aptitude test: (a) low scorers, who scored below 20; (b) moderate scorers, who scored between 21 and 40; and (c) high scorers, who scored 41 or higher.").

GENERAL FORMAT

In general, the major divisions of the research paper are designated by main headings: *Method, Results, Discussion*, and *References*. Student papers usually also include summaries. In a published manuscript, an *abstract* replaces a summary. The subdivisions of these main headings are indicated by side (Level 2) headings. All major sections begin immediately below the preceding section except the References section that begins on a new page. The discussion of the specific sections of a report will be presented in the order the sections follow.

ABSTRACT

An *abstract* is an accurate one-paragraph summary of the contents of a paper. It typically appears at the beginning, is fully indented at both margins, and is not labeled. It may be the only part of a paper many professionals read. Most journals require abstracts for indexing and retrieval purposes. An abstract must be comprehensive, concise, and self-contained. It should describe in 100 to 200 words a study problem, subjects, method, findings, conclusions, and implications. Evaluative remarks or commentaries are inappropriate in an abstract. It should report only the contents of the paper in the order presented. Do not introduce information in the abstract that is not in the paper. An abstract should be succinct, interesting, and easy to understand. Typically, the abstract is written after the paper is completed, but may be written after the first draft and used as a reference to structure the final draft.

INTRODUCTION

Purpose

The purpose of the *introduction section* is to inform readers of the specific problem under study and the research strategy. Because its function is obvious, this section usually is not labeled. In preparing the introduction, you should consider the point of the study, the rationale or logical link between the problem and the research design, the theoretical implications of the study, and its relationship to previous work in the area. From the introduction, your readers should gain a firm sense of what you are doing and why.

The introduction should state the theoretical propositions from which your test hypotheses are derived, give the logic of their derivation, summarize the relevant arguments and data, and formally state the hypotheses to be tested.

The literature should be discussed but not as an exhaustive historical review. Cite only selected studies pertinent to the specific issue, emphasizing major conclusions, findings, and relevant methodological issues. Refer to general surveys or reviews of the topic if they are available.

Crediting Sources

When material included in a report is taken in whole or in part from another source, the source must be cited. Failure to give credit to a source constitutes *plagiarism*—a serious violation of professional ethics. This even applies to your own work, in that you cannot present your previously published material as though it was new. A good rule of thumb is to use only the amount of your previously published document that is necessary to clarify your new contribution.

Several methods exist for citing sources in text. For example, some disciplines use a numbering system in which material from other sources is numbered consecutively in the manuscript and the numbers and sources are listed sequentially in the references section. The method most frequently used in the behavioral and social sciences cites sources on the basis of author names and publication years. What follows is generally consistent with the APA style manuals that contain far more detail and cover methods of citing additional kinds of sources.

Single Source

When an author is named in the text, the year of publication follows in parentheses; if his or her name is not mentioned in the text, it appears within the parentheses, separated from the date by a comma:

According to Parsons (2009) …
… was found to be true (Parsons, 2009).

In cases of multiple authors, separate their names by *and*; if both their names and dates are included in the parentheses, separate their names by an ampersand (&) rather than *and*:

According to Jones and Fitzmanns (2009) …
It has been theorized (Able & Hunt, 1997) …

Multiple Sources

When several different references are used together in text, they should be separated by a semicolon and included within a single set of parentheses. List the sources in alphabetical order: "It has been shown (Adams, 2001; Campbell, 1996; Sample, 2007) that …"

One exception is when you wish to separate a major citation from other citations within the same parentheses. List the secondary sources in alphabetical order: (Jackson, 2009; see also Adams, 2001; Campbell, 1986; Sample, 2007).

Several Publications by Author in Same Year

Several publications by one author within a year are cited by adding lower case letters (a, b, c, etc.) in the order cited in the text to both text mentions and the entry on the references list. If several publications are cited together, separate them by commas:

In a series of studies reported by Holloman (2008a, 2008b, 2008c) …
… results in a decrement in performance (Holloman, 2008a, 2008b, 2008c).

Citing Primary Source Through Secondary Source

Secondary sources should be used sparingly, such as when a primary source is out of print, unavailable through normal sources, or is in a foreign language. If you reference a primary source through a secondary source, name the primary source but give the citation for the secondary source: "Johnson (as cited in Williamson, 2009) has shown …"

Citing Specific Part of Source

To cite a specific part of a source, indicate the page, chapter, figure, or table, as appropriate:

(Morgan, 2008, p. 23)
(Jackson, 1998, Chapter 3)

Quoting Author

Always cite the page number for the quotation:

In summarizing his research on leadership, Williamson (1996) noted, "in addition to intelligence, the one trait that consistently distinguished high performing from low performing leaders was integrity" (p. 32). Williamson defined …

Leadership research has shown that, "in addition to intelligence, the one trait that distinguished high performing … was integrity" (Williamson, 1996, p. 32).

If quoting from online material that shows paragraph numbers rather than page numbers, cite the paragraph number. If neither page nor paragraph numbers are shown, cite the heading and the number of the paragraph following it: "In his study, Samuelson (2007) found that 'high performing students differed from low performers on …' (Discussion section, para. 4)."

Personal Communication

Personal communications are not included on a reference list and are cited in text only. List initials, surname, and as accurate a date as possible:

> A.B. Lawrence (personal communication, June 4, 1984)
> (A.B. Lawrence, personal communication, June 4, 1984)

METHOD

The *method section* describes how the study was conducted. It should enable readers to evaluate the appropriateness of the experimental procedures and estimate the reliability and the validity of the results. It should provide sufficient detail for investigators to replicate the study. Appropriate side headings in this section usually include (a) Participants, (b) Apparatus (or Materials), and (c) Procedures. Student reports often also include a Treatment of Data heading.

Participants

The participants subsection must describe the subjects of the study in sufficient detail to allow another researcher to replicate your sample. It should answer the following questions:

- Who were your subjects and how were they chosen? State the number of participants, the population from which they were drawn, and any significant demographic information such as gender, age, type of institutional affiliation, profession, geographical location, education, and training or experience that may be relevant to their performance in the study.
- How were participants assigned to your treatment groups (experimental conditions)? Explain the assignment procedure used and the number assigned to each treatment condition.
- Did any participants not complete the experiment? If so, give the number who dropped out and the reasons if known. Describe the demographic characteristics of the dropouts and the extent to which their collective demographic profile differed from the remaining participants (i.e., did the numbers or demographics of the dropouts unbalance the treatment groups sufficiently to affect the outcome of the experiment).

Apparatus

Each piece of apparatus should be described in sufficient detail for other researchers to build or purchase it. If a standard piece of research apparatus is used, mention of its name should be sufficient. Sometimes a picture, diagram, or electrical schematic of the apparatus should be included—particularly if a device is complex. Descriptions of minor items such as ordinary writing pencils are not included.

Procedure

The step-by-step procedure followed from the initial contacts with subjects to the completion of subjects' involvement in the study should be set down in detail. Any instructions given to the subjects should be included. Randomization, counterbalancing, and other methods of experimental control should be described. The administration of the IV and the recording of the DM should be clearly explained. In general, this subsection should give readers an understanding of what you did and how you did it.

Treatment of Data

The treatment of data subsection usually is included in student reports and may sometimes be found in refereed research reports in journals. All criterion-dependent measures should be listed in this section along with descriptions of how these data were treated (statistical procedures, designs, and critical significance levels). State which descriptive and inferential statistics were used, on what computing hardware and software, and indicate the criteria used for dealing with outliers (data that do not necessarily reflect the population's characteristics of interest). In refereed journal articles, this material often is included in the procedure subsection.

RESULTS

The *results section* graphically and statistically reports your findings. Your data should be reported in sufficient detail to support your conclusions. However, do not include individual scores or raw data. All relevant results should be reported, including those that do not support your experimental hypothesis. Typically, the results of a study are summarized in the first part of this section, usually with a table. Figures also are used if graphic presentations may facilitate reader understanding. Attention should be drawn to the important points depicted in the tables and figures. Following the summary of a set of results, any statistical tests conducted using the data are summarized, and the resultant conclusion stated. This section does not include discussion of the implications of the results; implications are covered in the discussion section. Overall organization varies, but it is often organized by independent variables, describing each dependent measure, and concluding with the interaction of variables.

DISCUSSION

Having presented the results, you now are in a position to evaluate and interpret their implications, particularly with respect to your original hypotheses in the *discussion section*. Limitations of the conclusions, correspondences or differences between the findings and widely accepted points of view, and any implications for theory or practice should be pointed out.

Reports of research resulting in negative or unexpected findings should not end with long discussions of possible reasons for the outcome. Study shortcomings should be noted and explained briefly, but do not dwell on every flaw. In general, a good discussion will convey to your readers what your study contributed, how it has helped resolve the original problem, and what conclusions and theoretical implications can be drawn. Remember to clearly state the practical importance or application of findings.

SUMMARY

Student research reports usually include a *summary section*. Summaries typically contain fewer than 100 words but may be longer. As noted earlier, an abstract usually replaces a summary in a published manuscript. An abstract is similar in form but may be more detailed. If a summary appears in a report that contains an abstract or executive summary, the summary should not simply restate the same information. A summary should succinctly answer the research question, identify immediate implications, and highlight the major contributions of the research.

REFERENCES

Every research paper concludes with a *references section* that lists all the resources cited in the text. Just as the data from a study document the conclusions, reference citations document statements made about the literature. Each entry on a references list must contain all data necessary for identification and library search. Not all journals use the precise reference style shown in the following. For that reason, you should consult recent issues of the journal for which the manuscript is being prepared. The references section should begin on a separate page. *References* should be the main or first order heading. In student reports, common practice is not to double space within individual reference entries and this practice has been followed here. When preparing a manuscript for publication, the references should be double-spaced throughout and, as in student reports, have a hanging indent. The references on the list below are fictitious.

Book References

Human Factors Society. (2010). *Directory and yearbook.* Santa Monica, CA: Author.

Macklin, T. M., & Kim, W., Jr. (1982). *Social psychology* (2nd ed.). New York: Macmillan.

Morgan, S. P. (2006). Motivation and selective perception. In T. A. Smith & R. Stone (Eds.), *Readings in social psychology* (Vol. 3, pp. 275–294). New York: Academic Press.

Rice, W., Thomas, P. R., & Hughes, L. T. (in press). *Reading and you.* Belmont, CA: Brooks/ Cole.

Thompson, S. L. (Ed.). (1998). *Readings in organizational psychology.* Boston: Allyn & Bacon.

Vroom, T. L. (2002–2008). *Annals of human factors* (Vols. 3–9). Chicago: Eldon Press.

Journal References

Baker, R. M., & Smith, A. L. (2003). The development of achievement motivation. *Journal of Applied Psychology, 287*(2), 67–75.

The Society of Safety Management. (2007). Biorhythms and accident control. *American Scientist, 67,* 132–139.

Proceedings References

Robertson, M. M. (2008). Evaluation of an ergonomics computer workstation training program for office workers. *Proceedings of the Human Factors and Ergonomics Society 52nd Annual Meeting* (pp. 148–153). Santa Monica, CA: Human Factors and Ergonomics Society.

Dissertation References

Unpublished Doctoral Dissertation (or Other Manuscript With University Cited)

Brogmus, G. E. (1991). *Effects of age and gender on speed and accuracy of hand movements, and the refinements they suggest for Fitt's law.* Unpublished master's thesis, University of Southern California (Human Factors Department), Los Angeles.

Jones, B. A. (2007). *Worker performance as a function of leader behavior.* Unpublished doctoral dissertation, Purdue University, West Lafayette, IN.

Reynolds, S. L. (1991). *Longitudinal analysis of age changes in speed of behavior.* Unpublished master's thesis, University of Southern California (Human Factors Department), Honolulu.

Ryan, A. M. (2006). *Age and sex differences in simulated collision avoidance driving.* Unpublished master's thesis, University of Hawaii (Kinesiology Department), Honolulu.

Siu, K. (1996). *Pedestrian estimates of time to vehicle arrival: Differences by age as a function of preview distance and approach speed.* Unpublished undergraduate thesis, University of Hawaii (Honors Program), Honolulu.

Williams, R. S. (2009). *A new theory of leadership.* Unpublished manuscript, Department of Sociology, University of Michigan, Ann Arbor, MI.

Dissertation Retrieved From Institutional Database

Halpin, G. W. (2007). *Development and validation of a motivation test for predicting job performance* (Doctoral dissertation). Retrieved from http://www.ohiolink.edu/etd/

Dissertation Abstracted in DAI

Walton, B. E. (2003). Factors affecting group problem solving behavior. *Dissertation Abstracts International: Section B, Science and Engineering, 57*(11), 4723.

Master's Thesis From Commercial Database

Johnstone, R. L. (2008). *A comparison of organizational climate in a high performing versus a low performing manufacturing plant* (Master's thesis). Available from ProQuest Dissertations and Theses database (UMI No. 146329).

Technical Report References

Horrocks, R. L. (1978). *Comparison of teaching methodologies in high school courses* (ERIC Document Reproduction Service No. ED 021 463). West Lafayette, IN: Purdue University.

U.S. Department of Labor, Occupational Health and Safety Administration. (1993). *Ergonomics program management guidelines for assembly plants* (OSHA 3174). Washington, DC: Government Printing Office.

Motion Picture References

Thomas, P. (Producer), & R. J. Samuels (Director). (1979). *If schools could talk* [Motion picture]. United States: Educational Films.

Newspaper and Magazine References

Adams, C. R. (2009, January 14). Gas explosion blamed on human error. *The New York Times*, pp. A1, A14–15.

Smith, R. S. (2007, April 16). New incentive system for teachers. *Denver Post*. Retrieved from http://www.denverpost.com

The brain game. (1979, August 24). *Time*, pp. 53–54.

Moody, H. L. (1979). Leadership and organizational effectiveness. *New Horizons, 4*, pp. 21–22; 72–74.

Other References

For less common reference types and explanatory comments on the types illustrated, see the American Psychological Association's *Concise Rules of APA Style* (6th ed.) and the *Publication Manual of the American Psychological Association* (6th ed.).

TABLES AND FIGURES

In student reports, it is common practice to include tables and figures on a separate sheet, immediately following the page of text on which they are first mentioned. Alternately, some instructors prefer to tables and figures placed on separate sheets at the end of a report. In that case the locations of tables and figures should be shown by clear breaks in the text, with instructions centered and set off by a line space above and below the instruction. The latter procedure is usually required for manuscripts being prepared for publication. Always consult the specific journal or book publisher for instructions regarding table and figure preparation. In text, tables and figures always should be cited by their numbers. Be specific; refer to Table 1 or Figure 1, not *the following table* or *the figure above*. Figures and tables are numbered consecutively (Arabic numbers) throughout a report. In books, tables and figures are cited by chapter number, period, and the consecutive number within the chapter (Table 7.3; Figure 7.4).

TABLE 7.3
Summary of Results on Heading Error

	Experienced Subjects	Inexperienced Subjects	Difference	*df*	*t*
Normal Control					
$\bar{X}$	8.44	10.78	2.34	18	1.67
Σ	2.13	3.56			
Reversed Control					
$\bar{X}$	60.25	35.00	25.25	18	5.79[a]
Σ	30.56	26.96			

[a] $p < 0.01$.

Tables

Tables are centered on a page and labeled with the appropriate number at the left margin. The table title begins one double-spaced line below the table number. The title is italicized with the first letter of each principal word capitalized. No punctuation follows the title. The Table 7.3 example follows these guidelines. Lines should denote the top of the table, separate columnar headings from body, and indicate the end of the table. Other lines are optional—a judgment call of the preparer.

Figures

The most common figures are (a) graphs, (b) charts, (c) maps, (d) drawings, and (e) photographs. A figure caption appears at the bottom and only the first letter of the first word in each sentence is capitalized. *Figure* and the appropriate number are italicized (*Figure 7.1*). As was the case for tables, figures are numbered consecutively in a research report or by chapter number and individual consecutive number within a book.

Generally in graphical presentations, the DM is plotted on the ordinate or Y-axis and the IV is plotted on the abscissa or X-axis. The plotting of data points on the graph should be precise and the points should be connected with straight lines. If more than one class is plotted on a single graph, the classes should be differentiated as solid lines, dashed lines, dotted lines, etc. Do not use color coding. Figures should be drawn only in blue or black ink. See Figure 7.1.

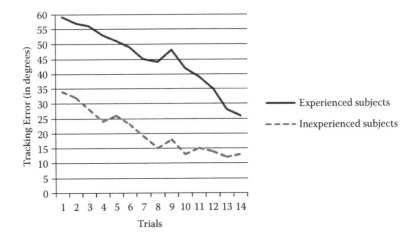

FIGURE 7.1   Effects of practice on reversed control tracking. Averaged data for 10 experienced and 10 inexperienced subjects over 14 trials.

Additional Comments About Writing Scientific Reports

Be specific — If you use generalizations, support them with concrete evidence. Write short, clear sentences. If a word adds no meaning to a sentence, omit it (e.g., small *in size*, summarize *briefly*).

Avoid beginning a sentence with an indefinite pronoun — *This showed* is less precise than *this study showed* ...

Use past tense — The literature cited has already been written and your study has been completed. Remember, you are reporting on work already done. An exception to this rule is using present tense for statements that continue to apply: *Thorndike defines* ... or *Behavior theory holds*....

Use commas freely — This rules applies even before *and* where three or more items are enumerated. Research writing should be "tight."

Spell out numbers below 10 — The exception is the use of numbers in a series: 2, 4, 9, 12, and 16.

Define and then use abbreviations — Use abbreviations for widely understood technical and statistical terms. Abbreviations for technical terms or organizations should appear in all capital letters, without periods or spaces. Examples of standard abbreviations are REM for rapid eye movement, ESP for extrasensory perception, and IQ for intelligence quotient. For less familiar technical terms or organizational names, spell them out in full when first used followed by their abbreviations in parentheses. Use only the abbreviation thereafter: The Human Factors and Ergonomics Society (HFES) published.... Five years later, HFES....

Data is plural — The singular form of *data* is *datum* but the singular is rarely used. *Data* thus requires a plural verb: *data are* (not *data is*). Similarly, *criteria* is plural and *criterion* is singular.

Statistical tests show significance not data — The results of a statistical analysis may be significant; the data are not significant. For example, the observed *t* is significant; the difference between the means is not.

Use non-sexist language — Do not use *man* as a generic noun or *his* as a generic pronoun. Avoid stereotyping such as *woman doctor, the teacher ... she, chairman* (use *chair* or *chairperson), policeman* (use *police officer), mothering* (use *parenting* or *nurturing*). Avoid prompting readers into irrelevant evaluation of roles (*man and wife, behavior was typically male*). Use *gender* rather than *sex*.

Avoid ethnic biases — Do not employ evaluative statements that imply one group (especially your group) is the standard against which others (e.g., *culturally deprived*) are compared.

Proofread — Proofread your writing to see whether you left out any words. Spelling can be corrected by most word processors but missing words must be checked by reading.

Eliminate ambiguity — In particular, avoid using pronouns such as *this, that, these*, and *those* when referring to a person or item mentioned in a previous sentence. Instead, use phrases such as *these participants, that trial, those reports*, and other specific descriptions. Avoid ambiguity also by using personal pronouns (*I* and *we*) rather than third person nouns (*the author*) when describing steps followed in an experiment and use active voice. *We reviewed the literature* is easier to read than *the authors reviewed the literature* or *the literature was reviewed*.

Restrict use of *we* and *I* — Use these pronouns to refer only to yourself and your co-authors: *Researchers often classify...* rather than *We usually classify...*

SUMMARY

1. The survey of the literature should be completed early in the planning of a study to (a) more clearly identify and formulate the problem, (b) identify and gain information about potentially contaminating variables, (c) gain insight into how best to design the study, and (d) more knowledgeably formulate the experimental hypothesis. An excellent place to begin a literature search is PsychINFO (http://www.apa.org/psychinfo).

2. The problem should be stated in a form that can be answered empirically in a positive or negative manner. The hypothesis should explicitly identify the IV and DM.

3. When operationally defining the IV and DM, consider all apparatus to be used in the study that will help the experimenter to (a) accurately manipulate the IV and (b) precisely measure the DM.

4. Information gained during a literature search along with the decisions made in the foregoing steps should enable a researcher to select an optimum design for a study. An estimate of the number of subjects required can be computed using the power formula and solving for n. However, the researcher must have a preliminary notion of what results he or she expects.

5. In planning a study, the experimental procedure including instructions should be written in detail.

6. Conduct a pilot study to test the procedure and instructions.

7. How you plan to statistically analyze the results should be carefully considered before you collect data. Failure to consider the analysis at that point could lead to an inability to perform the kinds of analyses necessary to test the hypothesis. Similarly, the kinds of conclusions and generalizations that may result also should be considered to prevent a possibly worthless study.

8. The techniques of extraneous variable control are (a) elimination, (b) holding conditions constant, (c) balancing, (d) counterbalancing, and (e) randomization. Generally, these techniques should be considered in the order listed. Balancing requires the use of a control group to balance out extraneous variable effects. Counterbalancing is used in conjunction with repeated measures designs to distribute the effects of going from one treatment condition to another equally among all conditions. Latin square design is one form of counterbalancing.

9. The ethical principles of conducting an experiment include protection of subjects from harm, undoing any harm, and protection of confidentiality. Informed consent is an essential element of working with human subjects. Participation is voluntary and subjects are free to withdraw at any time. The use of deception requires special safeguards to prevent psychological harm to the participants.

10. The basic format of a research report is almost universal. A report contains main headings, side headings, and paragraph headings. Main headings are used for all major sections except the introduction. The usual main headings are Method, Results, Discussion, and References. In student reports, a summary section often is included and follows the discussion section. In published reports, the summary is replaced by an abstract that appears at the beginning of the report. Side headings are used for subsections. The method section headings are usually Subjects, Apparatus (or materials), Procedure, and (in student reports) Treatment of Data.

11. When material included in a report is taken from another source, the source must be cited in the text and listed in the references section.

12. All charts, graphs, photographs, and schematics are called *figures*. On graphs, the DM is plotted on the ordinate, and the IV is plotted on the abscissa. Tables are used to summarize data. Both tables and figures are numbered consecutively in the text.

KEYWORD DEFINITIONS

Abscissa: X-axis.

Abstract Section: Accurate, concise summary of contents of a paper; appears fully indented at the beginning of the paper and normally has no heading.

Apparatus: Experimental aid for accurately manipulating the IV and precisely measuring the DM.

Balancing: Use of a control group to balance out extraneous variable effects.

Confounding Variable: Variable that varies systematically with the IV(s), thereby degrading internal validity of an experiment and clouding the cause–effect relationships of the independent and dependent variables. In short, confounding variables make it difficult or impossible to determine the effects of treatments on dependent measures. When a second variable varies simultaneously with the treatment, it is unclear whether a change in a dependent measure was produced by a manipulated variable (IV), a potential confounding variable, or both variables acting together.

Control Group: One of two subject groups is called the control group, and usually receives the normal or standard amount of treatment variable.

Counterbalancing: Approach used in conjunction with repeated measures designs to distribute equally the effects of going from one treatment condition to another among all conditions.

Design: General plan of an experiment including the number and arrangement of IVs and the methods of selecting and assigning subjects to conditions.

Discussion Section: Report section used to evaluate and interpret the implications of results, particularly with respect to the original hypotheses.

Extraneous Variable: A potential source of confounding or error variance that is of no interest to an investigator; it must be controlled to maintain internal validity.

Figure: General term covering charts, graphs, photographs, and schematics included in a report.

Informed Consent: An experimenter must give participants enough information to make a judgment about participating in an experiment. Informed consent requires a fair and clear agreement that explains all obligations and responsibilities of both the experimenters and the participants. The participants must understand that participation is voluntary and involves no coercion or pressure.

Intervening (Mediating) Variable: An unnoticed or unobservable process or entity that is hypothesized to explain relationships between antecedent conditions and consequent events (like a hypothetical construct). Such variables may have no effect on the experiment or may act as unwanted or extraneous variables in confounding the research.

Introduction Section: Report section that informs readers of the specific problem under study and the research strategy.

Latin Square Design: Form of counterbalancing.

Main Heading: First-order heading; used for all major sections of a report except the introduction.

Method Section: Report section describing study method; should enable readers to evaluate the appropriateness of experimental procedures and estimate the reliability and the validity of results; should provide sufficient detail for investigators to replicate the study.

Nuisance Variable: Unwanted, unplanned, or uncontrolled variable that makes interpretation of results more difficult. Extraneous, intervening, and confounding variables are nuisances and may compromise the research effort, integrity, or findings.

Ordinate: Y-axis.

Paragraph Heading: Heading subordinate to main heading; follows paragraph indentation, ends with period; usually only initial letter of first word is capitalized.

Problem: What an experimenter proposes to solve.

Procedure: Detailed description of conduct of an experiment.

Randomization: Selection process by which each person in a population has an equal probability of being selected. By randomly assigning subjects to treatment conditions, we expect differences from unidentified extraneous variables to average out. This technique is most effective when the subject groups are large.

Reference Section: List of all the literature sources cited in text.

Results Section: Report section that summarizes data collected and their statistical treatment by briefly stating the main results or findings and then reporting the data in sufficient detail to justify the conclusions.

Seriation: Arrangement of a report that helps readers understand the organization of key points in a section or paragraph. Separate paragraphs, sentences, or items in a series are indented and an Arabic numeral and period identifies each entry. Alternately, bullets are used instead of numerals.

Side Heading: Second-order headings; used for subsections of report.

Subjects: First subheading in the method section of a report; describes the subjects, how they were selected, and how they were assigned to treatment conditions. *Subject* can also refer to a study participant.

Summary Section: Report section that succinctly answers the research question, identifies immediate implications, highlights the major contributions of the study, and briefly notes shortcomings or limitations of the study.

Survey of Literature: A search of relevant literature in an effort to clearly identify and formulate a problem, obtain information about potentially contaminating variables, gain insight into how best to design a study, and knowledgeably formulate an experimental hypothesis.

Table: Tabular format for summarizing data.

Treatment of Data: Methods of data collection, analysis, and storage.

EXERCISES

1. An Air Force behavioral scientist is interested in knowing whether learning to fire an M-16 rifle will facilitate learning to accurately fire a 20 mm cannon at a moving aerial target from a combat aircraft. He identifies a group of air cadets who just completed their initial flight training and received their wings to serve as subjects in the study. Those who never learned to fire a gun are randomly divided into experimental and control groups. The experimental group undergoes M-16 rifle training while the control group does not. The behavioral scientist then assigns the experimental group to one combat aircraft and instructor and the control group to another. To hold flying conditions constant, each group also is assigned its own tow plane and target so that the two groups can run training and firing flights simultaneously. The procedure followed by both groups during the training missions is identical. Each cadet makes five firing runs at an airborne low target and the average accuracy score for the best three runs is used as the cadet's score. Results showed that the experimental group was significantly more accurate than the control group. The behavioral scientist concluded that learning to fire the M-16 rifle facilitated learning to accurately fire the 20 mm cannon at a moving aerial target.

2. A teacher wishes to determine whether reading out loud helps students understand the content of printed material. To test this hypothesis, a group of grade schoolers are assigned to individual reading booths and instructed to read an assigned passage silently. Each student then is given a second passage and told to read it out loud. After reading the second passage, each student takes a short comprehension exam on the two passages. One-half of the group read Passage A first and B second; the other half read B first and A second. Results indicated that comprehension on the oral reading was better than on the silent reading for the entire group. The teacher concluded that oral reading facilitated student comprehension for the type of material used in the study.

3. A corporate manager is interested in knowing whether learning to become a computer programmer using a hands-on method of working on real problems with an experienced computer programmer is superior to a traditional classroom approach. A new group of trainees is randomly divided into two groups. One group is sent to the South Dakota plant where the traditional classroom approach is used; the other group is assigned to the Los Angeles plant

where the new hands-on approach is tried. Both plants use the same computer equipment and software packages and both perform the same types of operations. Average scores for both groups on a common exam administered at the end of the training period showed no difference. The executive concluded that the two training methods are equally effective.

4. A researcher wishes to test the theory that children of managerial and professional people, because of parental exposure, will demonstrate greater analytical skills than children whose parents have other work backgrounds. To test this theory, all high school seniors in a city were administered an analytical skills test. Scores for the seniors from families having at least one parent in a managerial or professional position were significantly higher than the scores for the rest of the seniors. The researcher concludes that managerial and professional parental background facilitates the development of analytical skills of offspring to a greater extent than other parental backgrounds.

5. An industrial psychologist developed a new test for possible use in selecting sales staff for the ABC Widget Company. To concurrently validate the test, the psychologist administers it to all present company sales employees and correlates their test scores with their dollar volumes of sales for the past year. The resulting correlation is too low to show practical or statistical significance. The psychologist concludes that the new test is not a valid predictor of widget sales performance.

6. An ergonomist wishes to determine which of two aircraft instrument display arrangements results in more effective pilot performance in flying through a series of 10 programmed flight maneuvers typical of those most frequently used in combat. One display is installed in an F-16 fighter flight simulator at Air Force Base X; the other display is installed in an F-18 fighter flight simulator at Air Force Base Y. A sample of ten F-16 pilots is randomly selected from the two F-16 squadrons at Base X. Similarly, a sample of 10 is randomly selected from members of the two F-18 squadrons at Base Y. Each pilot flies the programmed maneuvers in his respective flight simulator 10 times for familiarization, and then 5 more times in which his error scores for each programmed flight maneuver are automatically recorded by the simulator. Each pilot's average error score for all the maneuvers across all five trials constitutes his performance score. The means and standard deviations are computed for both groups' performance scores. The average error scores for the F-16 pilots were higher than for the F-18 pilots, indicating that the F-18 pilot sample performed better. To test to see whether the result represented a real difference that would be found for the entire population of pilots using the two instrument panel arrangements, a t-test for independent samples was conducted. The test was found significant at the 0.05 point (one-tailed test). The null hypothesis thus was rejected. The ergonomist concluded that the display arrangement in the F-18 simulator was superior to the one in the F-16 simulator, and thus recommended that the Air Force adopt the one in the F-18 for use in future fighter aircraft.

EXERCISE ANSWERS

Answers to the exercises are to be discussed in class with your instructor. Prepare notes for a response to each exercise so you are ready to discuss your answers aloud when called upon to do so.

REFERENCES

American Psychological Association. (2009a). *Concise rules of APA style* (6th ed.). Washington, D.C.: Author.
American Psychological Association. (2009b). *Publication manual of the American Psychological Association* (6th ed.). Washington, D.C.: Author.
Strunk, W., Jr., & White, E. B. (1979). *The elements of style*. New York: Macmillan.

Appendix A: Statistical Tables

TABLE A
Squares, Square Roots, and Reciprocals

N	N^2	$\sqrt{N}$	$\sqrt{10N}$	$1/N$
1	1	1.0000000	3.1622777	1.0000000
2	4	1.4142136	4.4721360	0.5000000
3	9	1.7320508	5.4772256	0.3333333
4	16	1.0000000	6.3245553	0.2500000
5	25	2.2360680	7.0710678	0.2000000
6	36	2.4494897	7.7459667	0.1666667
7	49	2.6457513	8.3666003	0.1428571
8	64	2.8284271	8.9442719	0.1250000
9	81	3.0000000	9.4868330	0.1111111
10	100	3.1622777	10.0000000	0.1000000
11	121	3.3166248	10.4880885	0.0909091
12	144	3.4641016	10.9544512	0.0833333
13	169	3.6055513	11.4017543	0.0769231
14	196	3.7416574	11.8321596	0.0714286
15	225	3.8729833	12.2474487	0.0666667
16	256	4.0000000	12.6491106	0.0625000
17	289	4.1231056	13.0384048	0.0588235
18	324	4.2426407	13.4164079	0.0555556
19	361	4.3588989	13.7840488	0.0526316
20	400	4.4721360	14.1421356	0.0500000
21	441	4.5825757	14.4913767	0.0476190
22	484	4.6904158	14.8323970	0.0454545

(*continued*)

TABLE A (CONTINUED)
Squares, Square Roots, and Reciprocals

N	N^2	$\sqrt{N}$	$\sqrt{10N}$	$1/N$
23	529	4.7958315	15.1657509	0.0434783
24	576	4.8989795	15.4919334	0.0416667
25	625	5.0000000	15.8113883	0.0400000
26	676	5.0990195	16.1245155	0.0384615
27	729	5.1961524	16.4316767	0.0370370
28	784	5.2915026	16.7332005	0.0357143
29	841	5.3851648	17.0293864	0.0344828
30	900	5.4772256	17.3205081	0.0333333
31	961	5.5677644	17.6068169	0.0322581
32	1024	5.6568542	17.8885438	0.0312500
33	1089	5.7445626	18.1659021	0.0303030
34	1156	5.8309519	18.4390889	0.0294118
35	1225	5.9160798	18.7082869	0.0285714
36	1296	6.0000000	18.9736660	0.0277778
37	1369	6.0827625	19.2353841	0.0270270
38	1444	6.1644140	19.4935887	0.0263158
39	1521	6.2449980	19.7484177	0.0256410
40	1600	6.3245553	20.0000000	0.0250000
41	1681	6.4031242	20.2484567	0.0243902
42	1764	6.4807407	20.4939015	0.0238095
43	1849	6.5574385	20.7364414	0.0232558
44	1936	6.6332496	20.9761770	0.0227273
45	2025	6.7082039	21.2132034	0.0222222
46	2116	6.7823300	21.4476106	0.0217391
47	2209	6.8556546	21.6794834	0.0212766
48	2304	6.9282032	21.9089023	0.0208333
49	2401	7.0000000	22.1359436	0.0204082
50	2500	7.0710678	22.3606798	0.0200000
51	2601	7.1414284	22.5831796	0.0196078
52	2704	7.2111026	22.8035085	0.0192308
53	2809	7.2801099	23.0217289	0.0188679
54	2916	7.3484692	23.2379001	0.0185185
55	3025	7.4161985	23.4520788	0.0181818
56	3136	7.4833148	23.6643191	0.0178571
57	3249	7.5498344	23.8746728	0.0175439
58	3364	7.6157731	24.0831892	0.0172414
59	3481	7.6811457	24.2899156	0.0169492
60	3600	7.7459667	24.4948974	0.0166667
61	3721	7.8102497	24.6981781	0.0163934
62	3844	7.8740079	24.8997992	0.0161290
63	3969	7.9372539	25.0998008	0.0158730
64	4096	8.0000000	25.2982213	0.0156250

TABLE A (CONTINUED)
Squares, Square Roots, and Reciprocals

N	N^2	$\sqrt{N}$	$\sqrt{10N}$	$1/N$
65	4225	8.0622577	25.4950976	0.0153846
66	4356	8.1240384	25.6904652	0.0151515
67	4489	8.1853528	25.8843582	0.0149254
68	4624	8.2462113	26.0768096	0.0147059
69	4761	8.3066239	26.2678511	0.0144928
70	4900	8.3666003	26.4575131	0.0142857
71	5041	8.4261498	26.6458252	0.0140845
72	5184	8.4852814	26.8328157	0.0138889
73	5329	8.5440037	27.0185122	0.0136986
74	5476	8.6023253	27.2029410	0.0135135
75	5625	8.6602540	27.3861279	0.0133333
76	5776	8.7177979	27.5680975	0.0131579
77	5929	8.7749644	27.7488739	0.0129870
78	6084	8.8317609	27.9284801	0.0128205
79	6241	8.8881944	28.1069386	0.0126582
80	6400	8.9442719	28.2842712	0.0123500
81	6561	9.0000000	28.4604989	0.0123457
82	6724	9.0553851	28.6356421	0.0121951
83	6889	9.1104336	28.8097206	0.0120482
84	7056	9.1651514	28.9827535	0.0119048
85	7225	9.2195445	29.1547595	0.0117647
86	7396	9.2736185	29.3257566	0.0116279
87	7569	9.3273791	29.4957624	0.0114943
88	7744	9.3808315	29.6647939	0.0113636
89	7921	9.4339811	29.8328678	0.0112360
90	8100	9.4868330	30.0000000	0.0111111
91	8281	9.5393920	30.1662063	0.0109890
92	8464	9.5916630	30.3315018	0.0108696
93	8649	9.6436508	30.4959014	0.0107527
94	8836	9.6953597	30.6594194	0.0106383
95	9025	9.7467943	30.8220700	0.0105263
96	9216	9.7979590	30.9838668	0.0104167
97	9409	9.8488578	31.1448230	0.0103093
98	9604	9.8994949	31.3049517	0.0102041
99	9801	9.9498744	31.4642654	0.0101010
100	10000	10.0000000	31.6227766	0.0100000

Note: Table generated using Turbo Pascal software.

TABLE B
Areas and Ordinates of Standard Normal Curve

(1) Z	(2) A AREA FROM μ TO Z	(3) B AREA IN LARGER PORTION	(3) C AREA IN SMALLER PORTION	(5) y ORDINATE AT Z
0.00	0.0000	0.5000	0.5000	0.3989
0.01	0.0040	0.5040	0.4960	0.3989
0.02	0.0080	0.5080	0.4920	0.3989
0.03	0.0120	0.5120	0.4880	0.3988
0.04	0.0160	0.5160	0.4840	0.3986
0.05	0.0199	0.5199	0.4801	0.3984
0.06	0.0239	0.5239	0.4761	0.3982
0.07	0.0279	0.5279	0.4721	0.3980
0.08	0.0319	0.5319	0.4681	0.3977
0.09	0.0359	0.5359	0.4641	0.3973
0.10	0.0398	0.5398	0.4602	0.3970
0.11	0.0438	0.5438	0.4562	0.3965
0.12	0.0478	0.5478	0.4522	0.3961
0.13	0.0517	0.5517	0.4483	0.3956
0.14	0.0557	0.5557	0.4443	0.3951
0.15	0.0596	0.5596	0.4404	0.3945
0.16	0.0636	0.5636	0.4364	0.3939
0.17	0.0675	0.5675	0.4325	0.3932
0.18	0.0714	0.5714	0.4286	0.3925
0.19	0.0753	0.5753	0.4247	0.3918
0.20	0.0793	0.5793	0.4207	0.3910
0.21	0.0832	0.5832	0.4168	0.3902
0.22	0.0871	0.5871	0.4129	0.3894
0.23	0.0910	0.5910	0.4090	0.3885
0.24	0.0948	0.5948	0.4052	0.3876
0.25	0.0987	0.5987	0.4013	0.3867
0.26	0.1026	0.6026	0.3974	0.3857
0.27	0.1064	0.6064	0.3936	0.3847
0.28	0.1103	0.6103	0.3897	0.3836
0.29	0.1141	0.6141	0.3859	0.3825
0.30	0.1179	0.6179	0.3821	0.3814
0.31	0.1217	0.6217	0.3783	0.3802
0.32	0.1255	0.6255	0.3745	0.3790
0.33	0.1293	0.6293	0.3707	0.3778
0.34	0.1331	0.6331	0.3669	0.3765
0.35	0.1368	0.6368	0.3632	0.3752
0.36	0.1406	0.6406	0.3594	0.3739
0.37	0.1443	0.6443	0.3557	0.3725
0.38	0.1480	0.6480	0.3520	0.3712
0.39	0.1517	0.6517	0.3483	0.3697
0.40	0.1554	0.6554	0.3446	0.3683
0.41	0.1591	0.6591	0.3409	0.3668
0.42	0.1628	0.6628	0.3372	0.3653

TABLE B (CONTINUED)
Areas and Ordinates of Standard Normal Curve

(1) Z	(2) A AREA FROM μ TO Z	(3) B AREA IN LARGER PORTION	(3) C AREA IN SMALLER PORTION	(5) y ORDINATE AT Z
0.43	0.1664	0.6664	0.3336	0.3637
0.44	0.1700	0.6700	0.3300	0.3621
0.45	0.1736	0.6736	0.3264	0.3605
0.46	0.1772	0.6772	0.3228	0.3589
0.47	0.1808	0.6808	0.3192	0.3572
0.48	0.1844	0.6844	0.3156	0.3555
0.49	0.1879	0.6879	0.3121	0.3538
0.50	0.1915	0.6915	0.3085	0.3521
0.51	0.1950	0.6950	0.3050	0.3503
0.52	0.1985	0.6985	0.3015	0.3485
0.53	0.2019	0.7019	0.2981	0.3467
0.54	0.2054	0.7054	0.2946	0.3448
0.55	0.2088	0.7088	0.2912	0.3429
0.56	0.2123	0.7123	0.2877	0.3410
0.57	0.2157	0.7157	0.2843	0.3391
0.58	0.2190	0.7190	0.2810	0.3372
0.59	0.2224	0.7224	0.2776	0.3352
0.60	0.2257	0.7257	0.2743	0.3332
0.61	0.2291	0.7291	0.2709	0.3312
0.62	0.2324	0.7324	0.2676	0.3292
0.63	0.2357	0.7357	0.2643	0.3271
0.64	0.2389	0.7389	0.2611	0.3251
0.65	0.2422	0.7422	0.2578	0.3230
0.66	0.2454	0.7454	0.2546	0.3209
0.67	0.2486	0.7486	0.2514	0.3187
0.68	0.2517	0.7517	0.2483	0.3166
0.69	0.2549	0.7549	0.2451	0.3144
0.70	0.2580	0.7580	0.2420	0.3123
0.71	0.2611	0.7611	0.2389	0.3101
0.72	0.2642	0.7642	0.2358	0.3079
0.73	0.2673	0.7673	0.2327	0.3056
0.74	0.2703	0.7703	0.2297	0.3034
0.75	0.2734	0.7734	0.2266	0.3011
0.76	0.2764	0.7764	0.2236	0.2989
0.77	0.2793	0.7793	0.2207	0.2966
0.78	0.2823	0.7823	0.2177	0.2943
0.79	0.2852	0.7852	0.2148	0.2920
0.80	0.2881	0.7881	0.2119	0.2897
0.81	0.2910	0.7910	0.2090	0.2874
0.82	0.2939	0.7939	0.2061	0.2850
0.83	0.2967	0.7967	0.2033	0.2827
0.84	0.2995	0.7995	0.2005	0.2803

(continued)

TABLE B (CONTINUED)
Areas and Ordinates of Standard Normal Curve

(1) Z	(2) A AREA FROM μ TO Z	(3) B AREA IN LARGER PORTION	(3) C AREA IN SMALLER PORTION	(5) y ORDINATE AT Z
0.85	0.3023	0.8023	0.1977	0.2780
0.86	0.3051	0.8051	0.1949	0.2756
0.87	0.3078	0.8078	0.1922	0.2732
0.88	0.3106	0.8106	0.1894	0.2709
0.89	0.3133	0.8133	0.1867	0.2685
0.90	0.3159	0.8159	0.1841	0.2661
0.91	0.3186	0.8186	0.1814	0.2637
0.92	0.3212	0.8212	0.1788	0.2613
0.93	0.3238	0.8238	0.1762	0.2589
0.94	0.3264	0.8264	0.1736	0.2565
0.95	0.3289	0.8289	0.1711	0.2541
0.96	0.3315	0.8315	0.1685	0.2516
0.97	0.3340	0.8340	0.1660	0.2492
0.98	0.3365	0.8365	0.1635	0.2468
0.99	0.3389	0.8389	0.1611	0.2444
1.00	0.3413	0.8413	0.1587	0.2420
1.01	0.3438	0.8438	0.1562	0.2396
1.02	0.3461	0.8461	0.1539	0.2371
1.03	0.3485	0.8485	0.1515	0.2347
1.04	0.3508	0.8508	0.1492	0.2323
1.05	0.3531	0.8531	0.1469	0.2299
1.06	0.3554	0.8554	0.1446	0.2275
1.07	0.3577	0.8577	0.1423	0.2251
1.08	0.3599	0.8599	0.1401	0.2227
1.09	0.3621	0.8621	0.1379	0.2203
1.10	0.3643	0.8643	0.1357	0.2179
1.11	0.3665	0.8665	0.1335	0.2155
1.12	0.3686	0.8686	0.1314	0.2131
1.13	0.3708	0.8708	0.1292	0.2107
1.14	0.3729	0.8729	0.1271	0.2083
1.15	0.3749	0.8749	0.1251	0.2059
1.16	0.3770	0.8770	0.1230	0.2036
1.17	0.3790	0.8790	0.1210	0.2012
1.18	0.3810	0.8810	0.1190	0.1989
1.19	0.3830	0.8830	0.1170	0.1965
1.20	0.3849	0.8849	0.1151	0.1942
1.21	0.3869	0.8869	0.1131	0.1919
1.22	0.3888	0.8888	0.1112	0.1895
1.23	0.3906	0.8906	0.1094	0.1872
1.24	0.3925	0.8925	0.1075	0.1849
1.25	0.3943	0.8943	0.1057	0.1826
1.26	0.3962	0.8962	0.1038	0.1804
1.27	0.3980	0.8980	0.1020	0.1781

TABLE B (CONTINUED)
Areas and Ordinates of Standard Normal Curve

(1) Z	(2) A AREA FROM μ TO Z	(3) B AREA IN LARGER PORTION	(3) C AREA IN SMALLER PORTION	(5) y ORDINATE AT Z
1.28	0.3997	0.8997	0.1003	0.1758
1.29	0.4015	0.9015	0.0985	0.1736
1.30	0.4032	0.9032	0.0968	0.1714
1.31	0.4049	0.9049	0.0951	0.1691
1.32	0.4066	0.9066	0.0934	0.1669
1.33	0.4082	0.9082	0.0918	0.1647
1.34	0.4099	0.9099	0.0901	0.1626
1.35	0.4115	0.9115	0.0885	0.1604
1.36	0.4131	0.9131	0.0869	0.1582
1.37	0.4147	0.9147	0.0853	0.1561
1.38	0.4162	0.9162	0.0838	0.1539
1.39	0.4177	0.9177	0.0823	0.1518
1.40	0.4192	0.9192	0.0808	0.1497
1.41	0.4207	0.9207	0.0793	0.1476
1.42	0.4222	0.9222	0.0778	0.1456
1.43	0.4236	0.9236	0.0764	0.1435
1.44	0.4251	0.9251	0.0749	0.1415
1.45	0.4265	0.9265	0.0735	0.1394
1.46	0.4279	0.9279	0.0721	0.1374
1.47	0.4292	0.9292	0.0708	0.1354
1.48	0.4306	0.9306	0.0694	0.1334
1.49	0.4319	0.9319	0.0681	0.1315
1.50	0.4332	0.9332	0.0668	0.1295
1.51	0.4345	0.9345	0.0655	0.1276
1.52	0.4357	0.9357	0.0643	0.1257
1.53	0.4370	0.9370	0.0630	0.1238
1.54	0.4382	0.9382	0.0618	0.1219
1.55	0.4394	0.9394	0.0606	0.1200
1.56	0.4406	0.9406	0.0594	0.1182
1.57	0.4418	0.9418	0.0582	0.1163
1.58	0.4429	0.9429	0.0571	0.1145
1.59	0.4441	0.9441	0.0559	0.1127
1.60	0.4452	0.9452	0.0548	0.1109
1.61	0.4463	0.9463	0.0537	0.1092
1.62	0.4474	0.9474	0.0526	0.1074
1.63	0.4484	0.9484	0.0516	0.1057
1.64	0.4495	0.9495	0.0505	0.1040
1.65	0.4505	0.9505	0.0495	0.1023
1.66	0.4515	0.9515	0.0485	0.1006
1.67	0.4525	0.9525	0.0475	0.0989
1.68	0.4535	0.9535	0.0465	0.0973
1.69	0.4545	0.9545	0.0455	0.0957

(continued)

TABLE B (CONTINUED)
Areas and Ordinates of Standard Normal Curve

(1) Z	(2) A AREA FROM μ TO Z	(3) B AREA IN LARGER PORTION	(3) C AREA IN SMALLER PORTION	(5) y ORDINATE AT Z
1.70	0.4554	0.9554	0.0446	0.0940
1.71	0.4564	0.9564	0.0436	0.0925
1.72	0.4573	0.9573	0.0427	0.0909
1.73	0.4582	0.9582	0.0418	0.0893
1.74	0.4591	0.9591	0.0409	0.0878
1.75	0.4599	0.9599	0.0401	0.0863
1.76	0.4608	0.9608	0.0392	0.0848
1.77	0.4616	0.9616	0.0384	0.0833
1.78	0.4625	0.9625	0.0375	0.0818
1.79	0.4633	0.9633	0.0367	0.0804
1.80	0.4641	0.9641	0.0359	0.0790
1.81	0.4648	0.9648	0.0352	0.0775
1.82	0.4656	0.9656	0.0344	0.0761
1.83	0.4664	0.9664	0.0336	0.0748
1.84	0.4671	0.9671	0.0329	0.0734
1.85	0.4678	0.9678	0.0322	0.0721
1.86	0.4686	0.9686	0.0314	0.0707
1.87	0.4693	0.9693	0.0307	0.0694
1.88	0.4699	0.9699	0.0301	0.0681
1.89	0.4706	0.9706	0.0294	0.0669
1.90	0.4713	0.9713	0.0287	0.0656
1.91	0.4719	0.9719	0.0281	0.0644
1.92	0.4726	0.9726	0.0274	0.0632
1.93	0.4732	0.9732	0.0268	0.0620
1.94	0.4738	0.9738	0.0262	0.0608
1.95	0.4744	0.9744	0.0256	0.0596
1.96	0.4750	0.9750	0.0250	0.0584
1.97	0.4756	0.9756	0.0244	0.0573
1.98	0.4761	0.9761	0.0239	0.0562
1.99	0.4767	0.9767	0.0233	0.0551
2.00	0.4772	0.9772	0.0228	0.0540
2.01	0.4778	0.9778	0.0222	0.0529
2.02	0.4783	0.9783	0.0217	0.0519
2.03	0.4788	0.9788	0.0212	0.0508
2.04	0.4793	0.9793	0.0207	0.0498
2.05	0.4798	0.9798	0.0202	0.0488
2.06	0.4803	0.9803	0.0197	0.0478
2.07	0.4808	0.9808	0.0192	0.0468
2.08	0.4812	0.9812	0.0188	0.0459
2.09	0.4817	0.9817	0.0183	0.0449
2.10	0.4821	0.9821	0.0179	0.0440
2.11	0.4826	0.9826	0.0174	0.0431
2.12	0.4830	0.9830	0.0170	0.0422

TABLE B (CONTINUED)
Areas and Ordinates of Standard Normal Curve

(1) Z	(2) A AREA FROM μ TO Z	(3) B AREA IN LARGER PORTION	(3) C AREA IN SMALLER PORTION	(5) y ORDINATE AT Z
2.13	0.4834	0.9834	0.0166	0.0413
2.14	0.4838	0.9838	0.0162	0.0404
2.15	0.4842	0.9842	0.0158	0.0396
2.16	0.4846	0.9846	0.0154	0.0387
2.17	0.4850	0.9850	0.0150	0.0379
2.18	0.4854	0.9854	0.0146	0.0371
2.19	0.4857	0.9857	0.0143	0.0363
2.20	0.4861	0.9861	0.0139	0.0355
2.21	0.4864	0.9864	0.0136	0.0347
2.22	0.4868	0.9868	0.0132	0.0339
2.23	0.4871	0.9871	0.0129	0.0332
2.24	0.4875	0.9875	0.0125	0.0325
2.25	0.4878	0.9878	0.0122	0.0317
2.26	0.4881	0.9881	0.0119	0.0310
2.27	0.4884	0.9884	0.0116	0.0303
2.28	0.4887	0.9887	0.0113	0.0297
2.29	0.4890	0.9890	0.0110	0.0290
2.30	0.4893	0.9893	0.0107	0.0283
2.31	0.4896	0.9896	0.0104	0.0277
2.32	0.4898	0.9898	0.0102	0.0270
2.33	0.4901	0.9901	0.0099	0.0264
2.34	0.4904	0.9904	0.0096	0.0258
2.35	0.4906	0.9906	0.0094	0.0252
2.36	0.4909	0.9909	0.0091	0.0246
2.37	0.4911	0.9911	0.0089	0.0241
2.38	0.4913	0.9913	0.0087	0.0235
2.39	0.4916	0.9916	0.0084	0.0229
2.40	0.4918	0.9918	0.0082	0.0224
2.41	0.4920	0.9920	0.0080	0.0219
2.42	0.4922	0.9922	0.0078	0.0213
2.43	0.4924	0.9924	0.0076	0.0208
2.44	0.4927	0.9927	0.0073	0.0203
2.45	0.4929	0.9929	0.0071	0.0198
2.46	0.4931	0.9931	0.0069	0.0194
2.47	0.4932	0.9932	0.0068	0.0189
2.48	0.4934	0.9934	0.0066	0.0184
2.49	0.4936	0.9936	0.0064	0.0180
2.50	0.4938	0.9938	0.0062	0.0175
2.51	0.4940	0.9940	0.0060	0.0171
2.52	0.4941	0.9941	0.0059	0.0167
2.53	0.4943	0.9943	0.0057	0.0163
2.54	0.4945	0.9945	0.0055	0.0158

(continued)

TABLE B (CONTINUED)
Areas and Ordinates of Standard Normal Curve

(1) Z	(2) A AREA FROM μ TO Z	(3) B AREA IN LARGER PORTION	(3) C AREA IN SMALLER PORTION	(5) y ORDINATE AT Z
2.55	0.4946	0.9946	0.0054	0.0154
2.56	0.4948	0.9948	0.0052	0.0151
2.57	0.4949	0.9949	0.0051	0.0147
2.58	0.4951	0.9951	0.0049	0.0143
2.59	0.4952	0.9952	0.0048	0.0139
2.60	0.4953	0.9953	0.0047	0.0136
2.61	0.4955	0.9955	0.0045	0.0132
2.62	0.4956	0.9956	0.0044	0.0129
2.63	0.4957	0.9957	0.0043	0.0126
2.64	0.4959	0.9959	0.0041	0.0122
2.65	0.4960	0.9960	0.0040	0.0119
2.66	0.4961	0.9961	0.0039	0.0116
2.67	0.4962	0.9962	0.0038	0.0113
2.68	0.4963	0.9963	0.0037	0.0110
2.69	0.4964	0.9964	0.0036	0.0107
2.70	0.4965	0.9965	0.0035	0.0104
2.71	0.4966	0.9966	0.0034	0.0101
2.72	0.4967	0.9967	0.0033	0.0099
2.73	0.4968	0.9968	0.0032	0.0096
2.74	0.4969	0.9969	0.0031	0.0093
2.75	0.4970	0.9970	0.0030	0.0091
2.76	0.4971	0.9971	0.0029	0.0088
2.77	0.4972	0.9972	0.0028	0.0086
2.78	0.4973	0.9973	0.0027	0.0084
2.79	0.4974	0.9974	0.0026	0.0081
2.80	0.4974	0.9974	0.0026	0.0079
2.81	0.4975	0.9975	0.0025	0.0077
2.82	0.4976	0.9976	0.0024	0.0075
2.83	0.4977	0.9977	0.0023	0.0073
2.84	0.4977	0.9977	0.0023	0.0071
2.85	0.4978	0.9978	0.0022	0.0069
2.86	0.4979	0.9979	0.0021	0.0067
2.87	0.4979	0.9979	0.0021	0.0065
2.88	0.4980	0.9980	0.0020	0.0063
2.89	0.4981	0.9981	0.0019	0.0061
2.90	0.4981	0.9981	0.0019	0.0060
2.91	0.4982	0.9982	0.0018	0.0058
2.92	0.4982	0.9982	0.0018	0.0056
2.93	0.4983	0.9983	0.0017	0.0055
2.94	0.4984	0.9984	0.0016	0.0053
2.95	0.4984	0.9984	0.0016	0.0051
2.96	0.4985	0.9985	0.0015	0.0050
2.97	0.4985	0.9985	0.0015	0.0048

TABLE B (CONTINUED)
Areas and Ordinates of Standard Normal Curve

(1) Z	(2) A AREA FROM μ TO Z	(3) B AREA IN LARGER PORTION	(3) C AREA IN SMALLER PORTION	(5) y ORDINATE AT Z
2.98	0.4986	0.9986	0.0014	0.0047
2.99	0.4986	0.9986	0.0014	0.0046
3.00	0.4986	0.9986	0.0014	0.0044
3.01	0.4987	0.9987	0.0013	0.0043
3.02	0.4987	0.9987	0.0013	0.0042
3.03	0.4988	0.9988	0.0012	0.0040
3.04	0.4988	0.9988	0.0012	0.0039
3.05	0.4989	0.9989	0.0011	0.0038
3.06	0.4989	0.9989	0.0011	0.0037
3.07	0.4989	0.9989	0.0011	0.0036
3.08	0.4990	0.9990	0.0010	0.0035
3.09	0.4990	0.9990	0.0010	0.0034
3.10	0.4990	0.9990	0.0010	0.0033
3.11	0.4991	0.9991	0.0009	0.0032
3.12	0.4991	0.9991	0.0009	0.0031
3.13	0.4991	0.9991	0.0009	0.0030
3.14	0.4992	0.9992	0.0008	0.0029
3.15	0.4992	0.9992	0.0008	0.0028
3.16	0.4992	0.9992	0.0008	0.0027
3.17	0.4992	0.9992	0.0008	0.0026
3.18	0.4993	0.9993	0.0007	0.0025
3.19	0.4993	0.9993	0.0007	0.0025
3.20	0.4993	0.9993	0.0007	0.0024
3.21	0.4993	0.9993	0.0007	0.0023
3.22	0.4994	0.9994	0.0006	0.0022
3.23	0.4994	0.9994	0.0006	0.0022
3.24	0.4994	0.9994	0.0006	0.0021
3.25	0.4994	0.9994	0.0006	0.0020
3.26	0.4994	0.9994	0.0006	0.0020
3.27	0.4995	0.9995	0.0005	0.0019
3.28	0.4995	0.9995	0.0005	0.0018
3.29	0.4995	0.9995	0.0005	0.0018
3.30	0.4995	0.9995	0.0005	0.0017
3.31	0.4995	0.9995	0.0005	0.0017
3.32	0.4995	0.9995	0.0005	0.0016
3.33	0.4996	0.9996	0.0004	0.0016
3.34	0.4996	0.9996	0.0004	0.0015
3.35	0.4996	0.9996	0.0004	0.0015
3.36	0.4996	0.9996	0.0004	0.0014
3.37	0.4996	0.9996	0.0004	0.0014
3.38	0.4996	0.9996	0.0004	0.0013
3.39	0.4997	0.9997	0.0003	0.0013

(continued)

TABLE B (CONTINUED)
Areas and Ordinates of Standard Normal Curve

(1) Z	(2) A AREA FROM μ TO Z	(3) B AREA IN LARGER PORTION	(3) C AREA IN SMALLER PORTION	(5) y ORDINATE AT Z
3.40	0.4997	0.9997	0.0003	0.0012
3.41	0.4997	0.9997	0.0003	0.0012
3.42	0.4997	0.9997	0.0003	0.0012
3.43	0.4997	0.9997	0.0003	0.0011
3.44	0.4997	0.9997	0.0003	0.0011
3.45	0.4997	0.9997	0.0003	0.0010
3.46	0.4997	0.9997	0.0003	0.0010
3.47	0.4997	0.9997	0.0003	0.0010
3.48	0.4997	0.9997	0.0003	0.0009
3.49	0.4998	0.9998	0.0002	0.0009
3.50	0.4998	0.9998	0.0002	0.0008
3.51	0.4998	0.9998	0.0002	0.0008
3.52	0.4998	0.9998	0.0002	0.0008
3.53	0.4998	0.9998	0.0002	0.0008
3.54	0.4998	0.9998	0.0002	0.0008
3.55	0.4998	0.9998	0.0002	0.0007
3.56	0.4998	0.9998	0.0002	0.0007
3.57	0.4998	0.9998	0.0002	0.0007
3.58	0.4998	0.9998	0.0002	0.0007
3.59	0.4998	0.9998	0.0002	0.0006
3.60	0.4998	0.9998	0.0002	0.0006
3.61	0.4998	0.9998	0.0002	0.0006
3.62	0.4999	0.9999	0.0001	0.0006
3.63	0.4999	0.9999	0.0001	0.0005
3.64	0.4999	0.9999	0.0001	0.0005
3.65	0.4999	0.9999	0.0001	0.0005
3.66	0.4999	0.9999	0.0001	0.0005
3.67	0.4999	0.9999	0.0001	0.0005
3.68	0.4999	0.9999	0.0001	0.0005
3.69	0.4999	0.9999	0.0001	0.0004
3.70	0.4999	0.9999	0.0001	0.0004

Note: Table generated using Turbo Pascal software.

TABLE C
Distribution of *t* for Given Probability Levels

df	Level of Significance for One-Tailed Test					
	.10	.05	.025	.01	.005	.0005
	Level of Significance for Two-Tailed Test					
	.20	.10	.05	.02	.01	.001
1	3.078	6.314	12.706	31.821	63.657	636.619
2	1.886	2.920	4.303	6.965	9.925	31.598
3	1.638	2.353	3.182	4.541	5.841	12.941
4	1.533	2.132	2.776	3.747	4.604	8.610
5	1.476	2.015	2.571	3.365	4.032	6.859
6	1.440	1.943	2.447	3.143	3.707	5.959
7	1.415	1.895	2.365	2.998	3.499	5.405
8	1.397	1.860	2.306	2.896	3.355	2.041
9	1.383	1.833	2.262	2.821	3.250	4.781
10	1.372	1.812	2.228	2.764	3.169	4.587
11	1.363	1.796	2.201	2.718	3.106	4.437
12	1.356	1.782	2.179	2.681	3.055	4.314
13	1.350	1.771	2.160	2.650	3.012	4.221
14	1.345	1.761	2.145	2.642	2.977	4.140
15	1.341	1.753	2.131	2.602	2.947	4.073
16	1.337	1.746	2.120	2.583	2.921	4.015
17	1.333	1.740	2.110	2.567	2.898	3.965
18	1.330	1.734	2.101	2.552	2.878	3.992
19	1.328	1.729	2.093	2.539	2.861	3.883
20	1.325	1.725	2.086	2.528	2.845	3.850
21	1.323	1.721	2.080	2.518	2.831	3.819
22	1.321	1.717	2.074	2.508	2.879	3.792
23	1.319	1.714	2.069	2.500	2.807	3.767
24	1.318	1.711	2.064	2.492	2.797	3.745
25	1.316	1.708	2.060	2.485	2.787	3.725
26	1.315	1.706	2.056	2.479	2.779	3.707
27	1.314	1.703	2.052	2.473	2.771	3.690
28	1.313	1.701	2.048	2.467	2.763	3.674
29	1.311	1.699	2.045	2.462	2.750	3.659
30	1.310	1.697	2.042	2.457	2.750	3.646
40	1.303	1.684	2.021	2.423	2.704	3.551
60	1.296	1.671	2.000	2.390	2.660	3.460
120	1.289	1.658	1.980	2.358	2.167	3.373
∞	1.282	1.645	1.960	2.326	2.576	3.291

Source: Abridged from Table III of R. A. Fisher and F. Yates, *Statistical Tables for Biological, Agricultural, and Medical Research.* Edinburgh: Oliver and Boyd, Ltd. With permission from authors and publishers.

Note: Reject H_0 if $t_{observed} > t_{table}$, also called $t_{critical}$

TABLE D
Critical Values of U in Mann–Whitney Test

n_1	9	10	11	12	13	14	n_2 15	16	17	18	19	20

(a) Critical Values of U for One-Tailed Test at $\alpha = 0.001$ or Two-Tailed Test at $\alpha = 0.002$

n_1	9	10	11	12	13	14	n_2 15	16	17	18	19	20
1												
2												
3									0	0	0	0
4		0	0	0	1	1	1	2	2	3	3	3
5	1	1	2	2	3	3	4	5	5	6	7	7
6	2	3	4	4	5	6	7	8	9	10	11	12
7	3	5	6	7	8	9	10	11	13	14	15	16
8	5	6	8	9	11	12	14	15	17	18	20	21
9	7	8	10	12	14	15	17	19	21	23	25	26
10	8	10	12	14	17	19	21	23	25	27	29	32
11	10	12	15	17	20	22	24	27	29	32	34	37
12	12	14	17	20	23	25	28	31	34	37	40	42
13	14	17	20	23	26	29	32	35	38	42	45	48
14	15	19	22	25	29	32	36	39	43	46	50	54
15	17	21	24	28	32	36	40	43	47	51	55	59
16	19	23	27	31	35	39	43	48	52	56	60	65
17	21	25	29	34	38	43	47	52	57	60	66	70
18	23	27	32	37	42	36	51	56	61	66	71	76
19	25	29	34	40	45	50	55	60	66	71	77	82
20	26	32	37	42	48	54	59	65	70	76	82	88

n_1	9	10	11	12	13	14	n_2 15	16	17	18	19	20

(b) Critical Values of U for One-Tailed Test at $\alpha = 0.01$ or Two-Tailed Test at $\alpha = 0.02$

n_1	9	10	11	12	13	14	n_2 15	16	17	18	19	20
1												
2					0	0	0	0	0	0	1	1
3	1	1	1	2	2	2	3	3	4	4	4	5
4	3	3	4	5	5	6	7	7	8	9	9	10
5	5	6	7	8	9	10	11	12	13	14	15	16
6	7	8	9	11	12	13	15	16	18	19	20	22
7	9	11	12	14	16	17	19	21	23	24	26	28
8	11	13	15	17	20	22	24	26	28	30	32	34
9	14	16	18	21	23	26	28	31	33	36	38	40
10	16	19	22	24	27	30	33	36	38	41	44	47
11	18	22	25	28	31	34	37	41	44	47	50	53
12	21	24	28	31	35	38	42	46	49	53	56	60
13	23	27	31	35	39	43	47	51	55	59	63	67
14	26	30	34	38	43	47	51	56	60	65	69	73
15	28	33	37	42	47	50	56	61	66	70	72	80
16	31	36	41	46	51	56	61	66	71	76	82	87
17	33	28	44	49	55	61	66	71	77	82	88	93
18	36	41	47	53	59	66	70	76	82	88	94	100
19	38	44	50	56	63	69	75	82	88	94	101	107
20	40	47	53	60	67	73	80	87	93	100	107	114

TABLE D
Critical Values of U in Mann–Whitney Test

n_1	9	10	11	12	13	14	n_2 15	16	17	18	19	20
(c) Critical Values of U for One-Tailed Test at $\alpha = 0.025$ or Two-Tailed Test at $\alpha = 0.05$											
1												
2	0	0	0	1	1	1	1	1	2	2	2	2
3	2	3	3	4	4	5	5	6	6	7	7	8
4	4	5	6	7	8	9	10	11	11	12	13	13
5	7	8	9	11	12	13	14	15	17	18	19	20
6	10	11	13	14	16	17	19	21	22	24	25	27
7	12	14	16	18	20	22	24	26	28	30	32	34
8	15	17	19	22	24	26	29	31	34	36	38	41
9	17	20	23	26	28	31	34	37	39	42	45	48
10	20	23	26	29	33	36	39	42	45	48	52	55
11	23	26	30	33	37	40	44	47	51	55	58	62
12	26	29	33	37	41	45	49	53	57	61	65	69
13	28	33	37	41	45	50	54	59	63	67	72	76
14	31	36	40	45	50	55	59	64	67	74	78	83
15	34	39	44	49	54	59	64	70	75	80	85	90
16	37	42	47	53	59	64	70	75	81	86	92	98
17	39	45	51	57	63	67	75	81	87	93	99	105
18	42	48	55	61	67	74	80	86	93	99	106	112
19	45	52	58	65	72	78	85	92	99	106	113	119
20	48	55	62	69	76	83	90	98	105	112	119	127

n_1	9	10	11	12	13	14	n_2 15	16	17	18	19	20
(d) Critical Values of U for One-Tailed Test at $\alpha = 0.05$ or Two-Tailed Test at $\alpha = 0.10$											
1											0	0
2	1	1	1	2	2	2	3	3	3	4	4	4
3	3	4	5	5	6	7	7	8	9	9	10	11
4	6	7	8	9	10	11	12	14	15	16	17	18
5	9	11	12	13	15	16	18	19	20	22	23	25
6	12	14	16	17	19	21	23	25	26	28	30	32
7	15	17	19	21	24	26	28	30	33	35	37	39
8	18	20	23	26	28	31	33	36	39	41	44	47
9	21	24	27	30	33	36	39	42	45	48	51	54
10	24	27	31	34	37	41	44	48	51	55	58	62
11	27	31	34	38	42	46	50	54	57	61	65	69
12	30	34	38	42	47	51	55	60	64	68	72	77
13	33	37	42	47	51	56	61	65	70	75	80	84
14	36	41	46	51	56	61	66	71	77	82	87	92
15	39	44	50	55	61	66	72	77	83	88	94	100
16	42	48	54	60	65	71	77	83	89	95	101	107
17	45	51	57	64	70	77	83	89	96	102	109	115
18	48	55	61	68	75	82	88	95	102	109	113	123
19	51	58	65	72	80	87	94	101	109	116	123	130
20	54	62	69	77	84	92	100	107	115	123	130	138

Source: Adapted and abridged from Tables 1, 3, 5, and 7 of D. Aube, Extended tables for the Mann–Whitney statistic. *Bulletin of the Institute of Educational Research at Indiana University*, 1953, 1, No. 2. Reproduced from S. Siegal. *Nonparametric Statistics for the Behavioral Sciences.* New York: McGraw-Hill, 1956. Reprinted by permission of the author, Institute of Educational Research, and McGraw-Hill Book Company.

Note: Reject H_0 if $U_{observed} > U_{table}$.

TABLE E
Critical Values of _T_ (Wilcoxon Matched-Pairs Signed-Ranks Test)

	Level of Significance for One-Tailed Test		
	.025	.01	.005
	Level of Significance for Two-Tailed Test		
N	.05	.02	.01
6	1	—	—
7	2	0	—
8	4	2	0
9	6	3	2
10	8	5	3
11	11	7	5
12	14	10	7
13	17	13	10
14	21	16	13
15	25	20	16
16	30	24	19
17	35	28	23
18	40	33	28
19	46	38	32
20	52	43	37
21	59	49	43
22	66	56	49
23	73	62	55
24	81	69	61
25	90	77	68
26	98	85	76
27	107	93	84
28	117	102	92
29	127	111	100
30	137	120	109

Source: Adapted from Table 2 of F. Wilcoxon and R. A. Wilcox. _Some Rapid Approximate Statistical Procedures._ New York: American Cyanamid Company, 1964, p. 28. Reprinted by permission of Lederle Laboratories, a division of American Cyanamid Company.

Note: Reject H_0 if $T_{observed} \geq T_{table}$.

TABLE F
Distribution of χ^2

df	Levels of Significance				
	.10	.05	.02	.01	.001
1	2.706	3.841	5.412	6.635	10.827
2	4.605	5.991	7.824	9.210	13.815
3	6.251	7.815	9.837	11.345	16.268
4	7.779	9.488	11.668	13.277	18.465
5	9.236	11.070	13.338	15.086	20.517
6	10.645	12.592	15.033	16.812	22.457
7	12.017	14.067	16.622	18.475	24.322
8	13.362	15.507	18.168	20.090	26.125
9	14.684	16.919	19.679	21.666	27.877
10	15.987	18.307	21.161	23.209	29.588
11	17.275	19.675	22.618	24.725	31.264
12	18.549	21.026	24.054	26.217	32.909
13	19.812	22.362	25.472	27.688	34.528
14	21.064	23.685	26.873	29.141	36.123
15	22.307	24.996	28.259	30.578	37.697
16	23.542	26.296	29.633	32.000	39.252
17	24.769	27.587	30.995	33.409	40.790
18	25.989	28.869	32.346	34.805	42.312
19	27.204	30.144	33.687	36.191	43.820
20	28.412	31.410	35.020	37.566	45.315
21	29.615	32.671	36.343	38.932	46.797
22	30.813	33.924	37.659	40.289	48.268
23	31.007	35.172	38.968	41.638	49.728
24	33.196	36.415	40.270	42.980	51.179
25	34.382	37.652	41.566	44.314	52.620
26	35.563	38.885	42.856	45.642	54.052
27	36.741	40.113	44.140	46.963	55.472
28	37.916	41.337	45.419	48.278	56.893
29	39.087	42.557	46.693	49.588	58.302
30	40.256	43.773	47.962	50.892	59.703

Source: From Table IV of R. A. Fisher and F. Yates. *Statistical Tables for Biological, Agricultural, and Medical Research.* London: Longman Group Ltd. (previously published by Oliver and Boyd, Ltd., Edinburgh). Reprinted with permission of authors and publishers.

Note: Reject H_0 if $\chi^2_{observed} > \chi^2_{table}$.

TABLE G
Values of *r* for Different Levels of Significance

df	.1	.05	.02	.01	.001
1	.98769	.99692	.999507	.999877	.9999988
2	.90000	.95000	.98000	.990000	.99900
3	.8054	.8783	.93433	.95873	.99116
4	.7293	.8114	.8822	.91720	.97406
5	.6694	.7545	.8329	.8745	.95074
6	.6215	.7067	.7887	.8343	.92493
7	.5822	.6664	.7498	.7977	.8982
8	.5494	.6319	.7155	.7646	.8721
9	.5214	.6021	.6851	.7348	.8471
10	.4973	.5760	.6581	.7079	.8233
11	.4762	.5529	.6339	.6835	.8010
12	.4575	.5324	.6120	.6614	.7800
13	.4409	.5139	.5923	.6411	.7603
14	.4259	.4973	.5742	.6226	.7420
15	.4124	.4821	.5577	.6055	.7246
16	.4000	.4683	.5425	.5897	.7084
17	.3887	.4555	.5285	.5751	.6932
18	.3783	.4438	.5155	.5614	.6787
19	.3687	.4329	.5034	.5487	.6652
20	.3598	.4227	.4921	.5368	.6524
25	.3233	.3809	.4451	.4869	.5974
30	.2960	.3494	.4093	.4487	.5541
35	.2746	.3246	.3810	.4182	.5189
40	.2573	.3044	.3578	.3932	.4896
45	.2428	.2875	.3384	.3721	.4648
50	.2306	.2732	.3218	.3541	.4433
60	.2103	.2500	.2964	.3248	.4078
70	.1954	.2319	.2737	.3017	.3799
80	.1829	.2172	.2565	.2830	.3568
90	.1726	.2050	.2422	.2673	.3375
100	.1638	.1946	.2301	.2540	.3211

Source: Reprinted from Table VI of R. A. Fisher and F. Yates. *Statistical Tables for Biological, Agricultural, and Medical Research.* London: Longman Group Ltd. (previously by Oliver and Boyd, Ltd., Edinburgh). Reprinted with permission of authors and publishers.

Note: Based on two-tailed tests. Reject H_0 if $r_{observed} > r_{table}$.

TABLE H
Critical Values of Spearman Correlation Coefficient

N	Significance Level (One-Tailed Test)	
	.05	**.01**
4	1.000	
5	.900	1.000
6	.829	.943
7	.714	.893
8	.643	.833
9	.600	.783
10	.564	.746
12	.506	.712
14	.456	.645
16	.425	.601
18	.399	.564
20	.377	.534
22	.359	.508
24	.343	.485
26	.329	.465
28	.317	.448
30	.306	.432

Source: Adapted from E. G. Olds. (1949). Distributions of sums of squares of rank differences for small numbers of individuals, *Annals of Mathematical Statistics, 20,* 117–118. Reprinted with permission of author and publisher.

Note: Reject H_0 if $rho_{observed} \geq rho_{table}$.

TABLE I
F Distributions

α = 0.01 Level

df_1 df_2	1	2	3	4	5	6	7	8	9	10	12	15	20	24	30	40	60	120	∞
1	4052	4999	5403	5625	5764	5859	5028	5982	6022	6056	6106	6157	6209	6235	6261	6287	6313	6339	6366
2	98.5	99.00	99.17	99.25	99.30	99.33	99.36	99.37	99.39	99.40	99.42	99.43	99.45	99.46	99.47	99.47	99.48	99.49	99.50
3	34.12	30.82	29.46	28.71	28.24	27.91	27.67	27.49	27.35	27.23	27.05	26.87	26.69	26.60	26.50	26.41	26.32	26.22	26.13
4	21.20	18.00	16.69	15.98	15.52	15.21	14.98	14.80	14.66	14.55	14.37	14.20	14.02	13.93	13.84	13.75	13.65	13.56	13.14
5	16.26	13.27	12.06	11.39	10.97	10.67	10.46	10.29	10.16	10.05	9.89	9.72	9.55	9.47	9.38	9.29	9.20	9.11	9.02
6	13.75	10.92	9.78	9.15	8.75	8.47	8.26	8.10	7.98	7.87	7.72	7.56	7.40	7.31	7.23	7.14	7.06	6.97	6.88
7	12.25	9.55	8.45	7.85	7.46	7.19	6.99	6.81	6.72	6.62	6.47	6.31	6.16	6.07	5.99	5.91	5.82	5.74	5.65
8	11.26	8.65	7.50	7.01	6.63	6.37	6.18	6.03	5.91	5.81	5.67	5.52	5.36	5.28	5.20	5.12	5.03	4.95	4.86
9	10.56	8.02	6.99	6.42	6.06	5.80	5.61	5.47	5.35	5.26	5.11	4.96	4.81	4.73	4.65	4.57	4.48	4.40	4.31
10	10.04	7.56	6.55	5.99	5.64	5.39	5.20	2.06	4.94	4.85	4.71	4.56	4.41	4.33	4.25	4.17	4.08	4.00	3.91
11	9.65	7.21	6.22	5.67	5.32	5.07	4.89	4.74	4.63	4.54	4.40	4.25	4.10	4.02	3.95	3.86	3.78	3.69	3.60
12	9.33	6.93	5.95	5.41	5.06	4.82	4.61	4.50	4.39	4.30	4.16	4.01	3.86	3.78	3.70	3.62	3.54	3.45	3.36
13	9.07	6.70	5.74	5.21	4.86	4.62	4.44	5.30	4.19	4.10	3.96	3.82	3.66	3.59	3.51	3.43	3.31	3.25	3.17
14	8.86	6.51	5.56	5.04	4.69	4.46	4.28	4.14	4.03	3.94	3.80	3.66	3.51	3.43	3.35	3.27	3.18	3.09	3.00
15	8.68	6.36	5.42	4.89	4.56	4.32	4.14	4.00	3.89	3.80	3.67	3.52	3.37	3.29	3.21	3.13	3.05	2.96	2.87
16	8.53	6.23	5.29	4.77	4.44	4.20	4.03	3.89	3.78	3.69	3.55	3.41	3.26	3.18	3.10	3.02	2.93	2.84	2.75
17	8.40	6.11	5.19	4.67	4.34	4.10	3.93	3.79	3.68	3.59	3.46	3.31	3.16	3.08	3.00	2.92	2.83	2.75	2.65
18	8.20	6.01	5.09	4.58	4.25	4.01	3.84	3.71	3.60	3.51	3.37	3.23	3.08	3.00	2.92	2.81	2.75	2.66	2.57
19	8.18	5.93	5.01	4.50	4.17	3.94	3.77	3.63	3.52	3.43	3.30	3.15	3.00	2.92	2.84	2.76	2.67	2.58	2.49
20	8.10	5.85	4.94	4.43	4.10	3.87	3.70	3.56	3.46	3.37	3.23	3.09	2.94	2.86	2.78	2.69	2.61	2.52	2.42
21	8.02	5.78	4.87	4.37	4.04	3.81	3.64	3.51	3.40	3.31	3.17	3.03	2.88	2.80	2.72	2.64	2.55	2.46	2.36
22	7.95	5.72	4.82	4.31	3.99	3.76	3.59	3.45	3.35	3.26	3.12	2.98	2.83	2.75	2.67	2.58	2.50	2.40	2.31
23	7.88	5.66	4.76	4.26	3.94	3.71	3.54	3.41	3.30	3.21	3.07	2.93	2.78	2.70	2.62	2.54	2.45	2.35	2.26
24	7.82	5.61	4.72	4.22	3.90	3.67	3.50	3.36	3.26	3.17	3.03	2.89	2.74	2.66	2.58	2.49	2.40	2.31	2.21

(continuation of α = 0.01 Level table)

df_2 \ df_1	1	2	3	4	5	6	7	8	9	10	12	15	20	24	30	40	60	120	∞
25	7.77	5.57	4.68	4.18	3.85	3.63	3.46	3.32	3.22	3.13	2.99	2.85	2.70	2.62	2.54	2.45	2.36	2.27	2.17
26	7.72	5.53	4.64	4.14	3.82	3.59	3.42	3.29	3.18	3.09	2.96	2.81	2.66	2.58	2.50	2.42	2.33	2.23	2.13
27	7.68	5.49	4.60	4.11	3.78	3.56	3.39	3.26	3.15	3.06	2.93	2.78	2.63	2.55	2.47	2.38	2.29	2.20	2.10
28	7.62	5.45	4.57	4.07	3.75	3.53	3.36	3.23	3.12	3.03	2.90	2.75	2.60	2.52	2.44	2.35	2.26	2.17	2.06
29	7.60	5.42	4.54	4.04	3.73	3.50	3.33	3.20	3.09	3.00	2.87	2.73	2.57	2.49	2.41	2.33	2.23	2.14	2.03
30	7.56	5.39	4.51	4.02	3.70	3.47	3.30	3.17	3.07	2.98	2.84	2.70	2.55	2.47	2.39	2.30	2.21	2.11	2.01
40	7.31	5.19	4.31	3.83	3.51	3.29	3.12	2.99	2.89	2.80	2.66	2.52	2.37	2.29	2.20	2.14	2.02	1.92	1.80
60	7.08	4.98	4.13	3.65	3.34	3.12	2.95	2.82	2.72	2.63	2.50	2.35	2.20	2.12	2.03	1.94	1.81	1.73	1.60
120	6.85	4.79	3.95	3.48	3.17	2.96	2.79	2.66	2.56	2.47	2.34	2.19	2.03	1.95	1.86	1.76	1.66	1.53	1.38
∞	6.63	4.61	3.78	3.32	3.02	2.90	2.64	2.51	2.41	2.32	2.18	2.04	1.88	1.79	1.70	1.59	1.47	1.32	1.00

Source: Abridged from Table 18 of E. S. Pearson and H. O. Hartley. Eds. *Biometrika Tables for Statisticians*, Vol. 1. Reproduced with permission of E. S. Pearson and the trustees of Biometrika.

Notes: df_1 is from numerator F-ratio. df_2 is from denominator of F-ratio. Reject H_0 if $F_{observed} > F_{table}$.

α = 0.05 Level

df_2 \ df_1	1	2	3	4	5	6	7	8	9	10	12	15	20	24	30	40	60	120	∞
1	161.4	199.5	215.7	224.6	230.2	234.0	236.8	238.9	240.5	241.9	243.9	245.9	248.0	249.1	250.1	251.1	252.2	253.3	254.3
2	18.51	19.00	19.16	19.25	19.30	19.33	19.35	19.37	19.38	19.40	19.41	19.43	19.45	19.45	19.46	19.47	19.48	19.49	19.50
3	10.13	9.55	9.28	9.12	9.01	8.94	8.89	8.85	8.81	8.79	8.74	8.70	8.66	8.64	8.62	8.59	8.57	8.55	8.53
4	7.71	6.94	6.59	6.39	6.26	6.16	6.09	6.04	6.00	5.96	5.91	5.86	5.80	5.77	5.75	5.72	5.69	5.66	5.63
5	6.61	5.79	5.41	5.19	5.05	4.95	4.88	4.82	4.77	4.74	4.68	4.62	4.56	4.53	4.50	4.46	4.43	4.40	4.36
6	5.99	5.14	4.76	4.53	4.39	4.28	4.21	4.15	4.10	4.06	4.00	3.94	3.87	3.84	3.81	3.77	3.74	3.70	3.67
7	5.59	4.74	4.35	4.12	3.97	3.87	3.79	3.73	3.68	3.64	3.57	3.51	3.44	3.41	3.38	3.34	3.30	3.27	3.23
8	5.32	4.46	4.07	3.84	3.69	3.58	3.50	3.44	3.39	3.35	3.28	3.22	3.16	3.12	3.08	3.04	3.01	2.97	2.93
9	5.12	4.26	3.80	3.63	3.48	3.37	3.29	3.23	3.18	3.14	3.07	3.01	2.94	2.90	2.86	2.83	2.79	2.75	2.71
10	4.96	4.10	3.71	3.46	3.33	3.22	3.14	3.07	3.02	2.98	2.91	2.85	2.77	2.74	2.70	2.66	2.62	2.58	2.54
11	4.84	3.96	3.59	3.36	3.20	3.09	3.01	2.95	2.90	2.85	2.79	2.72	2.65	2.61	2.57	2.53	2.40	2.45	2.40
12	4.75	3.89	3.49	3.20	3.11	3.00	2.91	2.85	2.80	2.75	2.69	2.62	2.54	2.51	2.47	2.43	2.38	2.34	2.30
13	4.67	3.81	3.41	3.18	3.03	2.92	2.83	2.77	2.71	2.67	2.60	2.53	2.46	2.42	2.38	2.34	2.30	2.25	2.21

(continued)

TABLE I (CONTINUED)
F Distributions

α = 0.05 Level

df_2 \ df_1	1	2	3	4	5	6	7	8	9	10	12	15	20	24	30	40	60	120	∞
14	4.60	3.74	3.34	3.11	2.96	2.85	2.76	2.70	2.65	2.60	2.53	2.46	2.39	2.35	2.31	2.27	2.22	2.18	2.13
15	4.54	3.68	3.29	3.06	2.90	2.79	2.71	2.64	2.59	2.54	2.48	2.40	2.33	2.29	2.25	2.20	2.16	2.11	2.07
16	4.49	3.63	3.24	3.01	2.85	2.74	2.66	2.59	2.54	2.49	2.42	2.35	2.28	2.24	2.19	2.15	2.11	2.06	2.01
17	4.45	3.59	3.26	2.96	2.81	2.70	2.61	2.55	2.49	2.45	2.38	2.31	2.23	2.19	2.15	2.10	2.06	2.01	1.96
18	4.41	3.55	3.16	2.93	2.77	2.66	2.58	2.51	2.46	2.41	2.34	2.27	2.19	2.15	2.11	2.06	2.02	1.97	1.92
19	4.38	3.52	3.13	2.90	2.74	2.63	2.54	2.48	2.42	2.38	2.31	2.23	2.16	2.11	2.07	2.03	1.98	1.93	1.88
20	4.35	3.49	3.10	2.87	2.71	2.60	2.51	2.45	2.39	2.35	2.28	2.20	2.12	2.08	2.04	1.99	1.95	1.90	1.84
21	4.32	3.47	3.07	2.84	2.68	2.57	2.49	2.42	2.37	2.32	2.25	2.18	2.10	2.05	2.01	1.96	1.92	1.87	1.81
22	4.30	3.44	3.05	2.82	2.66	2.55	2.46	2.40	2.34	2.30	2.23	2.15	2.07	2.03	1.98	1.94	1.89	1.84	1.78
23	4.28	3.42	3.03	2.80	2.64	2.53	2.44	2.37	2.32	2.27	2.20	2.13	2.05	2.01	1.96	1.91	1.86	1.81	1.76
24	4.26	3.40	3.01	2.78	2.62	2.51	2.42	2.36	2.30	2.25	2.18	2.11	2.03	1.98	1.94	1.89	1.84	1.79	1.73
25	4.24	3.39	2.99	2.76	2.60	2.49	2.40	2.34	2.28	2.24	2.16	2.09	2.01	1.96	1.92	1.87	1.82	1.77	1.71
26	4.23	3.37	2.98	2.74	2.59	2.47	2.39	2.32	2.27	2.22	2.15	2.07	1.99	1.95	1.90	1.85	1.80	1.75	1.69
27	4.21	3.35	2.96	2.73	2.57	2.46	2.37	2.31	2.25	2.20	2.13	2.06	1.97	1.93	1.88	1.84	1.79	1.73	1.67
28	4.20	3.34	2.95	2.71	2.56	2.45	2.36	2.29	2.24	2.19	2.12	2.04	1.96	1.91	1.87	1.82	1.77	1.71	1.65
29	4.18	3.33	2.93	2.70	2.55	2.43	2.35	2.28	2.22	2.18	2.10	2.03	1.94	1.90	1.85	1.81	1.75	1.70	1.64
30	4.17	3.32	2.92	2.69	2.53	2.42	2.33	2.27	2.21	2.16	2.09	2.01	1.93	1.89	1.84	1.79	1.74	1.68	1.62
40	4.08	3.23	2.81	2.61	2.45	2.34	2.25	2.18	2.12	2.08	2.00	1.92	1.84	1.79	1.74	1.69	1.64	1.58	1.51
60	4.00	3.15	2.76	2.53	2.37	2.25	2.17	2.10	2.04	1.99	1.92	1.84	1.75	1.70	1.65	1.59	1.53	1.47	1.39
120	3.92	3.07	2.68	2.45	2.29	2.17	2.09	2.02	1.96	1.91	1.83	1.75	1.66	1.61	1.55	1.50	1.43	1.35	1.25
∞	3.84	3.00	2.60	2.37	2.21	2.10	2.01	1.94	1.88	1.83	1.75	1.67	1.57	1.52	1.46	1.39	1.32	1.22	1.00

Source: Abridged from Table 18 of E. S. Pearson and H. O. Hartley, Eds. *Biometrika Tables for Statisticians*, Vol. 1. Reproduced with permission of E. S. Pearson and the trustees of Biometrika.

Notes: df_1 is from numerator F-ratio. df_2 is from denominator of F-ratio. Reject H_0 if $F_{observed} > F_{table}$.

TABLE J
95th Percentile Studentized Range, q : Distribution by Number of Groups and Denominator df

q Table — 95% Percentiles

df									Number of Group Means Being Compared										
	2	3	4	5	6	7	8	9	10	11	12	13	14	15	16	17	18	19	20
1	17.970	26.980	32.820	37.080	40.410	43.120	45.400	47.360	49.070	50.590	51.960	53.200	54.330	55.360	56.320	57.220	58.040	58.830	59.56
2	6.085	8.331	9.798	10.880	11.740	12.440	13.030	13.540	13.990	14.390	14.750	15.080	15.380	15.650	15.910	16.140	16.370	16.570	16.77
3	4.501	5.910	6.825	7.502	8.037	8.478	8.853	9.177	9.462	9.717	9.964	10.150	10.350	10.530	10.690	10.840	10.980	11.110	11.24
4	3.927	5.040	5.757	6.287	6.707	7.053	7.347	7.602	7.826	8.027	8.208	8.373	8.525	8.664	8.794	8.914	9.028	9.134	9.23
5	3.635	4.602	5.218	5.673	6.033	6.330	6.582	6.800	6.995	7.168	7.324	7.466	7.596	7.717	7.828	7.932	8.030	8.122	8.21
6	3.461	4.339	4.896	5.305	5.628	5.895	6.122	6.319	6.493	6.649	6.789	6.917	7.034	7.143	7.244	7.338	7.426	7.508	7.59
7	3.344	4.165	4.681	5.060	5.359	5.606	5.815	5.998	6.158	6.302	6.431	6.550	6.658	6.759	6.852	6.939	7.020	7.097	7.17
8	3.261	4.041	4.529	4.886	5.167	5.399	5.597	5.767	5.918	6.054	6.175	6.287	6.389	6.483	6.571	6.653	6.729	6.802	6.87
9	3.199	3.949	4.415	4.756	5.024	5.244	5.432	5.595	5.739	5.867	5.983	6.089	6.186	6.276	6.359	6.437	6.510	6.579	6.64
10	3.151	3.877	4.327	4.654	4.912	5.124	5.305	5.461	5.599	5.722	5.833	5.935	6.028	6.114	6.194	6.269	6.339	6.405	6.47
11	3.113	3.820	4.256	4.574	4.823	5.028	5.202	5.353	5.487	5.605	5.713	5.811	5.901	5.984	6.062	6.134	6.202	6.265	6.33
12	3.082	3.773	4.199	4.508	4.751	4.950	5.119	5.265	5.395	5.511	5.615	5.710	5.798	5.878	5.953	6.023	6.089	6.151	6.21
13	3.055	3.735	4.151	4.453	4.690	4.885	5.049	5.192	5.308	5.431	5.533	5.625	5.711	5.789	5.862	5.931	5.995	6.055	6.11
14	3.033	3.702	4.111	4.407	4.639	4.829	4.990	5.131	5.254	5.364	5.463	5.554	5.637	5.714	5.786	5.852	5.915	5.974	6.03
15	3.014	3.674	4.076	4.367	4.595	4.782	4.940	5.077	5.198	5.306	5.404	5.493	5.574	5.649	5.720	5.785	5.846	5.904	5.96
16	2.998	3.649	4.046	4.333	4.557	4.741	4.897	5.031	5.150	5.256	5.352	5.439	5.520	5.593	5.662	5.727	5.786	5.843	5.90
17	2.984	3.628	4.020	4.303	4.524	4.705	4.858	4.991	5.108	5.212	5.307	5.392	5.471	5.544	5.612	5.675	5.734	5.790	5.84
18	2.971	3.609	3.997	4.277	4.495	4.673	4.824	4.956	5.071	5.174	5.267	5.352	5.429	5.501	5.568	5.630	5.688	5.743	5.79
19	2.960	3.593	3.977	4.253	4.469	4.645	4.794	4.924	5.038	5.140	5.231	5.305	5.391	5.462	5.528	5.589	5.647	5.701	5.73
20	2.950	3.578	3.958	4.232	4.445	4.620	4.768	4.896	5.008	5.108	5.199	5.282	5.357	5.427	5.493	5.533	5.610	5.663	5.71
24	2.919	3.532	3.901	4.166	4.373	4.541	4.684	4.807	4.915	5.012	5.099	5.179	5.251	5.319	5.381	5.439	5.494	5.545	5.59
30	2.888	3.486	3.845	4.102	4.302	4.464	4.602	4.720	4.824	4.917	5.001	5.077	5.147	5.211	5.271	5.327	5.379	5.429	5.48
40	2.858	3.442	3.791	4.039	4.232	4.389	4.521	4.635	4.735	4.824	4.904	4.977	5.044	5.106	5.163	5.216	5.266	5.313	5.36
60	2.829	3.399	3.737	3.977	4.163	4.314	4.441	4.550	4.646	4.732	4.808	4.878	4.942	5.001	5.056	5.107	5.154	5.199	5.24
120	2.800	3.356	3.685	3.917	4.096	4.241	4.363	4.468	4.560	4.641	4.714	4.781	4.842	4.898	4.950	4.98	5.044	5.086	5.13
240	2.772	3.314	3.633	3.858	4.030	4.170	4.286	4.387	4.474	4.552	4.622	4.685	4.743	4.796	4.845	4.891	4.934	4.974	5.01

Denominator df

Note: Select the column that describes the number of means being compared. Select the row that represents the *df* from the denominator of the *F*-ratio. If the actual *df* falls between two rows on this table, use the upper row, i.e., the smaller of the two *df*.

TABLE K
Table of Random Numbers

Column

Row	1	2	3	4	5	6	7	8	9	10	11	12	13	14	15	16	17	18	19	20	21	22	23	24	25	26	27	28	29	30	31	32	33	34	35
1	2	0	6	5	9	0	1	8	7	6	0	4	7	6	8	2	1	0	9	5	3	1	6	7	9	0	5	6	8	2	1	0	8	3	3
2	7	6	8	0	4	1	8	7	1	4	3	5	8	0	7	6	7	1	5	1	7	3	2	5	3	9	4	5	8	7	9	5	3	4	5
3	2	2	8	4	2	8	8	6	1	2	3	9	4	6	2	4	6	8	9	2	2	5	5	8	8	0	2	2	1	5	8	8	4	6	2
4	4	5	8	1	0	0	7	4	0	9	4	3	7	4	9	8	8	9	9	2	0	2	9	4	4	5	7	2	6	5	4	4	8	4	5
5	5	8	7	8	3	8	2	8	6	7	3	7	2	0	7	7	3	2	1	7	9	6	2	2	3	3	4	6	3	8	4	8	4	8	0
6	4	0	6	3	1	7	1	5	6	9	0	6	1	1	6	5	4	6	6	3	8	3	3	3	9	7	6	5	7	3	0	7	3	6	3
7	5	4	7	5	1	8	0	7	4	3	5	4	0	5	2	8	0	3	1	6	6	2	5	8	3	9	3	3	4	1	3	9	5	9	7
8	5	3	4	1	6	1	5	5	2	6	2	0	5	0	5	3	5	2	6	5	5	3	4	4	5	5	5	5	0	2	7	1	1	4	4
9	7	1	1	0	3	9	3	4	3	9	9	4	9	7	9	9	3	3	3	9	9	6	4	7	3	0	1	8	7	9	7	5	6	4	2
10	0	9	7	8	0	8	1	0	8	1	7	3	1	3	8	8	2	3	2	1	1	3	3	9	4	2	9	6	1	0	1	8	9	6	1
11	0	4	1	0	5	7	7	0	5	6	5	5	2	9	2	4	6	9	0	1	3	5	5	2	3	6	5	4	9	2	8	9	3	6	4
12	2	6	8	3	2	5	4	1	2	3	7	0	1	1	3	1	9	5	5	8	0	8	7	3	7	9	2	9	3	0	0	6	5	9	3
13	7	7	2	8	6	6	4	5	3	5	9	4	7	8	6	6	6	0	8	8	3	8	4	4	5	6	8	3	5	4	3	5	1	4	5
14	9	7	5	5	9	4	9	8	8	6	2	3	0	2	2	2	2	2	7	6	7	2	3	3	9	0	7	0	0	7	7	0	8	9	9
15	7	1	5	9	9	8	9	1	6	5	6	7	6	9	3	4	8	8	6	4	4	6	8	1	6	4	5	6	1	5	5	4	1	6	0
16	0	1	8	4	4	2	1	1	6	4	9	9	0	9	2	6	5	9	2	2	5	2	4	0	3	6	5	8	2	9	8	0	5	2	3
17	3	7	1	0	0	9	1	9	5	9	3	2	0	7	7	3	5	6	8	4	8	7	0	3	7	0	3	0	7	6	0	6	3	1	9
18	3	4	6	8	6	2	1	9	6	6	4	0	0	3	5	9	5	7	0	0	1	0	3	6	0	8	5	3	0	1	3	5	5	7	4
19	6	7	6	5	3	8	7	7	0	3	9	9	3	5	5	6	6	4	1	6	4	4	0	4	7	4	6	7	6	8	7	4	7	1	9
20	7	8	6	6	1	2	0	0	5	7	1	1	9	3	2	4	3	2	5	6	8	8	4	4	4	4	3	8	4	4	1	7	8	7	9
21	4	4	3	2	2	3	1	7	2	6	4	5	5	4	1	2	2	6	3	2	5	5	6	7	9	6	1	9	2	6	3	4	6	9	7
22	2	8	4	2	4	5	0	6	8	9	3	4	3	5	9	6	7	9	1	6	9	9	9	5	2	9	2	0	7	2	5	2	7	9	6
23	2	0	7	3	2	7	1	2	1	6	1	9	5	9	8	2	0	6	3	5	2	3	3	2	8	6	8	4	8	8	7	8	7	6	6
24	5	3	7	8	5	6	2	1	5	8	8	7	5	5	5	4	5	5	7	3	0	9	4	0	4	0	9	5	0	5	8	0	0	1	1
25	4	4	7	0	1	4	2	8	5	5	3	0	8	2	2	0	4	4	5	4	4	6	9	0	8	9	1	1	8	8	1	9	9	8	9

Note: Table generated using Turbo Pascal software.

Appendix B: Glossary of Statistical Terms, Equations, and Symbols

Abscissa (x-Axis): Horizontal axis of coordinate system. The variable in question (e.g., performance measure) is typically placed on the abscissa in a frequency distribution; the frequency of occurrence is plotted on the ordinate (y-axis).

Abstract Section: An accurate, concise summary of the contents of the paper, which appears fully indented at the beginning of the paper and normally has no heading.

Acceptance Region: Portion of a sampling distribution that would lead to acceptance of the null hypothesis if it contained the sample result.

Acceptance Sampling: Statistical quality control technique used to determine whether a completed lot meets specifications.

Action Limit: Value of boundary between the acceptance and rejection regions stated in the units of the original variables used in testing.

Addition Rule: $\Pr(A \text{ or } B) = \Pr(A) + \Pr(B)$; shows the probability for two or more sets of outcomes connected by *or*.

Additivity: Relationship in which responses are determined by the sum of the effects of the independent variables. In ANOVA, nonadditivity is usually an interaction of two or more IVs.

Alpha (α) Error (Type I): Probability of rejecting null hypothesis when it should not be rejected; when one should fail to reject it.

Alpha (α) Level: Probability of Type I error; incorrect rejection of null hypothesis; probability of obtaining an outcome that supports rejection of the null hypothesis when such rejection is unwarranted. See *Type I Error*.

Alpha (α) Risk: Probability assigned to Type I error in a testing problem.

Alternative Hypothesis (H_1): Opposite of null hypothesis. When testing the hypothesis of difference, the alternative hypothesis states that the differences between means or populations are not the result of chance alone (i.e., population differences exist). When testing the hypothesis of association, the alternative hypothesis states that a correlation exists in the population. See *Hypothesis*.

Analysis: Use of statistical techniques to describe a set of statistical data and make inferences therefrom.

Analysis of Covariance (ANCOVA): Inferential statistical procedure for testing the null hypothesis that the means of three or more populations are equal. Random samples taken from various populations are first adjusted to eliminate differences between their average values on one or more independent variables (called covariables). The covariance (cov_{xy}) calculation is

$$cov_{xy} = \frac{\Sigma XY - \frac{\Sigma X \Sigma Y}{N}}{N-1} = \frac{SP_{xy}}{N-1}$$

Analysis of Variance (ANOVA): Analysis of ratio of variability between the means to variability within treatment groups to determine whether a significant (nonchance) difference exists among several sample means; developed by Sir Ronald Fisher; also called *F*-ratio.

SS_{total}: $$\Sigma X^2 - \frac{(\Sigma X)^2}{N}$$

SS_{group} (one-way): $$\frac{\Sigma T_g^2}{n} - \frac{(\Sigma X)^2}{N}$$

SS_{error} (one-way): $$SS_{total} - SS_{group}$$

SS_{rows} (two-way): $$\frac{\Sigma T_{R_i}^2}{nc} - \frac{(\Sigma X)^2}{N}$$

SS_{col} (two-way): $$\frac{\Sigma T_{C_i}^2}{nr} - \frac{(\Sigma X)^2}{N}$$

SS_{cells} (two-way): $$\frac{\Sigma T_{cell}^2}{n} - \frac{(\Sigma X)^2}{N}$$

SS_{RxC} (two-way): $$SS_{cells} - SS_{rows} - SS_{col}$$

SS_{error} (two-way): $$SS_{total} - SS_{rows} - SS_{col} - SS_{RxC}$$

$$SS_{total} - SS_{cells}$$

Protected t (use only if F is significant): $$t = \frac{\bar{X}_i + \bar{X}_j}{\sqrt{\frac{MS_{error}}{n_i} + \frac{MS_{error}}{n_i}}}$$

Eta (η) squared: $$\eta^2 = \frac{SS_{group}}{SS_{total}}$$

Omega (ω) squared (one-way): $$\omega^2 = \frac{SS_{group} - (k-1)MS_{error}}{SS_{total} + MS_{error}}$$

A Posteriori (**Posterior**) **Probability:** Form of conditional probability determined by use of Bayes' theorem.

Apparatus: Used to aid the experimenter to accurately manipulate the IV and to precisely measure the DM.

Approaches to Science: Five basic approaches to scientific research: correlational, case history, field study, experimental, and quasi-experimental. The primary approach used by scientists is the experimental method because it has the advantage of determining causal relationships.

A Priori (**Prior**) **Probability:** Probability based on advance knowledge of the behavior of the process in question.

Archive: Extensive record or collection of data by an institution or society; facility for long-term storage of information.

Area Classification: Qualitative classification using the geographical origin of the data as the basis for classification.

Area in Smaller Portion: Probability of obtaining a sample mean. This is far from the population mean μ or further.

Arithmetic Mean: Calculated average; value of the sum of numerical values of a series of items or numbers divided by the number of items or numbers.

Array: Tabular presentation of quantitative data; data are arranged in order of magnitude.

Asymmetric Transfer Effect: Experimental design error caused by greater influence on behavior of having Condition A before B than the reverse. An unbalanced order or sequence effect in which the experience on the first condition that influences subsequent conditions is not the same across conditions.

Asymptotic: When the tails of the normal curve theoretically never quite touch the base line but extend infinitely in either direction.

Average: Typical value summarizing an array of data leading to the location of central tendency; value that purports to represent or summarize relevant features of a set of values that may include the median or mode; usually refers to arithmetic mean. See *Mean*.

Average Deviation: See *Mean Deviation*.

Average Squared Deviation: The average of the absolute deviation scores; also called the *mean deviation*.

Balancing: The use of a control group in order to balance out extraneous variable effects.

Bar Chart: Form of graph used for presenting data.

Bar Graph (Histogram): Data illustration format; rectangular columns (bars) represent values of the X or independent variable. The height of each bar indicates the amount of the Y or dependent variable associated with each value of X. Two or three kinds of information can be compared by groups of shaded or colored bars that aid identification. Figures involving increases and decreases can be shown by bars in opposite directions above and below a zero line. Bar graphs typically are used to present nominal or ordinal data or discrete variables. The heights of a series of bars (separated by gaps) reflect the frequencies for the various words or classes. Used with qualitative data.

Baseline: Typical level of a behavior that serves as a standard element of comparison in an experiment assessing treatment effects.

Bayes' Theorem: Theorem used to compute posterior probabilities, generally as a revision of prior probabilities in light of experimental evidence.

Bayesian Decision Theory: Modern method of statistical decision making using Bayes' theorem to revise probabilities.

Bell-Shaped Curve: Frequency curve characterized by a single peak and trailing off in both directions from the peak.

Beta Coefficient (β) or Slope: In the $y = a + bx$ equation that represents a straight line, the b indicates the slope of the line. Slope is a measure of the degree of change in the Y variable with each one-unit change in the X variable. A positive slope indicates that Y increases along with X; a negative slope means that Y decreases as X increases.

Beta (β) Error (Type II): Probability of retaining the null hypothesis when it should have been rejected.

Beta (β) Risk: Probability of making Type II error during hypothesis testing.

Between Groups: Completely independent selection of one group is not related to selection of the other groups; each subject receives only one treatment condition.

Between Groups ANOVA (*F*-Test): Procedure for testing the null hypothesis that two or more populations have equal variances; used in analysis of variance, analysis of covariance, and other applications to determine whether differences obtained are significant.

Bias: Nonrandom sampling error arising from selection of a sample that is not representative of the population. In biased samples, the difference between mean ($\bar{X}$) and mu (μ) is consistently in one direction.

Bimodal: Frequency distribution having two modes or peaks.

Binomial Distribution: Distribution over n independent trials of the probability of the number of occurrences of an event that may or may not occur with a constant probability on a single trial.

Biserial Correlation Coefficient (r_b): Correlation coefficient describing the relationship between two variables, one of which is continuous and the other dichotomous (usually a forced dichotomy).

Biserial *r*: Used when one variable is continuous, and the other is a forced dichotomy.

Bivariate: Frequency distribution describing simultaneously the distribution of scores on two variables.

Bivariate Frequency: Describes simultaneously the distribution of scores on two variables.

Bivariate Statistics: Descriptive and inferential statistical procedures designed for use when analysis involves two dependent measures of each subject or one measure on a matched pair of subjects.

Blind Procedure: Procedure designed to ensure that participants (subjects or experiments) are not aware of the true nature of the tests administered. It is designed to mitigate the effects of participant biases on test results. In a single-blind procedure, the subjects remain uninformed about the purpose of the experiment and the treatments received. In a double-blind procedure, neither the subjects nor the experimenters know the assignment of subjects to treatments or treatments to subjects.

Calculated Average (Mean): Average defined by an arithmetic operation.

Caption: Column heading of a table.

Carryover Effect: Temporary or permanent change in a participant's behavior; treatment in a prior test may have affected his or her behavior in a later test. (e.g., practice, fatigue, and drug use).

Case History Approach: To assist an individual in understanding and solving personal problems. It can also produce insight into possible causal relationships, which can be further investigated through experimentation.

Case Study: Intensive study of a single individual, event, or behavior. Data are collected from records, observations, and interviews for a descriptive study, not a cause-and-effect study.

Caterus Parabus: State, condition, or environment in which all variables are controlled except the one manipulated by the investigator.

Census: Complete enumeration of a statistical universe (population).

Centile: A point in an ordered set of scores in a distribution divided into 100 parts expressing the percentage of scores below it; a value expressing the percentile rank among all scores in the distribution.

Central Limit Theorem: Theorem or statistical theory stating that if relatively large (~30) randomly selected samples are taken from a population, the sampling distribution of the mean will continue to approach a normal distribution in shape even if the population distribution deviates from normality as the size of the random sample continues to increase.

Central Tendency: Statistic that describes the tendency of numerical data to fall about a middle value between extremes of a set of measures; measure of central tendency include the arithmetic mean, the median, and the mode.

Chain Index: Index number with a single fixed period as a base.

Chart: See *Graph*.

Chi-Square (χ^2): Statistic similar to F and t used to test hypotheses of differences usually on variances or about central tendencies. Although formerly used as a parametric statistic, recent use is usually as a nonparametric statistical test of significance used to determine whether differences occur by chance. This test of the null hypothesis is for categorical data, qualitatively expressed as frequencies. Chi-square data must be in nominal form or the number of cases (frequency of occurrence) must fall into two or more discrete categories; no population assumptions are required for its use. As a parametric test, the basic equation is:

$$\chi^2 = \frac{(n-1)\hat{\sigma}_j^2}{\sigma_0^2}$$

where σ_0^2 = the variance value under the null hypothesis. As a nonparametric test, the basic equation is

$$\chi^2 = \Sigma \frac{(O - E)^2}{E}$$

Chi-square: $\chi^2 = \Sigma \dfrac{(O - E)^2}{E}$

Chi-square for 2×2 table only: $\chi^2 = \dfrac{N(AD - BC)^2}{(A + B)(C + D)(A + C)(B + D)}$

Chi-Square Distribution: Continuous probability distribution used in single-sample variance problems and goodness of fit tests.

Chronological Classification: Classification of statistical data over time according to a set of consecutive historical dates usually called a time series.

Class Interval: Grouping used in construction of frequency tables.

Classification: Placing of statistical data into groups; preliminary step in analyzing statistical data.

Cluster Sample: Variation of multistage sampling; all items in the final stage are used.

Clustering Point: See *Average*.

Coefficient Correlation: Square root of coefficient of determination; alternative measure of the degree of correlation.

Coefficient of Contingency: Test requiring only nominal data sorted into any number of independent cells to determine tendency for certain characteristics in one distribution to be associated with characteristics in another distribution.

Coefficient of Determination (r^2): Coefficient indicating proportion of information about Y contained in X; calculated by squaring the Pearson r.

Coefficient of Nondetermination: Represents that portion of the variance that cannot be determined (predicted) by the other variable.

Coefficient of Regression: Slope coefficient of regression line; summary measure of pattern of a relationship.

Coefficient of Variation: Ratio of standard deviation to mean; relates the standard deviation to the absolute size of the objects measured and allows comparisons of different scales or units because the standard deviation is expressed as a percentage of the mean.

Cohort: Group of people (society) that undergoes a set of experiences at the same time.

Collection: Obtaining numerical data for a statistical investigation (study).

Column Diagram: Form of graph for presenting a frequency table.

Combination: Method of counting the number of possible outcomes in situations where order is immaterial and repetition of elements is not permitted.

Compensatory Equalization: Untreated (control) subjects obtain something positive from a study.

Compensatory Rivalry: A control (untreated) group works hard to ensure that the facilitative effect of the treatment is not demonstrated.

Competency-Based Evaluation: A desired level of competency is identified, and test subjects designated as competent must be able to meet or exceed this level.

Composite Index Number: Combining information from a number of related time series into a single series that attempts to show the general movement of the entire group.

Concurrent Validity: Validity measured by correlation of a new measurement instrument and an existing validated measure of the same concept; reflects degree to which a predictor test correlates with a criterion test. See *Validity*.

Conditional Probability: Probability whose value is dependent on the previous occurrence of another event.

Confidence Interval: Statistical procedure for making universal estimates from sample data.

Confidence Interval on μ: $\qquad CI = \bar{X} \pm t_{.025}(\sigma_{\bar{X}})$

Confidence Interval on $\mu_1 - \mu_2$: $\quad CI = (\bar{X} - \bar{X}) \pm t_{.025}(\sigma_{\bar{X}_1 - \bar{X}_2})$

Confirmatory Experiments: Experiments conducted to confirm or refute experimental hypotheses.

Confounding Variable: Variable that changes systematically with one or more independent variables, thus degrading internal validity of an experiment and clouding the cause–effect relationships of the independent and dependent variables. Confounding variables make it difficult or impossible to determine effects of treatments on dependent measures. When a second variable varies simultaneously with the treatment, whether a change noted in a dependent measure was produced by a manipulated variable (IV), a potential confounding variable, or both variables is unclear.

Conservatism: Analysis and interpretation which is cautious; bias toward maintaining status quo.

Constant: Property, characteristic, or numerical value that is the same over all members of a sample or population, at least within the context of a given problem.

Constant Seasonal: Set of seasonal indices that remain the same over a period of years.

Construct: Unobservable concept used to label a consistent set of observable variables; cannot be measured directly; inferred from observable events; examples are motivation, social class, and intelligence.

Construct Validity: Degree to which a test measures an attribute or trait of a psychological concept but is not proven conclusively; determined by complex inferential and hypothetical procedures. See *Validity*.

Content Validity: Determines whether adequate coverage and representation are contained in a test, for example spelling and history tests from which items are taken for analysis; degree to which a sample of test items, tasks, or questions are representative of a defined universe or domain of content. See *Validity*.

Contiguity of Events: The assumption that observed events are temporally or spatially contiguous (adjacent).

Contingency Coefficient: Method of determining validity of a mastery test.

Contingency Table: Data table showing levels of two or more variables by placing the numbers of observations in cells. The numbers take into consideration the intersections of the different levels of variables. For example, using a sample of adults by age and socioeconomic class ranking (high, middle, and low income), the table would have two rows (young, old) and three columns (income levels). Each cell would indicate the number of adults of that age and income.

Continuous Probability: Probability in which calculus is used to measure the occurrences of events.

Continuous Scale: Data that can be described in varying degrees of fineness are called continuous.

Continuous Variable: Variable having values at all points along its possible range of values; determined by measurement.

Contrasting Groups Method: Determination of valid cut-off points by comparing students expected to master an objective with those not expected to master it.

Control: Method of checking validity of an observation that holds irrelevant factors constant. It produces comparisons and allows observation of cause-and-effect relationships.

Control Chart: Chart showing measurements of some quality characteristic for consecutive samples over time; used in process control.

Control Group: Comparison group; group that receives zero magnitude of the independent variable. Control groups are critical for evaluating the pure effects of an independent variable on the measured responses of subjects.

Controlled Observation: To control or otherwise discount the effects of all other variables that might affect Y in order to make the descriptive statement that "a change in variable X produces a change in variable Y."

Convenience Sample: Sample selected amount items that are most readily available.

Convergent Validity: Relationship among two or more measures of the same theoretical construct. See *Validity.*

Correction for Guessing: Alteration of test scores based on number of items answered incorrectly.

Correlated Data: Data associated with one another; occurs in repeated measures (or within-subject) designs.

Correlated *F*-Ratio: Statistical test of the hypothesis of differences among several sample means where interval or ratio data are correlated.

Correlated Sample: One of two or more samples that are not selected independently. In matched group experiments, the selection of one sample determines how the other samples will be selected.

Correlated *t* Ratio (Dependent *t*): Statistical test of the hypothesis of difference between two sample means where interval or ratio data are correlated.

Correlation: Measurement of the extent or degree to which two or more variables are interrelated; indicates a relationship but not necessarily a causal one; is positive or negative, depending on whether one variable changes in the same way as another or in an opposite way. See *Partial Correlation, Point Biserial Correlation,* and *Squared Correlation Coefficient.*

Sum of Squares (SS_x):

$$SS_x = \Sigma X^2 - \frac{(\Sigma X)^2}{N}$$

Sum of Products (SP_{xy}):

$$SP_{xy} = \Sigma XY - \frac{\Sigma X \Sigma Y}{N}$$

Covariance (cov_{xy}):

$$cov_{xy} = \frac{\Sigma XY - \frac{\Sigma X \Sigma Y}{N}}{N-1} = \frac{SP_{xy}}{N-1}$$

Correlation (Pearson *r*):

$$r = \frac{cov_{xy}}{\sigma_x \sigma_y} = \frac{SP_{xy}}{\sqrt{SS_x SS_y}}$$

$$= \frac{N\Sigma XY - \Sigma X \Sigma Y}{(N\Sigma X^2 - (\Sigma X)^2)(N\Sigma Y^2 - (\Sigma Y)^2)}$$

Correlation Analysis: Statistical technique for studying the relative closeness of the relationship among several quantitatively classified variables.

Correlation Approach: To establish relationships for such purposes as prediction and subgrouping of homogeneous elements.

Correlation Coefficient: Number between –1 and +1 that indicates degree of relationship between two variables. When high scores on the first variable are associated with high scores on the second and low scores on the first variable are associated with low scores on the second, the correlation of the two variables is positive (+1 indicates a perfect positive relationship). Conversely, two variables are negatively correlated when low scores on one are associated with high scores on the other (–1 indicates a perfect negative correlation); 0 indicates that the two variables are not related. The correlation is usually computed for a bivariate sample. Squaring a correlation coefficient determines the proportion of total variability in one variable associated with variability in the other. See *Biserial Correlation Coefficient*, *Coefficient of Variation*, *Multiple Correlation*, *Pearson Product–Moment Correlation Coefficient*, and *Spearman Correlation Coefficient*.

Correlation Ratio η: Used to determine the maximum possible relationship regardless of the shape of the bivariate distribution.

Correlation Matrix: Table that shows computation of the correlations between all possible pairs of variables; usually three or more variables are the headings for both the rows and the columns; the corresponding cells show the correlations.

Cost of Uncertainty: Expected long-run opportunity loss from the best decision.

Counterbalancing: Systematic technique for varying the order of conditions to distribute various carryover effects so they are not confounded with different treatment conditions; balances the order in which treatments are experienced.

Covariance: See *Analysis of Covariance*.

Cramer's V: Convenient way to describe the apparent strength of an association (correlation) on nominal data in a sample of more than four independent cells.

Criterion: Standard for determining validity of an observation or test; set of scores or other measure of performance; standard of behavior.

Criterion Test: Highly valid test of an attribute to be measured; cannot be used in a practical setting because of complexity, expense, and need for specially trained testers; standard against which a practical test is compared.

Critical Distance: Minimum difference between two numbers required for significance.

Critical Ratio: Ratio in a testing procedure at the boundary between the acceptance and rejection regions.

Critical Region: Region of rejection; range of values of test statistic that leads to rejection of the null hypothesis. If a test statistic lies within the critical region, the null hypothesis is rejected. If the statistic lies outside the critical region, the null hypothesis cannot be rejected.

Critical z Score (z_{crit}): A z score that separates probable from improbable outcomes and dictates whether the null hypothesis should be retained or rejected.

Cross-Lagged Panel Technique: Statistical procedure that assists in drawing casual inferences from time-lagged correlation data; assumes that certain coefficients of correlation of two variables are available more than once.

Cross-Sectional Research: Nonexperimental research for examining selected characteristics of subjects of different ages, populations, and intelligence in a single test. For example, a researcher selects a sample (cross-section) at one age level (20 years old) and compares their data with data from a sample of older subjects (65 years old). Such comparisons may be misleading because modern 20-year-olds may have very different environmental backgrounds, societal pressures, and educational levels in comparison with 65-year-old subjects. This type of research contrasts with longitudinal studies.

Cumulative Frequency (c_f): Number of scores that fall at or below a particular interval or category; column in frequency distribution table obtained by adding frequencies from the bottom to the top of the column; the number at the top of the column (N) equals the number of scores; the cumulative frequency represents the number of subjects who score at or below a given interval.

Cumulative Frequency Distribution: Distribution of scores showing total frequency at or below each particular interval or category; obtained by summing frequencies of the columns starting with the lowest score and moving up through each successive score to the highest score in the distribution.

Cumulative Percent: Column in frequency distribution table obtained by dividing the cumulative frequency by N and multiplying by 100; represents percentage of subjects scoring at or below a given interval.

Curvilinear Relationship: Relationship between two variables; when plotted, it approximates a curved line instead of a straight line. If one variable increases and the amount of change of the other differs as opposed to both variables increasing at a constant rate, the relationship is curvilinear (nonlinear).

Cycle: Repetitive wave-like movement in a time series with a duration exceeding 1 year; generally 3 to 5 years in length.

Cyclical Relative: Measure of cyclical variation in a time series determined by dividing actual data by measure of trend.

Data: Plural form of *datum*. Raw materials of statistics; information that can be collected, stored, analyzed, evaluated, and interpreted. Data may be a series of numbers that provide numerical values for a characteristic of a sample or a population; may be obtained by experiment or collecting scores from a measured behavior.

Deciles: Result of dividing a distribution into tenths. The first decile corresponds to the tenth percentile; the fifth represents the fiftieth percentile (median).

Decision: Choice among alternative courses of action.

Decision Making Under Certainty: Decision situation in which knowledge of the states of the world exists.

Decision Making Under Uncertainty: Decision situation in which the states of the world are not known.

Decision Rule: Specification of conditions under which the null hypothesis will or will not be rejected. Criteria for choosing decision rules are the extent to which expected loss is minimized and the extent to which subjective loss is minimized.

Decision Validity: Accuracy of classification of masters and nonmasters.

Deduction: Reasoning or logical thinking that proceeds from the general to the specific.

Degrees of Freedom (*df*): Number of independent observations on which a statistic is based, less the number of restrictions placed on the freedom of those observations to vary. In inferential statistics, the larger the size of the sample, the larger the number of degrees of freedom. For interval data, *df* represents the number of values (or scores) that are free to vary. For example, with six observations and a restriction that the sum of these scores must equal a specified value, there would be five degrees of freedom. Stated another way, five of these scores are free to be any value, whereas the sixth score is fixed (is not free to vary). For nominal data, *df* represents the number of frequency values that are free to vary after the sum of the frequencies from all cells has been determined. For this category of data, degrees of freedom depend not on sample size, but on the number of categories in which the observations are allocated.

Demand Characteristic: Hint or cue available to a subject in a research situation that leads him or her to determine the purpose of an experiment and influence his or her perception of what is expected. The particular demand characteristic can be a source of experimental bias if the subject behaves in accordance with the perceived purpose of the experiment.

Dependent Event: The probability that a second event will occur is conditional on the occurrence of a first event.

Dependent (Paired) Sample: The choice of an item for one sample automatically includes a corresponding or related item for the other sample.

Dependent Variable or Measure (DM): Variable to be measured; aspect of behavior that depends on changes arising from manipulation of one or more independent variables. In the social sciences, the DM is typically a response measure; in experimental research, it is the effect half of the cause-and-effect relationship; in correlational research, the DM is the measure predicted.

Descriptive Statistics: Method of summarizing, tabulating, organizing, condensing, or graphing to describe large amounts of data; describes a measured sample of objects or individuals; includes measures of central tendency, variability, relationship, skewness, kurtosis, and standard scores.

Design: The general plan of an experiment, including the number and arrangement of independent variables and the way subjects are selected and assigned to conditions.

Determinism: The philosophy that every event has a cause.

Development Trend: Progressive and consistent change in behavior resulting from aging.

Deviation Measure of Variability: Technique that measures variability as the average amount by which the values of a set differ from their own average.

Deviation Score (x): Difference between a single score and the mean of the distribution: $x = X - \bar{X}$.

Dichotomous: Dividable or divided into parts.

Differential Attrition: Differences in the number of subjects who drop out of a study, presumably due to treatment received.

Differential Maturation: Changes in differences between groups over time because groups mature at different rates.

Diffusion: Transfer of treatment effects from experimental (treated) to control (untreated) groups.

Direction: Correlations may also vary in direction—they may be positive or negative. For positive correlations, the higher a score is on one variable, the higher the score is likely to be on the second variable. Conversely, for negative correlations, the higher a score is on one variable, the lower the score is on the second variable.

Directional Test: Statistical test specifying location of critical region for the test statistic. In a one-tailed test, the critical region is at only one end of the distribution; in a two-tailed test, the critical region is located in both tails of the distribution. See *One-Tailed Test* and *Two-Tailed Test*.

Discrete: Data that can only be expressed as whole units.

Discrete Probability: Probability in which the occurrences of events are measured by counting.

Discrete Variable: Variable taking on values only at certain points along its possible range of values; usually obtained by counting.

Discriminant Validity: Correlation among two or more measurable constructs. If the measure to be validated has little or no correlation with other measures, it has discriminant validity (measuring a different construct). See *Validity*.

Discussion Section: Evaluation and interpretation of the implications of the results, particularly with respect to your original hypotheses.

Dispersion (Variability): Frequency curve property concerned with horizontal spread of a set of data; measured as an interval.

Distance Measure of Variability: Measure of variability that uses the difference (interval) between two points.

Distribution: Set of measured scores arranged by magnitude; the advantage of listing scores this way instead of using unordered raw scores is that trends can be noticed more readily. See *Frequency Distribution*.

Domain-Referenced Validity: Extent to which tasks sampled by a test adequately represent the total domain of tasks.

Double-Blind Study: Method for eliminating experimental error. The researchers and subjects do not know which group is experimental and which group serves as a control. The procedure prevents unconscious bias of researchers and contaminating motivational sets on the parts of the subjects.

Ecological Validity: Extent to which research can match all environmental contexts of interest in an investigation. See *Validity*.

Editing: Careful check of returned questionnaires or schedules for obvious errors.

Effect: Difference between true and hypothesized population means; any difference between two (or more) population means.

Empirical: Derived from experimentation or direct observation.

Empirical Approach: Direct observation and experience are the only firm bases for obtaining knowledge; based on experiments, surveys, and facts.

Empirical Confirmation: Explains the descriptive statement from which it was derived, while all other statements that can be derived from it are also found to be true.

Empirical Probability: Probability based on experimental evidence.

Empirical Verification: States that a descriptive statement can be regarded as true only if it is found to correspond with reality.

Error Variance: Variability due to uncontrolled factors despite the use of appropriate controls.

Estimation: Inferential statistics method concerned with finding procedures for point estimates or interval estimates for parameters such as the mean of a population. In point estimation, the value of a sample statistic is used as a guess of the value of a corresponding population parameter. In interval estimation, sample statistics are used to form a confidence interval.

Estimator: Formula used to estimate a population parameter from a sample; numerical value obtained is the estimate. The formula used to compute a sample mean is an example of an estimator.

Eta (η): Nonlinear correlation test. To the extent that eta exceeds the Pearson r, assumptions of linearity and homoscedasticity are not met. The equation is

$$\eta^2 = \frac{SS_{group}}{SS_{total}}$$

Exclusion Area: Extreme area under the normal curve. Because of curve symmetry, the extreme areas at both the top and bottom are excluded by two z-scores that are equidistant from the mean.

Expected Frequency (f_e): Frequency with which a specific type of response will occur according to some theoretical or empirical basis (relevant to the Chi-square test). The theoretical frequency for each category of the qualitative variable expected when the null hypothesis is true.

Expected Monetary Value (EMV): Anticipated long-run average gain of selecting a certain course of action (making a particular decision).

Expected Net Gain From Sampling (ENGS): Difference between expected value of a specific sample and its cost.

Expected Opportunity Loss (EOL): Anticipated long-run average opportunity loss from making a particular decision.

Expected Value of Certainty (EVC): Expected value of always making a correct decision.

Expected Value of Perfect Information (EVPI): Difference between expected value of certainty and expected monetary value of optimal decision; equal to cost of uncertainty; the maximum management should be willing to pay for additional information before making a decision.

Expected Value of Sample Information (EVSI): Expected reduction in cost of uncertainty that a particular sample will produce.

Experimental Approach: The primary method for determining causal relationships. It can control the variables and induce stimulus conditions at will, enabling the experimenter to efficiently and clearly determine the effect of independent variables and to replicate experiments to determine validity.

Experimental Design: Plan of an experiment (numbers and levels of independent variables) and other factors. Design dictates how subjects are selected and assigned to conditions and may include provisions for controlling for confounding variables (e.g., using a Latin square design reduces the probability of order effects). Designs may be one-way, factorial, or nested, with respect to the number and arrangement of IVs. Classifications for selection and assignment of subjects to treatments include between-groups, within-subjects (repeated measures), or mixed designs.

Experimental Error: Differences due to the effect of extraneous variables.

Experimental Group: The primary method for determining causal relationships. It can control the variables and induce stimulus conditions at will, enabling the experimenter to efficiently and clearly determine the effect of independent variables and to replicate experiments to determine validity.

Experimental Hypothesis: Also called the research hypothesis or working hypothesis; statement that describes the predicted relationship between levels of an independent variable and a dependent measure; same as alternative hypothesis H_1.

Experimental Research: Research conducted via experimental method involving manipulation of one or more independent variables while measuring resulting changes in the dependent variables; using experimental method allows scientists to discern cause-and-effect relationships.

Experimenter Bias: Extent to which an experimenter allows his or her ideas or expectations about the results of an experiment to interfere with his or her objectivity when collecting, analyzing, and interpreting experimental data.

Experimenter Role: Making observations, giving tests, and administering treatments; he or she must avoid behaviors that may bias experimental results despite direct contacts with subjects.

Exploratory Experiments: Experiments conducted to learn more about relationships of variables in order to formulate hypotheses. Confirmatory experiments are conducted to confirm or refute experimental hypotheses.

External Validity: Generalizability of experimental results; representative of "real life" without distorting the question under investigation. See *Validity*.

Extraneous Variable: Variable representing a potential source of confounding or error variance but is of no interest to the investigator; must be controlled to maintain internal validity. An extraneous variable is not necessarily a confounding variable, but may become one if appropriate measures are not taken to keep its effects constant across all treatment conditions.

***F*-Distribution:** Continuous probability distribution used to compare two sample variances to see whether they represent the same universe.

Face Validity: Having the appearance of validity but not formally validated.

Factor Analysis: Reduction of a large number of variables in a correlation matrix to a few relatively independent variables called factors; intent is to achieve parsimony and discover the main variables that underlie all of the information in a large number of variables.

Factorial ANOVA: Analysis technique for employing more than one independent variable; determines effects of the individual variables (main effects) and interactions among them; requires data in interval form.

Factorial Design: Studies in which two or more independent variables are manipulated simultaneously.

Fatigue Effect: Performance impairment attributed only to boredom or tiredness.

Field Study Approach: The observation of events as they actually occur in their natural environment. It has the advantages of realism and of reducing the number of possible causes of real life events.

Field Test: Practical but valid substitute for a more complex laboratory test.

Figure: All charts, graphs, photographs, and schematics.

File Drawer Problem: Tendency to file away unpublished null results that creates a bias toward Type I errors in the published literature.

Finite Universe Correction: Also known as finite population correlation; common factor used to adjust standard error formulas when sampling from a finite universe (population).

Fisher, Sir Ronald (1890–1962): English mathematician and statistician who developed the F (for Fisher)-ratio that serves as the basis for the analysis of variance (ANOVA) technique.

Follow-Up Testing: See *Post Hoc Test*.

Forced Dichotomy: A continuous variable that has been arbitrarily divided into an upper and lower proportion (e.g., pass-fail on an examination).

Forecasting: Predicting future behavior through time series analysis; involves analysis of patterns underlying previously observed behaviors.

Frame: Means for identifying items making up a statistical universe.

Freedman ANOVA by Ranks: Nonparametric test for differences among three or more correlated groups.

Frequency: Number of scores in an interval; sum of tallies in an interval.

Frequency Curve: Smooth curve representation of a conceptual or theoretical frequency distribution.

Frequency Distribution: Set of raw scores arranged in order from lowest to highest and the frequency with which each score occurs; technique organizes raw data into mutually exclusive intervals. When graphing frequency distributions, scores are typically plotted on the horizontal axis (abscissa), with the frequency of occurrence on the vertical axis (ordinate).

Frequency Polygon (Line Graph): Illustration of frequency distribution; scores (or class intervals) are placed on the x-axis and frequencies on the y-axis; points are then connected by a series of straight lines.

Frequency Table: Tabular presentation of quantitative data; data are classified (distributed) into intervals.

Friedman's Two-Way ANOVA for Ranked Data: Nonparametric test of the hypothesis of differences among levels of an independent variable presented as a repeated measure or randomized-block design; used on nominal data; is an alternative to matched group ANOVA; ordinal counterpart of the correlated F.

F-Test: Hypothesis testing procedure for testing null hypothesis for two or more population means; may be used in the analysis of variance and covariance. If the null hypothesis is true, the numerator and denominator of the F-ratio will be similar or the same; if null hypothesis is false, the numerator will be larger.

Fully Crossed Design: Characteristic of designs in which each level of every IV is crossed with every level of all other IVs.

Galton, Sir Francis (1822–1911): The "father of intelligence testing" and creator of the concept of individual differences; also introduced the concepts of regression and correlation; Karl Pearson, his friend and colleague, devised the mathematical equations.

Gambler's Fallacy: Erroneous assumption that independent events are related. If a coin is flipped 10 times and comes up heads each time, the fallacy predicts virtual certainty that the result on the next toss will be tails. In reality, each coin flip is independent of all preceding flips and the probability remains the same (0.50) for each flip, regardless of past results. The gambler remembers the past, but the coin does not.

Gauss, Karl Friedrich (1777–1855): German mathematician credited with originating the normal curve, often called the Gaussian curve.

General Purpose (Reference) Table: Table arranged to allow easy location of material.

Geographic Classification: See *Area Classification*.

Geometric Mean: Calculated average defined as nth root of the product of a set of n items.

Gompertz Curve: Growth curve used in trend analysis.

Goodman and Kraskalls Gamma: Test for association of two correlated groups when ordered ranks are not treated as interval scales.

Gossett, William Sealy (1876–1937): While working for the Guinness Brewing Company in Ireland, Gossett developed (under the pen name of "Student") the technique of using sample data to predict population parameters that led to the development of the t-test.

Graph: Illustration of relationship of two or more variables; x-axis usually denotes the independent variable; y-axis denotes the dependent variable; types include pie charts, histographs, and frequency polygons.

Group: Set of subjects or items.

Grouped Data: Frequency distribution in which the scores are organized into classes of more than one value.

Grouped Score Distribution: Two-column table used to summarize a set of data. First column lists score intervals, score classes, or score groups (all synonyms). Second column lists the number of scores falling into each score interval. Construction of grouped frequency distributions is based on rules of experience rather than theory so instructions in statistics texts are similar, but not necessarily identical.

Growth Curves: Mathematical curves used to represent trends that have mathematical properties logically corresponding to the processes of economic growth.

Halo Effect: Bias error arising when subjects viewed positively on one trait are also thought to have many other positive traits; occurs when a rater is unduly influenced by previous performance of a subject; halo effect contaminates the independent variable.

Hawthorne Effect: Performance increase in response to study treatment because the participants know that they are participating in an experiment; a change in a dependent measure results from flattery or attention from the experimenter—not from manipulation of the independent variable; type of error likely to appear in before-and-after experimental studies without adequate control groups.

Histogram (Bar Chart): Illustration of a grouped frequency distribution of qualitative data; graphic representation of data by drawing a series of rectangles (bars) above the measure of performance; heights of bars indicate frequencies for various class intervals.

History: Environmental event other than the treatment of interest that occurs between successive measurements or tests.

Homogeneity of Variance: Underlying assumption of t- and F-ratios that presumes that the variability within each compared sample group is similar.

Homoscedasticity: Standard deviation of Y scores along the regression line should be fairly equal; otherwise the standard error of estimate is not a valid index of accuracy.

Hypergeometric Distribution: Distribution that shows probabilities of the number of occurrences of an event that may or may not occur on a single trial, but whose probability of occurrence on a single trial is conditional on events from other trials.

Hypothesis: A tentative statement that proposes an explanation for a phenomenon, observation, or scientific problem. These statements are most useful if testable (empirical).

Hypothesis of Association: Testing whether a correlation exists between different groups or subjects.

Hypothesis of Difference: Method for determining whether sample groups are different from each other.

Hypothesis Testing: Procedure for choosing between two hypotheses on the basis of sample data and decision rules to determine whether an effect is present.

Hypothetical Population: Statistical population that is not real and is imagined to be generated by repetitions of events of a certain type; all potential observations are not accessible at the time of sampling.

Idiographic Approach: Study of each individual as a unique person; psychological view for assessing behavior derived from an individual with emphasis on individual differences.

Inclusion Area: Middle-most area of normal curve included between two z-scores equidistant from the mean. Because of normal curve symmetry, the middle-most area includes the area immediately to the left and to the right of the mean in equal proportions.

Independent Data: Type of data required for between-groups deign; selection of one group or subject must not affect selection of other groups of subjects.

Independent Event: Event that does not affect the probability that a second event will occur.

Independent (Nonpaired) Sample: Sample chosen to ensure that selection of an item for one sample in no way determines the selection of an item for the second sample.

Independent Variable (IV): Variable experimental factor; may be manipulated (researcher deliberately alters environmental conditions to which subjects are subjected) or assigned (researcher categorizes subjects based on a pre-existing trait); the causal half of the cause-and-effect relationship in experimental research; the measure from which predictions are made in correlational research.

Index Number: Relative (percentage) statement of relationship between two numbers of different time periods.

Index of Discrimination: Ratio of high to low scorers who answer a given test item correctly.

Individually Based Norm: Norm based on the distribution of scores of individual subjects; used in AAHPER Youth Fitness Test.

Induction: Reasoning based on facts that leads to a general conclusion; method of accumulating specific data, observing it, and making inductions about principles involved.

Inductive Fallacy: Logic error in logic resulting from overgeneralizing on the basis of too few observations; occurs when one assumes that all members of a class have a certain characteristic because one member has it; for example, it is fallacious to assume that all members of an ethnic group are liars on the basis of having met one member who was a liar.

Inferential Statistics: Statistical method used to make general statements (inferences) about a population on the basis of information obtained in a sample from the population; hypothesis testing and estimation of population parameters are the most common forms of inferential statistical procedures.

Infinite Universe: Statistical universe that is conceptually unbounded.

Informed Consent: Ethical requirement; potential subjects must be able to decide whether to participate in an experiment only when they have sufficient information about the experiment to make informed decisions.

Insignificant Difference: Difference between sample value and hypothesized population value associated with acceptance of the null hypothesis; difference that may have arisen from random sampling errors.

Instrument Decay: Changes in characteristics of measurement equipment or procedures over time; also known as instrumentation error.

Interaction: Experimental result occurring when the effect of one independent variable differs at different levels of the other independent variables. In an education experiment, the effects of various curricula on student achievement may depend on the level of student motivation. Catalytic interaction results when two or more treatments must occur together to be effective; antagonistic and terminative interactions are two other forms.

Interaction Effect: Circumstance in which the combination of two or more IVs produces an effect which is not simply an additive representation of each factor's effects singly.

Intercept (a):

$$a = \frac{\Sigma Y - b\Sigma X}{N} = \overline{X}_y - b\overline{X}_x$$

Interdecile Range: Scores that include the middle-most 80% of a distribution; difference between first and ninth deciles.

Interjudge (Interrater) Reliability: Consistency in scoring of two or more independent judgments of the same performance.

Internal Consistency: When test reliability (consistency or conformity) is established using the split half method.

Internal Validity: Degree to which a test measures what it was intended to measure; allows straightforward statements about causality; degree to which an experiment is free from confounding variables. See *Validity*.

Interpercentile Range: Measure of variability used in conjunction with median; may be represented by a variety of ranges as long as equal portions of each end of the distribution are eliminated.

Interquartile Range: Scores included in middle-most 50% of a distribution; difference between first and third quartiles.

Interval: Range of scores identified to form categories of a frequency distribution.

Interval Data: Data in equal units of measure with a zero point whose position varies. The order of the numbers and the intervals between them are known; in addition to knowing whether one value is greater than or less than a second value, one also can determine the difference between the two values.

Interval Measurement: Measure that locates observations along a scale of equal intervals.

Interval Scale: Scale on which equal differences between measurements indicate equal differences in the amount of attribute measured but lacks an absolute zero point.

Intervening (Mediating) Variable: Variable representing an unnoticed or unobservable process or entity hypothesized to explain relationships between antecedent conditions and consequent events (like a hypothetical construct); may have no effect or may act as an unwanted or extraneous variable by confounding the research may be postulated as a predictor of one or more dependent variables and simultaneously predicted by one or more independent variables.

Intraclass Reliability Coefficient: Method of estimating reliability (rXX') using analysis of variance.

Intrajudge Objectivity: Consistency in scoring the same test two or more times.

Intratask Analysis: Identification of stages of development of a task from the first attempt to its performance at a mature or adult level.

Introduction: To inform the reader of the specific problem under study and the research strategy.

Introspection: Examination or observation of one's own thoughts; formal method of obtaining psychological data by studying one's own conscious experience and analyzing it into areas such as sensations, images, and feelings.

Investigator Role: See *Experimenter Role*.

Irregular: Movements in a time series that follow no definite pattern.

Judgment (Selective) Sample: Sample whose subjects are deliberately chosen by an investigator who is an expert in the study field.

Kappa (κ) Coefficient: Method of determining reliability of a mastery test that accounts for changes.

Kendall's Coefficient of Concordance: Describes the relationship between three or more sets of ranks, therefore requiring ordinal or higher level data. It often is used to determine interjudge reliability.

Kendall Tau (τ): Test for association between two correlated groups; ordered ranks are not treated as interval scales.

Kendall W: Correlation of concordance; test for association of three or more correlated ranks.

Kim's D: Test for association between two correlated groups; ordered ranks are not treated as interval scales.

Kind Classification: General qualitative classification based on any attribute of data other than geographic source.

Kruskal–Wallis _H_ Test: Nonparametric test of differences among three or more independent groups; ordinal counterpart of one-way ANOVA; variable must be distributed continuously.

$$H = \frac{12}{n(n+1)}\left[\Sigma\frac{R_i^2}{n_i} - 3(n+1)\right]$$

Kurtosis: Degree to which a frequency or probability distribution differs from a normal shape (mesokurtosis) by being more peaked (leptokurtosis) or flatter (platykurtosis).

Latin Square Design: The method of incomplete counterbalancing used to counter the order effects of the experiment.

Laws of Science: Experimental hypotheses that have been confirmed so many times that they are generally taken to be true.

Learning Effect: Treatment or experience that results in a relatively permanent change in behavior.

Least Squares Prediction Method: Technique for obtaining estimates of the parameters of a mathematical model; fitting a regression line such that the sum of the squares of the deviations from the regression line is less than from any other straight line.

Leptokurtic: Frequency distribution that is tall and thin, marked by more peaks than normal distribution; occurs when samples of data produce relative degrees of peakedness in a frequency distribution. See *Kurtosis*.

Leptokurtic Distribution: Unimodal frequency distribution in which the curve is relatively peaked; most scores occur in the middle of the distribution and very few occur in the tails; usually a small standard deviation indicates a leptokurtic distribution.

Less Than Cumulative Frequencies: Number of frequencies below the upper limit of interval in question.

Level: The different treatment values for a given IV or factor.

Level of Confidence: Range of numerical values that, considering all possible samples, contains the true value of a population parameter or shows a probability of locating true population value; indicates the percent of times (usually 95%) that a series of confidence intervals includes the unknown population mean.

Level of Measurement: Relationship assumed to exist between an underlying variable (construct) and an observed variable thought to be a measure of the construct. The four most common assumptions about such relationships are nominal level, ordinal level, interval level, and ratio level measurements.

Level of Significance (α): Proportion of total area of sampling distribution identified with improbable outcomes; alpha (α) value set for a significance test may be expressed as a percentage; normal value is 5% (0.05).

Limited Causality: The assumption (by science) that the number of things which cause another thing to occur or change, is, for the most part, limited to relatively few variables.

Linearity: A relationship between two variables which, as the value of one variable increases, the value of the other variable increases by a constant amount. It is graphically depicted as a dot cluster that approximates a straight line.

Line Graph: Graph form used to present chronological data.

Linear Relationship: Relationship between two variables; when the value of one variable increases, the value of the other increases a constant amount; depicted by a dot cluster that approximates a straight line.

Link Relative Index: Index number with a changing base, usually the value of the preceding time period.

Local History: Environmental event other than the treatment of interest that occurred for only some groups.

Logical Validity: Extent to which a test measures the most important components of skill required to perform a motor task adequately.

Logistic Curve: Growth curve used in trend analysis.

Longitudinal Study: Procedure used in developmental psychology to study age-related changes in groups of individuals over time. Measuring the same individuals at different times of their lives allows identification of possible trends in growth and development. Longitudinal research requires great patience but the resulting data is considered to be more valid than data from a cross-sectional approach.

Lower-Tailed Test: One-tailed test with the rejection region in the lower tail of the sampling distribution.

Lurking Variables: Nuisance or intervening variables that confound the study and make straightforward interpretation of the results difficult.

Machine Formula: The formula most readily used when solving equations with a calculator.

Magnitude: The amount or quantity of a variable.

Mail Questionnaire: Set of questions used to collect primary data; forms are transmitted by U.S. Postal Service.

Main Effect: Overall influence of an independent variable; effect of one independent variable is the same at all levels of another independent variable. In a test of the main effect, it compares marginal means for all scores for a particular level of the variable.

Main Heading: First order heading; used for all major sections of the report except the Introduction.

Mann–Whitney U Test: Nonparametric test of null hypothesis; uses ranks of various observations for two independent samples; serves as alternative to the T test for equality of means.

MANOVA (Multivariate Analysis of Variance): Analysis of variance on more than two dependent measures.

Matched Groups Design: Experimental design in which pairs of participants are matched on an extraneous variable known to be correlated with the DM. Participants are then randomly assigned to the treatment and control groups.

Matching Variable: Dimension on which participants are paired/matched.

Maturation: Development resulting from biological processes; anatomical, physiological, and neurological development of an organism distinguished from development resulting from learning or exercise.

McNemar Test: Test of symmetry developed by Quinn McNemar; uses correlated proportions and Chi-square to analyze nominal data from correlated samples.

Mean ($\bar{X}$): Measure of central tendency specifying arithmetic average of a distribution of scores; arithmetic mean is the most common form, and may be called the average of scores; scores are added and the total divided by the number of cases:

$$\bar{X} = \frac{(\Sigma X)}{n}$$

In normal distribution, the mean coincides with median and mode. When the entire population of scores is used, the mean is designated by the Greek mu (μ). See *Average and Central Tendency*.

Mean Deviation: Measure of variability for a set of measurements; equal to the mean of the sum of absolute deviations from the mean; may be used in place of the standard deviation since it is less sensitive to large deviations.

Mean Square (MS): Sum of squares of differences between observations and arithmetic mean divided by its degrees of freedom; used in the analysis of variance.

Meaningful Difference: Measured difference that is beyond significant; useful and relevant in a real-world context.

Meaningful Problems: Those problems that can be solved through direct observation.

Measure: As a noun, an instrument or technique for obtaining information (usually a score) about an attribute of a person or an object; as a verb, refers to obtaining a score.

Measurement: Quantifying observations by assigning a number to an attribute of a person or object on the basis of specific rules; obtaining test scores; rules chosen determine whether measurement scale used is nominal, ordinal, interval, or ratio.

Measurement Error: Difference between obtained value and theoretically true value; portion of a score arising from random or irrelevant factors; usually assumed to be normally distributed.

Measure of Association: Statistic whose magnitude indicates degree of correspondence or strength of relationship of two variables. An example is the Pearson product–moment correlation coefficient. Measures of association are different from statistical tests of association (Pearson Chi-square, F-test) whose primary purpose is assessing the probability that the strength of a relationship is different from a preselected value (usually zero).

Measures of Location: See *Average*.

Median (Mdn): One of the three most common measures of central tendency; middle-most value of an ordered set (numbers arranged in order of magnitude that divides the scale into two equal parts coinciding with the 50th percentile); middle value of an odd number of scores; halfway between the two middle values of an even number of scores; most valid measure of central tendency when a distribution is skewed. See *Central Tendency*.

Mesokurtic: Unimodal normal or medium (meso-) frequency distribution; curve without too much or too little peakedness. See *Kurtosis* and *Normal Curve*.

Meta-Analysis: Combining and examining a collection of independent or partial replications.

Method of Least Squares: Technique used in fitting regression lines; has the mathematical property of making the sum of the squared deviations of the points from the regression line a minimum.

Method of Selected Points: Technique used in fitting trend lines; forces the trend line to pass through certain predetermined (selected) points.

Methods Section: Describes how the study was conducted. It should enable readers to evaluate the appropriateness of your experimental procedures and estimate the reliability and the validity of your results. It should provide sufficient detail for investigators to replicate the study.

Mixed Design: Experimental design with two or more independent variables including one between-subject variable and one within-subject variable. Subjects are randomly assigned under different levels or combinations of variables and tested with every level or combination.

Mixed Group: Between- and within-groups designs combined in the same matrix.

Mode (Mo): One of the three measures of central tendency; most frequent number in a distribution. A distribution can be bi-modal if two separate values have the same maximum frequency. A distribution that has two most common points is bimodal. See *Central Tendency*.

Moments of Distribution: Statistics describing frequency distribution by average value, relative scatter of observations, symmetry, etc.; the four moments are central tendency, variability, skewness, and kurtosis.

Monotonic Relationship: When increases in one variable are accompanied by either systematic increases or decreases in the other variable.

Moving Averages: Set of averages over time obtained by dropping the first item in a set and replacing it by the next item before calculating the next average.

Moving Seasonal: Set of seasonal indices that exhibit a systematic change in timing, amplitude, or both over a period of years.

MS **(Mean Square):** Sum of squares of the differences between the observations and the mean divided by its degrees of freedom.

Multifactor ANOVA: Test of difference for experimental designs that include two or more independent variables.

Multifactor Mixed ANOVA: Test of difference used for multifactor designs that include at least one repeated-measures variable and one between-groups factor.

Multifactor Repeated Measures ANOVA: Similar to repeated measures ANOVA; analyzes data from experiments involving two or more independent measures. See *Repeated Measures ANOVA*.

Multimodal: Similar to modal; distribution that has three or more modes.

Multiple Correlation: Correlation describing relationship between two or more predictor (independent) variables and a criterion (dependent) variable.

Multiple IV Experiment: Study which employs more than one factor to examine the interaction of experimental variable on the dependent measure(s).

Multiple Regression: Statistical procedure for predicting value of dependent variable from the values of at least two independent variables; technique for weighing, combining, and ordering two or more IVs nonredundantly into a mathematical equation.

Multiplication Rule: Rule for finding the probability for two or more sets of outcomes connected by *and*. $Pr (A \text{ and } B) = [Pr(A)][Pr(B)]$.

Multistage Sample: Probability sample based on random selection via several steps (stages) to select final sample.

Mutually Exclusive Event: Occurrence of one event precludes the occurrence of one or more alternative events.

Negative Relationship: Direction of association between X and Y variables; relatively high values are paired with relatively low values and relatively low values are paired with relatively high values.

Negatively Accelerated Curve: A curve in which increases on variable X are accompanied by progressively smaller increases or decreases on variable Y.

Negative Skew: Lopsided shape of a frequency distribution based on a few observations with extremely small values. See *Skewness*.

Nominal Data: Data (measurements) in which numbers are used to label discrete, mutually exclusive categories; nose-counting data focusing on the frequency of occurrence within independent categories.

Nominal Measurement: Sorting of observations into different classes or categories.

Nominal Scale: Scaling technique placing items, objects, characteristics, or individuals into mutually exclusive categories; numbers are assigned to represent the categories but the numbers cannot be mathematically manipulated; data presented in classes or categories such as brands of coffee.

Nomothetic Approach: Belief that behavioral science should try to formulate general laws to explain behaviors of populations; system for assessing behavior based on averages of performance by many individuals; often contrasted with idiographic approach.

Nondirectional Test: Test that does not specify a positive or negative direction of a comparison of two means; two-tailed test, so named for location of the critical region for the statistic; shows that a true population value deviates in either direction from the hypothesized population value. See *Two-Tailed Test*.

Nonequivalent Control Group: A common type of quasi-experimental design.

Nonnumeric Classification: See *Qualitative Classification*.

Nonparametric: Technique requiring only nominal or ordinal levels of measurement or for use where interval or ratio data are not normally distributed.

Nonparametric Statistics: Tests that do not predict the population parameter (mu or μ) or make assumptions about the normality of the underlying population distribution; require only nominal or ordinal data and typically have less power than parametric tests.

Nonsampling Error: Systematic or unsystematic error.

Normal Curve (Distribution): Theoretical curve or distribution of a large number of data points plotted as unimodal distribution; perfectly symmetrical and bell-shaped; mean, median, and mode are superimposed.

Normal Deviate: Standard scale value for normal curve.

Normality: Degree (measured or assumed) to which the distribution of a DM is normal in shape.

Norm-Referenced Test: Test used to compare a subjects score with scores of similar subjects.

Null Hypothesis: Hypothesis evaluated via testing; assumption that the results are due only to chance; statistical statement of no relationship between two variables (population from which samples were drawn) so that its rejection provides support for a working (alternative) hypothesis. In testing the hypothesis of association, the null hypothesis states that the correlation in the population is equal to zero (does not exist). See *Hypothesis.*

Nuisance Variable: Potential independent variable that is not to be manipulated in an experiment and must be neutralized to prevent confounding with the treatment variable or variables.

Numeric Classification: See *Quantitative Classification.*

Objective Probability: Probability statement based on process knowledge or experimental evidence.

Objective Test: Test with highly precise scoring system that yields little error.

Objectivity: Precision with which test is scored.

Observed Frequency: Number of observations within a sample that fall into a response category; used with Chi-square test.

Odds: Chances that a specific event will not occur. For example, 5 to 1 odds indicate that an event will *not* occur five times for every one time it will occur.

Omega Squared (One-Way):

$$\omega^2 = \frac{SS_{group} - (k-1)MS_{error}}{SS_{total} + MS_{error}}$$

One-Sample Chi-Square: Test used to determine whether an observed frequency distribution is significantly different from a hypothesized or expected frequency distribution.

One-Sample Test of Runs: Nonparametric test for randomness for one group of correlated data.

One-Tailed (or Directional) Test: Decision rule reflecting concern that the true population value deviates in a particular direction from the hypothesized population value.

Open End Interval: Interval that is not closed on one end, for example, $10,000 and over.

Operating Characteristic (OC) Curve: Plot of probability of accepting the null hypothesis against various alternative values of parameter tested.

Operational Definition: Unique specification of the meaning of a term as equivalent to a single set of operations.

Operationalization: Choosing, or creating concrete procedures to represent a theoretical construct or concept.

Opportunity Loss: Difference between payoff of the best act that could have been selected based on the state of the world and the payoff of any other act for the same state; profit lost by failing to make the best decision.

Ordinal Data: Rank-ordered data derived only from the order of the numbers, not differences between them; provide information about greater-than or less-than status but not quantity.

Ordinal Measurement: Arrangement of observations in order.

Ordinal Scale: Scale of measurement in which numbers represent a rank ordering.

Ordinate: Vertical (or y-) axis on a graph or coordinate system; frequency of occurrence on a frequency distribution.

Orthogonal: Statistically independent.

Overestimate: Calculated value exceeds actual or "real" value.

Parabola (Second-Degree Curve): Mathematical curve used to represent trends characterized by squared values of independent variable; used in both natural number and logarithmic forms.

Paragraph Headings: Typed with paragraph indentation, end in a period, and usually have only the initial letter of the first word capitalized.

Parameter: Measure from entire population; usually inferred rather than directly measured.

Parametric Research: Research designed to establish a functional relationship between one or more quantitative IVs and a DM by testing at several levels of each IV. Testing multiple levels of a treatment is necessary to determine the parameters of the prediction equation relating the variables.

Parametric Statistics: Statistics for interval or higher level data that require normality and homogeneity of variance.

Parametric Test: Significance test in which the null hypothesis is a statement about the parameter(s) of the population(s) from which the samples were drawn; typically does not include distribution-free contrasts and is based on the assumption of sampling from binomially and normally distributed populations.

Partial Correlation: Correlation technique for ruling out possible effects of one or more variables on the relationship among the remaining variables. In three-variable situations, partial correlation rules out the influence of the third variable on the correlation between the remaining two. See *Correlation*.

Partial Regression and/or Correlation: Regression and correlation analysis of relationship of two variables as part of multiple analysis.

Pascal, Blaise (1623–1662): French mathematician and philosopher who introduced the concepts of probability and random events.

Path Analysis: Complex mathematical procedure for explicating and assessing patterns of causal relationships that underlie correlational data.

Payoff: Gain or loss associated with a particular decision and state of the world.

Pearson, Karl (1857–1936): English mathematician and colleague of Sir Francis Galton who translated Galton's ideas about correlation into precise mathematical terms; created the equation for the product–moment correlation coefficient (Pearson r).

Pearson Product–Moment Correlation Coefficient (r): Most widely used correlation coefficient, developed by Karl Pearson; measures degree of linear relationship between two variables. Computed correlation values range from +1.00 (perfect positive correlation) through 0 to –1.00 (perfect negative correlation). The farther the Pearson r is from 0 in a positive or negative direction, the stronger the relationship between the two variables. The Pearson r can be used for making better-than-chance predictions but cannot isolate causal factors; assumes linearity, homoscedasticity, and interval or higher level data. See *Correlation Coefficient*. Calculation is

$$r = \frac{\Sigma Z_x Z_y}{n} = \frac{n\Sigma xy - (\Sigma x)(\Sigma y)}{\sqrt{(n\Sigma X^2 - (\Sigma X)^2)(n\Sigma y^2 - (\Sigma y)^2)}}$$

Percentage Method: Method of grading based on assignment of a percentage to each grade category; norm-referenced approach to grading.

Percentiles (Centiles): Percentage of cases falling below a given score. If an individual scores at the 95th percentile, his or her results exceeded those of 95% of all persons taking the same test. If test scores are normally distributed and the standard deviation of the distribution is known, percentile scores can easily be converted to z-scores.

Percentile Rank: Percentage of observations with similar or smaller values in the entire distribution. Calculation for a percentile rank is

$$\text{Lower\%} + \frac{\text{Score} + \text{RLL}}{\text{Width}} (\text{Interval \%})$$

Score for a Percentile (Score_p): $RLL + \dfrac{Width}{Interval\ \%} (p - \text{Lower \%})$

Period: Frequency or rhythmicity with which a behavior repeats itself.

Permutation: Counting the number of possible outcomes in situations where order is important and repetition of elements is not permitted.

Personal Investigation: Direct collection of primary data through personal contacts.

Phi (Φ) Coefficient: Correlation test of nominal data when the number of independent cells is 4 (2×2 Chi-square analysis).

Phi Correlation: An extension of the Pearson r to the case where data on both dimensions are in the form of dichotomies.

Pie Chart: Illustration of percentages or proportions as sizes of slices in a complete circle ("pie").

Pilot Studies: Studies done for the purpose of serving as a dress rehearsal or dry run of a full-scale confirmatory experiment.

Pilot Test: Trial run of research procedures or hypotheses, usually undertaken to determine the feasibility of undertaking a larger or more formal study.

Placebo Effect: Change produced by an inert substance.

Platykurtic: Wide and flat frequency distribution; distribution that is markedly less peaked than a normal distribution. See *Kurtosis*.

Platykurtic Distribution: Unimodal frequency distribution whose curve is relatively flat; large numbers of scores appear in both tails; distribution of scores yields a relatively large standard deviation.

Plausible Rival Hypothesis: Rival hypothesis that has a more than remote chance of being correct.

Point Biserial Correlation: Index of linear correlation used where one variable is continuous and the other dichotomous (has only two values). See *Correlation*.

Point biserial r: A special case of the Pearson r, used when one of the variables is continuous but the other is believed to be a true dichotomy.

Point Estimate: Single value estimate.

Point of Intercept: In a scatter plot, location where the regression line crosses the ordinate or Y value when X equals zero. In the regression equation, $Y = bX + a$, the a represents the intercept.

Point of Central Tendency: See *Average*.

Poisson Distribution: If an event can occur at random over a large area or long period, the Poisson distribution shows the probability that an event will occur over a small area or short interval; occurrences are defined; nonoccurrences are not.

Pooling: Combining samples together to obtain an unbiased estimate of the common population variance.

Population: Set of potential observations having at least one limited (finite) or unlimited (infinite) trait in common.

Population Validity: Degree to which results can be generalized to the population of ultimate interest. See *Validity*.

Positionary Average: Average defined by describing a specific place or location in the data.

Positive Relationship: Relatively high values are paired with relatively high values and relatively low values are paired with relatively low values.

Positively Accelerated Curve: A curve in which increases on variable X are accompanied by progressively larger increases or decreases on variable Y.

Post Facto Research: Research that does not allow cause-and-effect conclusions but allows better-than-chance predictions. Subjects are measured on one response dimension and the measurements compared with different response measures. Responses are compared with responses, as in comparing SAT scores with grade-point averages. Since the experimenter does not treat the subjects differently (no manipulation of independent variable), cause-and-effect conclusions may not be drawn.

Post Hoc Comparison: Comparison conducted after data have been examined.

Protected t (only if F is significant): $\quad t = \dfrac{\bar{X}_i + \bar{X}_j}{\sqrt{\frac{MS_{error}}{n_i} + \frac{MS_{error}}{n_i}}}$

Omega squared (one-way): $\quad \omega^2 = \dfrac{SS_{group} - (k-1)MS_{error}}{SS_{total} + MS_{error}}$

Eta squared: $\quad \eta^2 = \dfrac{SS_{group}}{SS_{total}}$

Post-Hoc Tests: Pair-wise comparisons on greater than two means, conducted after the omnibus F-test has shown significance, to determine whether differences are significant.

Posterior Analysis: Analysis of a decision problem using Bayes' theorem after sampling.

Posterior (*A Posteriori*) Probability: Revised set of probabilities in a Bayesian decision problem.

Power (1 – Beta or 1 – β): Sensitivity of a statistical test. The more powerful a test, the less is the likelihood of committing a beta error (accepting null hypothesis when it should have been rejected). The higher a test power, the higher the probability of finding a small difference or small correlation significant.

Power Efficiency: See *Relative Power*.

Practice Effect: Improvement resulting merely from experience with a task or procedure.

Prediction: Accurate anticipation of future or still unobserved events.

Prediction Interval: Range of values that covers the unknown Y score with a known degree of confidence.

Predictive Validity: Appropriateness of a test as a predictor of behavior. See *Validity*.

Predictor: Test or variable that predicts a criterion behavior.

Presentation: Use of tables, graphs, or words to present results of a statistical analysis.

Price Index: Index number that measures changes in prices.

Primary Data: Study data collected under the control and supervision of the person conducting the study.

Primary Source: Source of secondary data that serves as its own collection agency.

Prior Analysis: Analysis of a decision problem before sampling.

Prior (*A Priori*) Probability: Probability based on advance knowledge; also refers to probabilities in a Bayesian decision problem before their revision based on sample evidence; may be objective or subjective.

Probability (*P or p*): Proportion or fraction of times that a certain outcome will occur; usually expressed as a decimal, for example a 1/20 probability is written as 0.05. For significance level, probability appears in some journals as a lower case or upper case letter in italics.

Probability Distribution: Distribution based on probabilities of all possible outcomes for a set of mutually exclusive events.

Probability Sample: Sample items are chosen by a chance (random) procedure; each universe item must have a known chance of appearing in the sample.

Problem: What the experimenter proposes to solve.

Procedure: Detailed description of how the experiment was conducted.

Process Control: Statistical quality control techniques that attempt to control the manufacturing during operation, thereby preventing poor quality products.

Prospective Study: Longitudinal study in which subjects are followed forward in time so that each important event or circumstance can be noted and recorded as it occurs.

Protected t: Formula for use only if F is significant:

$$t = \frac{\overline{X}_i + \overline{X}_j}{\sqrt{\frac{MS_{error}}{n_i} + \frac{MS_{error}}{n_i}}}$$

p Value: Degree of rarity of a test result if the null hypothesis is true.

Qualitative Classification: Point in time classification of statistical data wherein the data are classified according to some unmeasurable nonnumeric attribute; also known as nonnumeric classification; subdivided into kind and area (geographic) classifications.

Qualitative Data: Observations consisting of words, labels, or numerical codes that indicate differences in kind.

Qualitative Measurement: Judgmental approach to measurement.

Quantitative Classification: Point in time classification of statistical data wherein data are classified by size or magnitude of their numerical values; also known as numeric classification.

Quantitative Data: Numerical observations that indicate differences in amount or count.

Quantitative Measurement: Objective approach to measurement in which a number is assigned to the attribute of interest.

Quantity Index: Measurement limited to changes in physical volumes.

Quantity of Information: Weighting factor in continuous posterior analysis; reciprocal of appropriate variance.

Quartile: Division of a distribution into four (quarter) parts; the first quartile (Q_1) represents the 25th percentile, the second quartile the 50th percentile (or median), and the third quartile (Q_3) the 75th percentile. Q_1 has no more than one-fourth of the values smaller and no more than three-fourths larger. Q_3 has no more than three-fourths and no more than one-fourth larger.

Quartile Deviation: See *Semi-Interquartile Range*.

Quasi-Experiment: Study that employs experimental units of analysis and manipulated treatments without comparing groups formed by random assignment.

Quota Sampling: Selecting a sample that directly reflects population characteristics. If 45% of a population is composed of males and gender is a relevant research variable, a sample must contain 45% male subjects.

Random Assignment: Assigning subjects to treatments or groups so that each subject has an equal likelihood of assignment to any treatment or group; the assignment of any one individual does not influence the assignment of any other individual.

Random Digits: Set of digits used to draw random samples. Tables of random digits are constructed to have no predictable patterns; every digit has an equal chance (probability) of appearing at any position.

Random Error: Combined effects (on scores of individual subjects) of all uncontrolled factors such as individual differences among subjects, slight variations in experiment conditions, and measurement errors.

Random Sample: Sample selection method; all subjects in a population have equal chances of being included in a sample at each stage of sampling. When samples are selected randomly, sampling error should also be random, increasing the probability that the samples are representative of the population.

Random Selection: Selecting a sample randomly from the population of ultimate interest.

Randomization: A selection process by which each person in the population has an equal probability of being selected. By randomly assigning subjects to treatment conditions we expect differences on unidentified extraneous variables to average out. This technique is most effective when the subject groups are large.

Range (R): Measure of variability reflecting the spread of scores in distribution; the bottom score subtracted from the top score plus one. Most definitions state the range as the difference between the two most extreme scores in a distribution, equal to the highest value minus the lowest value. Our convention is to add 1.0 to this value to account for all intervals. For instance, if 1, 2, and 3 are used, the typical definition would dictate the range as 2 ($3 - 1 = 2$). We argue that the range is 3 ($3 - 1 + 1 = 3$), namely, 1, 2, and 3.

Rank: Position of a case (person) relative to other cases on a defined scale—as in first place, second place, etc. When actual values of the numbers designating relative positions (ranks) are used in analysis, they are treated as an interval scale, not an ordinal scale.

Rank Correlation: Technique for measuring relative closeness of a relationship between two sets of data in the form of rankings only.

Rank Difference Correlation Coefficient: Statistical technique used to calculate correlation coefficient when the scores are ranked.

Rank Order: Sequential order from highest to lowest or vice versa.

Rate of Acceleration: Measure of relative (percent) change in the rate of change in the trend of a time series when the latter rate is not constant; measure of the direction and degree of curvature of a trend line.

Rate of Change: Measure of relative (percentage) growth in a time series as shown by the trend line.

Rating Scale: Scale used to subjectively assess performance.

Ratio Data: Data in equal units that provide meaningful comparisons, for example, feet, gallons, pounds; data (measurements) that provide information about the order of numbers, the differences between numbers, and an absolute zero point; permits comparisons (A is three times B or A is half of B).

Ratio Measurement: One of four widely used sets of assumptions about the relationship between an underlying variable (construct) and an observed variable (measure of the construct). If the observed variable measures the construct on a ratio scale, a given interval on the scale of measurement corresponds to the same amount of the construct, no matter where that interval occurs on the scale. Zero on the scale of measurement corresponds to complete absence of a construct. Height measured in inches is an example of measurement on a ratio scale. The 4 inches between 48 and 52 inches represent the same difference in height as the difference between 58 and 62 inches. A person who measured 0 inches from the bottom of his or her feet to the top of his or her head has no height.

Ratio Scale: The highest scale of measurement, features equal units with an absolute zero.

Ratio to Moving Average Method: Most common procedure for obtaining seasonal indexes.

Raw Score: Piece of data that has not been "cooked." To be "raw," a score cannot be transformed to a derived scale such as a percentile rank or standard score.

Reactive Effect: Change in affect resulting from heightened awareness arising from the experience of taking an inventory measuring affective behavior.

Reactivity: Change in subject behavior merely as a reaction to being observed.

Real Limits: Upper and lower limits of intervals in a frequency distribution representing the entire area between intervals; do not represent score units obtained on a test.

Realism: Going out into the field to systematically observe events as they occur naturally in real life.

Reference Section: A list of all the resources cited in the text.

Reference Table: See *General Purpose Table.*

Region of Rejection: See Rejection Region.

Regression: Numerical measure of association between (or among) variables in which predictor and criterion variables are specified. Regression toward the mean refers to a tendency for individuals with unusually high or low scores to go back toward the mean on subsequent testing; also called statistical regression.

Regression:
$$Y = a + bx$$

Slope (b):
$$b = \frac{cov_{xy}}{S_x^2} = \frac{SP_{xy}}{SS_x}$$

Intercept (a):
$$a = \frac{\Sigma Y - b\Sigma X}{N} = \bar{X}_y - b\bar{X}_x$$

Standard error of estimate:
$$S_{y-\hat{y}} = \sigma_y \sqrt{\frac{\Sigma(Y - \hat{Y})^2}{N-2}} = \sqrt{\frac{SS_{error}}{N-2}}$$
$$= \sqrt{(1 - r^2)\frac{N-1}{N-2}}$$

SSy:
$$\Sigma Y^2 - \frac{(\Sigma Y)^2}{N}$$

$SS_{\hat{y}}$:
$$\Sigma \hat{Y}^2 - \frac{(\Sigma \hat{Y})^2}{N}$$

$SS_{y-\hat{y}}$:
$$SS_y - SS_{\hat{y}} = SS_{error}$$

SS_{error}:
$$SS_y(1 - r^2)$$

Regression Analysis: Testing a hypothesis of association through the use of prediction; $y = a + bx$.

Regression Line: $Y = bX + a$; single straight line lying closest to all points in a scatter plot; may be used to make correlational predictions when three important pieces of information are known: (1) how much the scatter points deviate from the line, (2) the slope of the line, and (3) the point of intercept.

Rejection Region: Portion of a sampling distribution that would lead to rejection of the null hypothesis if it contained the sample result.

Relative Frequency Distribution: Part or fraction of total frequency that occupies each class interval or category.

Relative Power: Ratio of sample size required to reject a given null hypothesis using a t- or F-test to the sample size required using the statistic under consideration; also known as *power efficiency.*

Reliability: Dependability of scores; relative freedom from error; consistency or repeatability of an individual's performance on a test.

Reliability (Criterion-Referenced): Consistency of classification of masters and nonmasters.

Reminiscence: Improvement in performance due simply to the passage of time.

Repeated Measures ANOVA: Analysis in which each subject performs experimental tasks under all possible treatment conditions (all levels of all factors); also called a completely within-subjects design; produces F-ratios and probabilities that determine whether performance on a given dependent measure differed significantly among experimental conditions.

Representative Sample: Sample that reflects the characteristics of an entire population; random sampling is assumed to result in representative samples.

Research Hypothesis: Informal hypothesis or hunch that inspires an investigation; lacks necessary precision to be tested directly. See *Hypothesis*.

Response Set: *A priori* tendency to respond in certain ways to test questions or items, including response acquiescence ("yeah"), response deviation (unusual responses), and social desirability (answering in the most socially accepted or approved direction).

Retrospective Study: Study in which subjects or those who knew earlier are asked to "look back" and recall what was said or done in the past.

Rival Hypothesis: Alternative explanation of unobserved relationship. See *Hypothesis*.

Root Mean Square Deviation: The square root of the average squared deviation.

Sample: Subject of observations; sample size selection charts (or power curves) identify a sample size that is not unduly small or excessively large to produce hypothesis test with the proper sensitivity.

Sampling Distribution: Distribution of measures (statistics) taken on successive random samples; when all samples in an entire population are measured, the resulting distributions are expected to be normal. Two important types are the distribution of means and the distribution of differences. See *Central Limit Theorem*.

Sampling Distribution of Mean: Probability distribution of means for all possible random samples of a given size from a population.

Sampling Error: Expected difference between mean of a sample and mean of a population (mean = mu or μ). Under random sampling, the probability of obtaining a sample mean greater than a population mean is identical to the probability of obtaining a sample mean less than the population mean ($P = 0.50$).

Sampling With Replacement: In selecting a representative sample, to keep the chance of being selected from a given population constant, each selected subject is placed back into the population before making the next random selection.

Scatter Diagram: Graph using correlation coefficients. Each plotted point represents the values of an individual observation on the two variables correlated. Values of one variable are shown on the horizontal axis (bottom) and values of the other are shown on the vertical axis (side).

Scattergram: The graph of a bivariate frequency distribution.

Scatter Plot: Graphic format in which each point depicts the relationship of a pair of X and Y scores. The array of points in a scatter plot typically forms an elliptical shape (result of the central tendency usually present in both X and Y distributions.)

Scheffe's Test: Test for all possible comparisons between population means after the overall null hypothesis is rejected in ANOVA.

Science: The study of meaningful problems through the use of the scientific method.

Scientific Method: A systematic process of investigation involving a sequential series of steps that the researcher follows in identifying and answering a problem.

Seasonal: Repetitive fluctuating movement in a time series with a duration of 1 year.

Secondary Data: Data available for use in a study but not originally collected for that study.

Secondary Source: Source of secondary data; agency that compiles already collected data.

Secular Trend Analysis: Use of regression technique to predict trends across time. Historical data are used to predict future results based on the assumption that a past trend will continue.

Selection Bias: Systematic difference between comparison groups not derived from experiencing the treatment of interest.

Selective Sample: See *Judgment Sample*.

Self-Fulfilling Prophecy: Biased experimenter behavior that cues subjects to produce the results wanted or expected.

Self-Selection: Subjects decide whether they will participate in research or decide which group to join.

Semi-Interquartile Range (Quartile Deviation): Spread of scores in a distribution; high quartile deviation indicates that the scores are spread out to a large degree. This population parameter can be used on ordinal level data. As the equation indicates, semi-interquartile range equals half the distance between the 75th percentile and the 25th percentile on the score scale.

$$Q = \frac{(Q_3 - Q_1)}{2}$$

Semilogarithmic Graph Paper: Graph paper printed with an arithmetically ruled horizontal scale and a logarithmically ruled vertical scale; used to study relative (percentage) changes in a time series.

Sequence Effect: See *Practice Effect*.

Sequential Sample: Probability sampling technique wherein items are selected individually.

Serendipitous Finding: Good luck; beneficial result that was not the original object of a search.

Seriation: The purpose of seriation is to help the reader understand the organization of key points within a section or paragraph. Separate paragraphs, sentences, or items within a series are indented and an Arabic numeral, followed by a period, identifies each. Alternately, bullets are used instead of numerals.

Side Heading: Second order headings; used for sub-sections of the report.

Significance: Indication that study results did not occur by chance. The assumption surrounding significant differences and significant correlations is that the past trend will continue; describes results if the null hypothesis is rejected. Test results are *not statistically significant* if the null hypothesis is retained.

Significant Difference: Difference between sample value and hypothesized population value sufficient to lead to rejection of the null hypothesis; a difference so great that it is improbable that it resulted from random sampling errors.

Sign Test: Test for difference using signs rather than quantitative values.

Simple Index Number: Conversion of consecutive values of a single series into percentages of the value for a time period chosen as a base.

Simple Regression and/or Correlation: Regression and correlation analysis involving only two variables.

Single IV Experiment: Study which employs only one experimental factor with multiple levels.

Skewed Distribution: Unbalanced distribution caused by having a few outliers (extreme scores) in one direction. The median is the best measure of central tendency for a skewed distribution.

Skewness: Shape of curve when most scores fall at one end of the distribution and the remainder taper off as reflected in a long tail. Negative skewness is a lopsided frequency distribution caused by a few observations with extremely small values. Positive skewness is a lopsided frequency distribution caused by a few observations with extremely large values.

Slope (*b*):

$$b = \frac{cov_{xy}}{S_x^2} = \frac{SP_{xy}}{SS_x}$$

Soloman Four Group Design: Experimental design that factorially combines the presence or absence of a pre-test with the presence or absence of a treatment.

Somer's *d*: Test for association between two ranks when a distinction is made between the independent and dependent variables.

Spearman, Charles E. (1863–1945): English psychologist and test expert who conducted research to identify the factors of intelligence and developed a correlation technique (Spearman rho or ρ) to analyze ordinal data.

Spearman Correlation Coefficient (ρ): Sample statistic or population parameter indicating degree of linear relationship between two variables measured on an ordinal scale; provides correlation between the ranks of two variables; can assume values between +1.0 and –1.0. See *Correlation Coefficient.* Calculation is

$$\rho = 1 - \frac{6\sum D^2}{N(N^2 - 1)}$$

Spearman Rank Order Correlation Coefficient: An adaptation of the Pearson *r* for use with ordinal data.

Special Purpose Table: Table designed to emphasize certain statistical information.

Split Half Reliability Estimate: Reliability estimate determined by dividing a test into halves and correlating scores from the halves.

Spurious Results or Findings: Experimental results that are not genuine; false findings.

Squared Correlation Coefficient (*r²*): Proportion of total variability in one variable associated with variability in the other variable.

SS **(Sum of Squares):** Sum of deviation squared of each individual raw score from the mean of the group.

Standard Deviation: Measure of variability that roughly indicates the average amount by which observations deviate from their mean; has a constant relationship with the area under the normal curve. See *Normal Curve.* The sample standard deviation is calculated with the following equation:

$$s = sd = \sqrt{s^2} = \sigma_{n-1} = \sqrt{\frac{\sum X^2 - \frac{(\sum X)^2}{n}}{n-1}}$$

The estimated standard deviation of the population is calculated with the following equation:

$$S = SD = \sigma = \sigma_n = \sqrt{\frac{\sum X^2 - \frac{(\sum X)^2}{n}}{n}}$$

When the standard deviation is calculated on the basis of all scores in an entire population, it is designated by the Greek letter sigma (σ).

Standard Deviation Method: Method of grading based on standard deviation of the distribution of scores; norm-referenced approach to grading. See *z-Score.*

Standard Error of Arithmetic Mean: Standard deviation of random sampling distribution of arithmetic means; measure of sampling error.

Standard Error of Difference: Standard deviation of a random sampling distribution of the differences between values from independent samples; measure of sampling error.

Standard Error of Difference Between Means: Measure of variability indicating average amount by which differences between sample means deviate from the difference between population means.

Standard Error of Estimate (SE$_{est}$): Technique for determining accuracy of a predicted Y value by using the regression equation.

$$SE_{est} = \sqrt{\Sigma \frac{(Y - \hat{Y})^2}{N - 2}} = \sqrt{\frac{\Sigma Y^2}{N - 2}} * \sqrt{1 - r^2}$$

Standard Error of Estimate $(S_{y-\hat{y}})$: $S_{y-\hat{y}} = \Sigma_y \sqrt{\frac{\Sigma (Y - \hat{Y})^2}{N - 2}} = \sqrt{\frac{SS_{error}}{N - 2}}$

$$= \sqrt{(1 - r^2)\frac{N - 1}{N - 2}}$$

Standard Error of Estimate of Individual Y: Measure for qualifying estimates from a regression line; allows for sampling error and unexplained variability.

Standard Error of Mean ($S_{\bar{x}}$ or $\sigma_{\bar{x}}$): Measure of variability that roughly indicates the average amount by which sample means deviate from the population mean.

$$\sigma_{\bar{x}} = \frac{\sigma}{\sqrt{n}}$$

Standard Error of Measurement: Estimate of absolute error of an individual score.

Standard Error of Percentage: Standard deviation of random sampling distribution of percentages; measure of sampling error.

Standard Error of Prediction: Measure of variability indicating average amount of predictive error (average amount by which known Y values deviate from their predicted Y′ values).

Standard Normal Curve: Normal curve that is actually tabled; always has a mean of 0 and a standard deviation of 1.

Standard of Performance: Degree to which a participant is expected to meet an objective.

Standard Score: Score standardized by dividing the deviation of the score from the mean by the standard deviation; score expressed relative to a known mean and standard deviation.

Standard Score Method: Same as the standard deviation method of grading; use of *z*-scores.

State of Control: Subject to chance variations only.

States of the World: Possible conditions (values) of the universe to be sampled.

Statistic: Measure obtained from a sample as opposed to an entire population. See *Statistics.*

Statistical Description: Summary of essential characteristics of a limited set of data, usually a sample.

Statistical Generalization: The results shown to hold for an adequate sample of a given population before attempting to generalize the results of an experiment to that population.

Statistical Inference: Generalization about a statistical universe on the basis of sample information.

Statistical Quality Control: Control of quality level of a manufactured product by statistical techniques, usually by sampling.

Statistical Significance: Probability that a difference occurred by chance; sometimes expressed as *P* value.

Statistical Universe (Population): All items that may be surveyed in a statistical study if a complete enumeration were made.

Statistics: Plural: statistical measures of samples; set of numbers. Singular: body of methods for obtaining and analyzing numerical data to make better decisions in an uncertain world.

Stratified Random Sample: Probability sample. The universe is divided into relatively homogeneous subuniverses (strata) and a simple random sample is taken from each stratum.

Stub: Row heading of a table.

Sturges' Rule: Formula for estimating approximate size of class interval to use in constructing a frequency table.

Subjective Probability: Probability statement based on personal assessment of the likelihood that an event will occur.

Subject: Individual taking part in an experiment; subjects may be independent or related.

Sum of Products (SP_{xy}):

$$SP_{xy} = \Sigma XY - \frac{\Sigma X \Sigma Y}{N}$$

Sum of Squares (SS_x): Sum of squared deviations for a set of scores about their mean; important concept for ANOVA.

$$SS_x = \Sigma X^2 - \frac{(\Sigma X)^2}{N}$$

$$SS_y = \Sigma Y^2 - \frac{(\Sigma Y)^2}{N}\sqrt{2}$$

$$SS_{y-\hat{y}} = SS_y - SS_{\hat{y}} = SS_{error}$$

$$SS_{error} = SS_y(1 - r^2)$$

$$SS_{total} = \Sigma X^2 - \frac{(\Sigma X)^2}{N}$$

$$SS_{group} \text{ (one-way)} = \frac{\Sigma T_g^2}{n} - \frac{(\Sigma X)^2}{N}$$

$$SS_{error} \text{ (one-way)} = SS_{total} - SS_{group}$$

$$SS_{rows} \text{ (two-way)} = \frac{\Sigma T_{R_i}^2}{nc} - \frac{(\Sigma X)^2}{N}$$

$$SS_{col} \text{ (two-way)} = \frac{\Sigma T_{C_i}^2}{nr} - \frac{(\Sigma X)^2}{N}$$

$$SS_{cells} \text{ (two-way)} = \frac{\Sigma T_{cell}^2}{n} - \frac{(\Sigma X)^2}{N}$$

$$SS_{RxC} \text{ (two-way)} = SS_{cells} - SS_{rows} - SS_{col}$$

$$SS_{error} \text{ (two-way)} = SS_{total} - SS_{rows} - SS_{col} - SS_{RxC} \text{ or } SS_{total} - SS_{cells}$$

Summary Section: A section that succinctly answers the research question, identifies immediate implications, highlights the major contributions of the research, and briefly notes any shortcomings or limitations of the study.

Survey of Literature: Done to more clearly identify and formulate the problem, to identify and gain information about potentially contaminating variables, to gain insight into how best to design the study, and to more knowledgeably formulate the experimental hypothesis.

Systematic Decrease in Scores: Gradual decrease in test scores from trial to trial, often because of fatigue or loss of motivation.

Systematic Error (Bias): Persistent one-directional error made during collection process.

Systematic Increase in Scores: Gradual increase in test scores from trial to trial, often because of practice or learning.

Systematic Selection: Selection of items by a regular pattern (e.g., every 10th item).

Systematic Variance: Measure of treatment effect.

***t* Distribution:** Continuous probability distribution used in problems involving estimated standard errors and small samples.

***t*-Score:** Distribution of standard scores ranging from 20 to 80, with a mean of 50 and standard deviation of 10; conversion of *z*-score distribution.

***t*-Test:** Also *t*-test for matched pairs; test of differences between groups; ratio of the difference between two sample means to an estimate of the standard deviation of the distribution of differences. A different *t*-test is done for independent and correlated data:

$$t = \frac{\bar{X} - \mu}{S_{\bar{X}}} = \frac{\bar{X} - \mu}{\frac{\sigma}{\sqrt{N}}}$$

Protected t (only if F is significant):
$$t = \frac{\bar{X}_i + \bar{X}_j}{\sqrt{\frac{MS_{error}}{n_i} + \frac{MS_{error}}{n_i}}}$$

Omega (ω) squared (one-way):
$$\omega^2 = \frac{SS_{group} - (k-1)MS_{error}}{SS_{total} + MS_{error}}$$

Eta (η) squared:
$$\eta^2 = \frac{SS_{group}}{SS_{total}}$$

***t*-Test for Population Correlation Coefficient:** Test to determine whether a sample correlation coefficient qualifies as a probable or improbable outcome under the null hypothesis.

***t*-Test for Population Mean:** Test to determine whether a sample mean qualifies as a probable or improbable outcome under the null hypothesis; requires estimation of the population standard deviation from the sample.

***t*-Test for Two Population Means (Dependent Samples):** Test to determine whether a sample mean difference qualifies as a probable or improbable outcome under the null hypothesis when the two samples are dependent. Dependent samples occur if the observations in one sample are paired with those in the other sample.

$$t = \frac{\bar{X}_D}{\sigma_D} = \frac{\bar{X}_D}{\frac{\sigma_D}{\sqrt{N}}}$$

***t*-Test for Two Population Means (Independent Samples):** Test to determine whether a sample mean difference qualifies as a probable or improbable outcome under the null hypothesis when the two samples are independent. Independent samples occur if the observations in one sample are not paired with those in the other sample.

t for two independent samples (unpooled):
$$t = \frac{\bar{X}_1 - \bar{X}_2}{\sigma_{\bar{X}_1 - \bar{X}_2}} = \frac{\bar{X}_1 - \bar{X}_2}{\sqrt{\frac{s_1^2}{N_1} + \frac{s_2^2}{N_2}}}$$

t for two independent samples (pooled):
$$t = \frac{\bar{X}_1 - \bar{X}_2}{\sigma_{\bar{X}_1 - \bar{X}_2}} = \frac{\bar{X}_1 - \bar{X}_2}{\sqrt{\frac{s_p^2}{N_1} + \frac{s_p^2}{N_2}}}$$

Table: Arrangement of numerical data in columns and rows.

Taxonomy: Classification scheme.

Temporal Precedence: Changes in one variable are known to precede changes in one or more other variables.

Temporal Validity: Generalizability of research results across time. See *Validity*.

Terminal Behavior: Behavior a student is expected to display upon successful attainment of an objective.

Terminative Interaction: Result when two or more individually effective treatments are combined and their effectiveness is not increased or enhanced by the combination. See *Interaction*.

Test: Instrument or technique for obtaining information (usually a score) about an attribute of a person or object.

Test Battery: Group of tests designed to measure complex abilities.

Tetracorich r: Correlation coefficient when both variables are dichotomous.

Theoretical Construct: Entity posited to be meaningful by a theory, often simply called a construct.

Theoretical Validity: Also known as construct validity; degree to which a research result is correctly interpreted or degree to which the formulation of a construct is true. See *Validity*.

Third Variable: Variable that may explain a relationship between X and Y.

Time Series: Procedure that examines behavior over time.

Transfer Effect: Change in an organism that carries from one condition, treatment, or experience to another to influence behavior on subsequent testing.

Treatment: How the data were collected, analyzed, and stored.

Treatment Effect: At least one difference appearing among population means for various experimental conditions.

Treatment Group: Sample of subjects receiving condition involving the experimental manipulation.

Trend: Long-term underlying growth movement of a time series; usually smooth, continuous, and irreversible.

True Score: Portion of a score that is real and not due to measurement error.

Tukey's HSD (Honestly Significant Difference): Technique developed by J. W. Tukey for establishing whether differences among sample means are significant.

Two-Tailed (or Nondirectional) Test: Decision rule reflecting concern that the true population value deviates in either direction from the hypothesized population value.

Type I Error: Rejection of true null hypothesis. See *Alpha Level*.

Type II Error: Failing to reject a false null hypothesis. See *Beta Error*.

Underestimate: A calculated value that is lower than the true (or actual) value.

Ungrouped Data: Observations organized into classes of single values.

Unimodal Distribution: Distribution of scores with only one mode.

Univariate (one variable): Refers to the distribution of scores for a single variable.

Unobtrusive Observation: Observing without interfering with the ongoing flow of behavior.

Unsystematic Error: Human and mechanical survey errors that have no set pattern of occurrence.

Upper-Tailed Error: One-tailed test with the rejection region in the upper tail of the sampling distribution.

Validity: Soundness of interpretation of a test; extent to which a test measures what it is supposed to measure. *See Concurrent Validity, Construct Validity, Content Validity, Convergent Validity, Discriminant Validity, Ecological Validity, External Validity, Internal Validity, Population Validity, Predictive Validity, Temporal Validity, and Theoretical Validity.*

Validity Coefficient: Correlation coefficient representing relationship between a test and a criterion measure.

Value Index: Measurement of change in total value of related series whether the result of a price change, quantity change, or combination of both.

Variability: Spread of scores in a distribution.

Variability Between Groups (ANOVA): Indication of treatment effect and random error.

Variability Within Groups (ANOVA): Indication of random error only.

Variable: Item that varies and can be measured. In experimental research, the variables manipulated by the experimenter are called the independent variables. The dependent variables measure subject responses or effects of experimental treatment. The independent variables represent the causes and the dependent variables represent the effects in cause-and-effect relationships.

Variance (s^2 or σ^2): Sum of squared deviations of scores from the mean divided by the number of scores; degree of spread of a set of scores or amount of variation of all scores in a distribution from the mean; standard deviation squared:

$$s^2 = \frac{\sum X^2 - \frac{(\sum X)^2}{n}}{n-1} = \Sigma_{n-1}^2$$

Pooled variance: $\qquad s_p^2 = \frac{(N_1 - 1)S_1^2 + (N_1 - 1)S_2^2}{N_1 + N_2 - 2}$

Variance Estimate (ANOVA): See *Mean Square*.

Volunteer Bias: Volunteers behave, respond, or perform differently from the general population of ultimate interest.

Weighted Mean: Special form of arithmetic mean; values of the items averaged are weighted before averaging.

Wilcoxon Matched-Pairs Signed-Ranks: Nonparametric alternative to *t*-test for correlated data.

Wilcoxon *T* Test: Nonparametric test (actually the ordinal counterpart of the dependent *t*-test) of the null hypothesis to determine whether two correlated groups of ranked data (e.g., as in the matched group design) are significantly different.

Within-Groups: Variance indicates how the groups vary about their own means.

Within-Subjects Design: Design in which each subject receives every treatment.

x-Axis: Abscissa or horizontal axis of graph.

y-Axis: Ordinate or vertical axis of graph.

z-Score: Number of standard deviations of an observation from its mean; standard score specifying how far above or below the mean a given score is in the standard deviation units. Scores above the mean convert to positive *z*-scores; scores below the mean convert to negative *z*-scores; the mean of the *z*-score distribution is zero.

General formula: $\qquad \dfrac{\text{Score} - \text{Mean}}{\text{Standard Deviation}} = z = \dfrac{X - \bar{X}_x}{\sigma}$

z for X given $\sigma_{\bar{X}}$: $\qquad \dfrac{\text{Statistic} - \text{Parameter}}{\text{Std Error of Statistic}} = z = \dfrac{(\bar{X} - \mu)}{\sigma_{\bar{X}}}$

REFERENCES

Kokoska, S., & Nevision, C. (1989). *Statistical tables and formulae.* New York: Springer.

Last, J. M. (1983). *A dictionary of epidemiology.* New York: Oxford University Press.

Marriott, F. H. C. (1990). *A dictionary of statistical terms.* New York: Longman.

Yaremko, R. M., Harari, H., Harrison, R. C., & Lynn, E. (1986). *Handbook of research and quantitative methods in psychology for students and professionals.* Hillsdale, NJ: Erlbaum.

Appendix C: Statistical Equations
Basic Statistical Formulae

DESCRIPTIVE STATISTICS

Arithmetic mean

$$\bar{X} = \frac{\sum X}{n}$$

Chp 2

Grand mean

$$\bar{\bar{X}} = \frac{(n_1 \bar{X}_2) + (n_2 \bar{X}_2)}{n_1 + n_2}$$

Chp 2

Semi-interquartile range

$$Q = \frac{(Q_3 - Q_1)}{2}$$

Chp 2

Variance (S^2)

$$S^2 = \frac{\sum X^2 - \frac{(\sum X)^2}{n}}{n-1} = \frac{SS}{n}$$

Chp 2

Population variance

$$\frac{\sum x^2}{n} = \frac{\sum X^2 - \frac{(\sum X)^2}{n}}{n} = \sigma_n^2 = \sigma^2$$

Chp 2

Sample variance

$$\frac{\sum x^2}{n-1} = \frac{\sum X^2 - \frac{(\sum X)^2}{n}}{n-1} = \sigma_{n-1}^2 = s^2$$

Chp 2

Standard deviation (σ)

$$\sigma = \sqrt{S^2}$$

Chp 2

Sample SD = $\sigma_{n-1} = \sqrt{\dfrac{\sum X^2 - \dfrac{(\sum X)^2}{n}}{n-1}}$

Chp 2

Population SD = $\sigma = \sqrt{\dfrac{\sum X^2 - \frac{(\sum X)^2}{n}}{n}}$

Chp 2

Averaging standard deviation $\qquad \sigma_T = \sqrt{\left(\dfrac{n_A\left(\bar{X}_A^2 + \sigma_A^2\right) + n_B\left(\bar{X}_B^2 + \sigma_B^2\right)}{(n_A + n_B)}\right) - \bar{X}_T^2}$ Chp 2

Normal curve $\qquad Y = \dfrac{N}{\sigma\sqrt{2\pi}}\,e - (x-m)\,{}^2\!\big/\!{}_{2\sigma^2}$ Chp 2

Linear transformation of z-scores $\qquad T = z(\sigma) + \bar{X}$ Chp 2

Standard score formula $\qquad r = \dfrac{\sum z_x z_y}{n}$ Chp 3

Raw score formula $\qquad r = \dfrac{n\sum XY - (\sum X)(\sum Y)}{\sqrt{[n\sum X^2 - (\sum X)^2][n\sum Y^2 - (\sum Y)^2]}}$ Chp 3

Raw score deviation formula $\qquad r = \dfrac{\sum xy}{n\sigma_x \sigma_y}$ Chp 3

OR $\qquad r = \dfrac{\sum xy}{\sqrt{(\sum x^2)(\sum y^2)}}$ Chp 3

Point biserial $\qquad r_{pb} = \dfrac{\bar{X}_p - \bar{X}_t}{\sigma_t}\sqrt{\dfrac{p}{q}}$ Chp 3

Biserial r_b $\qquad r_b = \dfrac{\bar{X}_p - \bar{X}_t}{\sigma_t}\left(\dfrac{p}{y}\right)$ Chp 3

Calculation of Spearman rho $\qquad \rho = 1 - \dfrac{6\sum D^2}{N(N^2 - 1)}$ Chp 3

Computation of W $\qquad W = \dfrac{12\sum D^2}{m^2(n)(n^2 - 1)}$ Chp 3

Computation of phi $\qquad \phi = \dfrac{(ad - bc)}{\sqrt{(k)(l)(m)(n)}}$ Chp 3

Correlation ratio (ETA) $\qquad relationship = \dfrac{\sigma_y'}{\sigma_y}$ Chp 3

Equation for straight line $\qquad Y = a + bX$ Chp 3

Regression line	$\hat{Y} = a + b(X)$	Chp 3
	$\hat{Y} = \left(r\left(\dfrac{\sigma_y}{\sigma_x}\right) \right)(X - \overline{X}_x) + \overline{X}_y$	Chp 3
	$\hat{X} = \left(r\left(\dfrac{\sigma_x}{\sigma_y}\right) \right)(Y - \overline{X}_y) + \overline{X}_x$	Chp 3
Computation of linear regression line	$b_{yx} = \dfrac{\sum XY - \frac{(\sum X)(\sum Y)}{N}}{\sum X^2 - \frac{(\sum X)^2}{N}} = \dfrac{\sum xy}{\sum x^2}$	Chp 3
a coefficient	$a_{yx} = \overline{X}_y - \overline{X}_x(b_{yx})$	Chp 3
Relation of b_{yx} and b_{xy} to r	$(b_{yx})(b_{xy}) = r^2$	Chp 3
Computation of SE_{est}	$SE_{est} = \sigma_y \sqrt{1 - r^2}$	Chp 3
Where:	$\sigma_y = \sqrt{\dfrac{\sum Y^2}{N - 2}}$	Chp 3
Relationship of sample size to $\sigma_{\overline{X}}$	$\sigma_{\overline{X}} = SEM = \dfrac{\sigma}{\sqrt{n}}$	Chp 3
Sampling distribution of difference between two means $\sigma_{D_{\overline{x}}}$		
	$z = \dfrac{\overline{X}_1 - \overline{X}_2}{\sigma_{D_{\overline{x}}}}$	Chp 4
Computing $\sigma_{D_{\overline{x}}}$	$S_{D_{\overline{x}}} \approx \sigma_{D_{\overline{x}}} = \sqrt{\dfrac{\sigma_{x_1}^2}{n} + \dfrac{\sigma_{x_2}^2}{n}}$	Chp 4
Statistical hypothesis testing		
Null hypothesis	$H_0 : \mu_1 - \mu_2 \leq 0$	Chp 4
Alternative hypothesis	$H_1 : \mu_1 - \mu_2 > 0$	Chp 4
Computation of t for correlated data	$t = \dfrac{\overline{X}_1 - \overline{X}_2}{S_{D_{\overline{x}}}} = \dfrac{\overline{D}}{\sqrt{\dfrac{\sum d^2}{n(n-1)}}} = \dfrac{\frac{\sum D}{n}}{\sqrt{\dfrac{\sum D^2}{n(n-1)}}}$	Chp 4
Where:	$\sum d^2 = \sum D^2 - \dfrac{(\sum D)^2}{n}$	Chp 4

Percentile rank $\qquad$ $\text{Lower\%} + \dfrac{\text{Score} + \text{RLL}}{\text{Width}}(\text{Interval \%})$

Correlation and the two matched groups design

$$\sigma_{D_{\overline{X}}} = \sqrt{\sigma_{\overline{X}_1}^2 + \sigma_{\overline{X}_2}^2 - 2(r_{12})(\sigma_{\overline{X}_1})(\sigma_{\overline{X}_2})} \qquad \text{Chp 4}$$

t Value $\qquad$ $t = \dfrac{\overline{D}}{\sqrt{\frac{\Sigma d^2}{n(n-1)}}} \qquad \text{Chp 4}$

Computation of Mann–Whitney U $\qquad U_a = n_a n_b + \dfrac{n_b(n_b+1)}{2} - \Sigma R_b \qquad \text{Chp 4}$

$$U_b = n_a n_b + \dfrac{n_a(n_a+1)}{2} - \Sigma R_a \qquad \text{Chp 4}$$

Test for significance of phi (Φ) $\qquad \chi^2 = N\phi^2 \qquad \text{Chp 4}$

Rationale for F–test $\qquad F = \dfrac{\text{between groups variance}}{\text{within groups variance}} \qquad \text{Chp 5}$

$$= \dfrac{\text{effect of the IV+random error}}{\text{random error}} \qquad \text{Chp 5}$$

Sample variance $\qquad \sigma^2 = \dfrac{SS}{n-1} \qquad \text{Chp 5}$

Mean square $\qquad MS = \dfrac{SS}{df} \qquad \text{Chp 5}$

Tukey's WSD $\qquad WSD = q_{\alpha,k,df_{denom}}\sqrt{\dfrac{MS_{error}}{n}} \qquad \text{Chp 5}$

Critical distance $\qquad WSD = q_{(\alpha,\ \#\ \text{of means},\ df_{MS_{error}})}\sqrt{\dfrac{MS_{error}}{\#\ \text{obs per mean}}} \qquad \text{Chp 5}$

Bonferroni t-test $\qquad t = \dfrac{\overline{X}_1 - \overline{X}_2}{\sqrt{\frac{2MS_{error}}{n}}} \qquad \text{Chp 5}$

Scheffe test $\qquad F = \dfrac{(\overline{X}_1 - \overline{X}_2)^2}{MS_{error}\left(\frac{1}{n_1} + \frac{1}{n2}\right)(k-1)} \qquad \text{Chp 5}$

Critical distance between means $\qquad d = \sqrt{\dfrac{2(k-1)(F_{critical})(MS_{error})}{n}} \qquad \text{Chp 5}$

Degrees of freedom for four additional component parts

$$df_S = n - 1$$ Chp 6

$$df_{K/S} = (k-1)(n-1)$$ Chp 6

$$df_{J/S} = (j-1)(n-1)$$ Chp 6

$$df_{JK/S} = (j-1)(k-1)(n-1)$$ Chp 6

Percentile rank (ungrouped frequency distribution) $$p = \left(\frac{f/2 + \text{Cum freq below}}{n} \right) 100$$ Chp 2

Score for a percentile (score$_p$) $$RLL + \frac{Width}{Interval\ \%}(p - Lower\ \%)$$ Chp 2

General formula for z-score $$\frac{\text{Score} - \text{Mean}}{\text{Standard Deviation}}$$ Chp 4

OR $$\frac{\text{Statistic} - \text{Parameter}}{\text{Std Error of Statistic}}$$

z-score for an observation $$z = \frac{X - \overline{X}_x}{\sigma}$$ Chp 4

TESTS ON SAMPLE MEANS

Standard error of the mean $S_{\overline{X}}$ or $\sigma \overline{x}$ $$\frac{\sigma}{\sqrt{n}}$$ Chp 4

z for X given σ $$z = \frac{(\overline{X} - \mu)}{\mu_{\overline{X}}}$$ Chp 4

t for one sample $$t = \frac{\overline{X} - \mu}{S_{\overline{X}}} = \frac{\overline{X} - \mu}{\frac{\sigma}{\sqrt{N}}}$$ Chp 4

Coincidence interval on μ $$CI = \overline{X} \pm t_{.025}(\sigma_{\overline{X}})$$ Chp 4

t for two related samples $$t = \frac{\overline{X}_D}{\sigma_D} = \frac{\overline{X}_D}{\frac{\sigma_D}{\sqrt{N}}}$$

t for two independent samples (unpooled) $$t = \frac{\overline{X}_1 - \overline{X}_2}{\sigma_{\overline{X}_1 - \overline{X}_2}} = \frac{\overline{X}_1 - \overline{X}_2}{\sqrt{\frac{s_1^2}{N_1} + \frac{s_2^2}{N_2}}}$$ Chp 4

Pooled variance ($S_p{}^2$) $$s_p^2 = \frac{(N_1 - 1)S_1^2 + (N_1 - 1)S_2^2}{N_1 + N_2 - 2}$$ Chp 4

t for two independent samples (pooled) $$t = \frac{\overline{X}_1 - \overline{X}_2}{\sigma_{\overline{X}_1 - \overline{X}_2}} = \frac{\overline{X}_1 - \overline{X}_2}{\sqrt{\frac{s_p^2}{N_1} + \frac{s_p^2}{N_2}}}$$ Chp 4

Confidence interval on $\mu_1 - \mu_2$ $$CI = (\overline{X} - \overline{X}) \pm t_{.025}(\sigma_{\overline{X}_1 - \overline{X}_2})$$ Chp 4

STATISTICAL POWER

Effect size (one sample) $$ES = \frac{(\mu_1 - \mu_2)}{\sigma} = \delta$$

Effect size (two samples) $$ES = \frac{\overline{X}_1 - \overline{X}_2}{\sqrt{\frac{SS_1 + SS_2}{n_1 + n_2}}}$$

Effect size (correlation) $$ES = \rho_1 - \rho_0$$

Delta (one-sample t) $$\delta = Y\sqrt{N}$$

Delta (two-sample t) $$\delta = Y\sqrt{\frac{N}{2}}$$

Delta (correlation) $$\delta = Y\sqrt{N - 1}$$

CORRELATION AND REGRESSION

Sum of squares $$SS_x = \sum X^2 - \frac{(\sum X)^2}{N}$$ Chp 2

SS_y $$\sum Y^2 - \frac{(\sum Y)^2}{N} SS_y$$

$SS_{\hat{y}}$ $$\sum \hat{Y}^2 - \frac{(\sum \hat{Y})^2}{N} SS_{\hat{Y}}$$

$SS_{y-\hat{y}}$ $$SS_y - SS_{\hat{y}} = SS_{error} SS_{y-\hat{y}}$$

SS_{error} $$SS_y(1 - r^2) SS_{error}$$

Sum of products $$SP_{xy} = \sum XY - \frac{\sum X \sum Y}{N}$$ Chp 3

Covariance $$cov_{xy} = \frac{\sum XY - \frac{\sum X \sum Y}{N}}{N-1} = \frac{SP_{xy}}{N-1}$$ Chp 3

Correlation (Pearson) $$r = \frac{cov_{xy}}{\sigma_x \sigma_y} = \frac{SP_{xy}}{\sqrt{SS_x SS_y}}$$ Chp 3

$$= \frac{N \sum XY - \sum X \sum Y}{(N \sum X^2 - (\sum X)^2)(N \sum Y^2 - (\sum Y)^2)}$$

Slope $$b = \frac{cov_{xy}}{S_x^2} = \frac{SP_{xy}}{SS_x}$$ Chp 3

Intercept $$\alpha = \frac{\sum Y - b \sum X}{N} = \overline{X}_y - b \overline{X}_x$$ Chp 3

Standard error of estimate $$S_{y-\hat{y}} = \sigma_y \sqrt{\frac{\sum (Y - \hat{y})^2}{N-2}} = \sqrt{\frac{SS_{error}}{N-2}}$$ Chp 3

$$= \sqrt{(1 - r^2) \frac{N-1}{N-2}}$$ Chp 3

ANALYSIS OF VARIANCE

SS_{total} $$\sum X^2 - \frac{(\sum X)^2}{N}$$ Chp 5

$SS_{between}$

$$\frac{(\sum X_1)^2}{n_1} + \frac{(\sum X_2)^2}{n_2} + \frac{(\sum X_3)^2}{n_3} \ldots + \frac{(\sum X_k)^2}{n_k} - \frac{(\sum X_1 + \sum X_2 + \sum X_3 \ldots \sum X_k)^2}{N}$$ Chp 5

$SS_{treatment}$

$$\frac{(\sum X_1)^2}{n_1} + \frac{(\sum X_2)^2}{n_2} + \frac{(\sum X_3)^2}{n_3} \ldots \frac{(\sum X_k)^2}{n_k} - \frac{(\sum X)^2}{N}$$ Chp 5

$SS_{subjects}$

$$\frac{\sum(\sum \text{rows})^2}{k} - \frac{(\sum X)^2}{N}$$

Chp 5

SS_{SXT}

$$SS_{\text{total}} - SS_{\text{treatment}} - SS_{\text{subjects}}$$

Chp 5

SS_{within}

$$SS_{total} - SS_{between}$$

Chp 5

SS_{group} (one-way)

$$\frac{\sum T_g^2}{n} - \frac{(\sum X)^2}{N}$$

Chp 5

SS_{error} (one-way)

$$SS_{total} - SS_{group}$$

Chp 5

SS_{rows} (two-way)

$$\frac{\sum T_{R_i}^2}{nc} - \frac{(\sum X)^2}{N}$$

Chp 6

SS_{col} (two-way)

$$\frac{\sum T_{C_i}^2}{nr} - \frac{(\sum X)^2}{N}$$

Chp 6

SS_{cells} (two-way)

$$\frac{\sum T_{cell}^2}{n} - \frac{(\sum X)^2}{N}$$

Chp 6

SS_{RxC} (two-way)

$$SS_{cells} - SS_{rows} - SS_{col}$$

Chp 6

SS_{error} (two-way)

$$SS_{total} - SS_{rows} - SS_{col} - SS_{RxC}$$

Chp 6

OR

$$SS_{total} - SS_{cells}$$

Chp 6

Protected t (use only if F is significant)

$$t = \frac{\overline{X}_i + \overline{X}_j}{\sqrt{\frac{MS_{error}}{n_i} + \frac{MS_{error}}{n_i}}}$$

Chp 5

Eta squared
$$\eta^2 = \frac{SS_{group}}{SS_{total}}$$
Chp 6

Omega squared (one-way)
$$\omega^2 = \frac{SS_{group} - (k-1)MS_{error}}{SS_{total} + MS_{error}}$$
Chp 5

SS_A
$$\frac{(\sum X_1 + \sum X_2)^2}{n_1 + n_2} + \frac{(\sum X_3 + \sum X_4)^2}{n_3 + n_4} + \frac{(\sum X_1 + \sum X_2 + \sum X_3 + \sum X_4)^2}{N}$$
Chp 6

SS_B
$$\frac{(\sum X_1 + \sum X_3)^2}{n_1 + n_3} + \frac{(\sum X_2 + \sum X_4)^2}{n_2 + n_4} - \frac{(\sum X_1 + \sum X_2 + \sum X_3 + \sum X_4)^2}{N}$$
Chp 6

SS_{AB}
$$SS_{between} - (SS_A + SS_B)$$
Chp 6

SS_J
$$\frac{(\sum X_1 + \sum X_2)^2 + (\sum X_3 + \sum X_4)^2}{nk} - C$$
Chp 6

SS_K
$$\frac{(\sum X_1 + \sum X_3)^2 + (\sum X_2 + \sum X_4)^2}{nj} - C$$
Chp 6

SS_S
$$\frac{(\sum S_1)^2 + (\sum S_2)^2 \ldots (\sum S_{10})^2}{jk} - C$$
Chp 6

SS_{JS}
$$\frac{\sum (\text{Each Subject's Row Total})^2}{k} - C - SS_J - SS_S$$
Chp 6

SS_{KS}
$$\frac{\sum \left(\text{Each Subject's Column Total}\right)^2}{j} - C - SS_K - SS_S$$
Chp 6

SS_{JKS}
$$SS_{total} - SS_J - SS_K - SS_S - SS_{JK} - SS_{JS} - SS_{KS}$$
Chp 6

CHI-SQUARE

Chi-square $\qquad$ $X^2 = \sum \dfrac{(O - E)^2}{E}$ $\qquad$ Chp 4

Degrees of freedom for chi-square tests $\qquad$ $df = (r - 1)\,(c - 1)$ $\qquad$ Chp 4

Yates's correction $\qquad$ $X^2 = \sum \dfrac{\left(\left|O - E\right| - .5\right)^2}{E}$ $\qquad$ Chp 4

Chi-square for 2×2 table only $\qquad$ $X^2 = \dfrac{N(AD - BC)^2}{(A + B)(C + D)(A + C)(B + D)}$ $\qquad$ Chp 4

Index

Printed and bound by CPI Group (UK) Ltd, Croydon, CR0 4YY

18/10/2024

01776252-0010